Semiconductor Physics: Devices and Technology

Karla Hodges

Clanrye International
www.clanryeinternational.com

Clanrye International,
750 Third Avenue, 9th Floor,
New York, NY 10017, USA

ISBN: 978-1-63240-914-0

Cataloging-in-Publication Data

Semiconductor physics : devices and technology / Karla Hodges.
p. cm.
Includes bibliographical references and index.
ISBN 978-1-63240-914-0
1. Semiconductors. 2. Electrical engineering--Materials.
3. Electronics--Materials. I. Hodges, Karla.
TK7871.85 .S46 2019
621.381 52--dc23

For information on all Clanrye International publications visit our website at www.clanryeinternational.com

CLANRYE INTERNATIONAL

Printed in China.

Contents

Preface

A semiconductor is a material that has the value of the electrical conductivity between that of a conductor and an insulator. Unlike metals, the electrical resistance decreases as the temperature increases in semiconductors. Silicon, germanium and compounds of gallium are commonly used semiconducting materials. The charge carriers in such materials are electrons, ions and electron holes. By controlling the introduction of impurities into their crystal structures, their conducting properties can be altered to suit requirements. This textbook is a compilation of chapters that discuss the most vital concepts in the field of semiconductor physics. It elucidates new techniques and applications of semiconductors in a multidisciplinary manner. It will serve as a valuable source of reference for physicists, engineers and students associated with this important discipline.

A detailed account of the significant topics covered in this book is provided below:

Chapter 1- Any material with the electrical conductivity value between a conductor and an insulator is known as a semiconductor. This chapter has been carefully written to provide insights into semiconductors and semiconducting materials. Various topics related to the different types of semiconductors, such as elemental, compound, doping, band gap and wide band gap semiconductors, have been discussed in detail in this chapter.

Chapter 2- Semiconductor devices exhibit properties of easy conductance of electric current, variable resistance and sensitivity to heat and light. These are therefore utilized for the design of logic gates and digital circuits. Science and technology have undergone rapid developments in the past decade, which has resulted in the design of innovative semiconductor devices and junctions, in the field of semiconductor technology, which have been extensively detailed in this chapter.

Chapter 3- Semiconductor device fabrication is the process that is used to build integrated circuits. The core concepts of crystal growth techniques and thin film growth techniques have been elucidated in great detail in this chapter. It also discusses in detail the theories and methodologies related to semiconductor device fabrication and the growth techniques of fabricating a p-n junction.

Chapter 4- A two-terminal electronic component, which has low electrical resistance in one direction of flow and high resistance in the other, is known as a diode. They are of different types, some of which are discussed in this chapter, like Zener diode, tunnel diode, Impatt diode, Light Emitting Diode (LED), and Gunn diode.

Chapter 5- A transistor is a semiconductor device that is used for the purpose of amplification or switching of electrical power and electronic signals. It is made of a semiconductor material and has three terminals that connect to an external circuit. The aim of this chapter is to explore the different types of transistors, such as bipolar junction transistors and field effect transistors. These topics are crucial for a complete understanding of transistor technology.

I would like to make a special mention of my publisher who considered me worthy of this opportunity and also supported me throughout the process. I would also like to thank the editing team at the back-end who extended their help whenever required.

Karla Hodges

1

An Introduction to Semiconductor and Semiconductor Materials

Any material with the electrical conductivity value between a conductor and an insulator is known as a semiconductor. This chapter has been carefully written to provide insights into semiconductors and semiconducting materials. Various topics related to the different types of semiconductors, such as elemental, compound, doping, band gap and wide band gap semiconductors, have been discussed in detail in this chapter.

Semiconductor

A semiconductor is generally defined as a material with electrical conductivity or resistivity lying between that of a metal and an insulator. The best known semiconductor is silicon (Si) with a band gap of around 1.1 eV. Semiconductors can be elemental or compounds. Elemental semiconductors, such as silicon (Si) and Germanium (Ge) consist of single type of atoms and have similar tetrahedrally bonded crystal structures. Compound semiconductors are formed from elements of group III and V or II and VI of periodic table. Pure semiconductors are usually single crystals of pure silicon or germanium. They are insulators at very low temperature. Upon increasing the temperature, some of the electrons gain thermal energy and are free to move and conduct electricity. These electrons leave behind vacancies or holes which carry a positive charge equal in magnitude to charge of an electron. These holes can also act as current carriers. So, a semiconductor consists of two types of charge carriers- electrons and holes.Semiconductors can be doped by pentavalent or trivalent elements to increase the concentration of electrons and holes. These elements can act as electron donors or acceptors. Silicon, being a tetravalent atom, contains four electrons in its outer most shell. A pentavalent atom (such as phosphorous, arsenic, antimony) will replace a silicon atom and donate one extra electron. A trivalent atom (such as aluminum, boron, gallium) will act as electron acceptor giving rise an extra hole. Semiconductors having electrons as the majority charge carriers are known as n-type semiconductors whereas those having holes as the majority carriers are known as p-type semiconductors.

The ability of semiconductors to conduct electricity can be greatly improved by replacing or adding certain donor or acceptor atoms to this crystalline structure thereby, producing more free electrons than holes or vice versa. That is by adding a small percentage of another element to the base material, either silicon or germanium.

On their own Silicon and Germanium are classed as intrinsic semiconductors, that is they are chemically pure, containing nothing but semi-conductive material. But by controlling the amount of impurities added to this intrinsic semiconductor material it is possible to control its conductivity.

Various impurities called donors or acceptors can be added to this intrinsic material to produce free electrons or holes respectively.

This process of adding donor or acceptor atoms to semiconductor atoms (the order of 1 impurity atom per 10 million (or more) atoms of the semiconductor) is called Doping. The as the doped silicon is no longer pure, these donor and acceptor atoms are collectively referred to as "impurities", and by doping these silicon material with a sufficient number of impurities, we can turn it into a *semi-conductor*.

The most commonly used semiconductor basics material by far is silicon. Silicon has four valence electrons in its outermost shell which it shares with its neighbouring silicon atoms to form full orbital's of eight electrons. The structure of the bond between the two silicon atoms is such that each atom shares one electron with its neighbour making the bond very stable.

As there are very few free electrons available to move around the silicon crystal, crystals of pure silicon (or germanium) are therefore good insulators, or at the very least very high value resistors.

Silicon atoms are arranged in a definite symmetrical pattern making them a crystalline solid structure. A crystal of pure silica (silicon dioxide or glass) is generally said to be an intrinsic crystal (it has no impurities) and therefore has no free electrons.

But simply connecting a silicon crystal to a battery supply is not enough to extract an electric current from it. To do that we need to create a "positive" and a "negative" pole within the silicon allowing electrons and therefore electric current to flow out of the silicon. These poles are created by doping the silicon with certain impurities.

Properties

Variable conductivity

> Semiconductors in their natural state are poor conductors because a current requires the flow of electrons, and semiconductors have their valence bands filled, preventing the entry flow of new electrons. There are several developed techniques that allow semiconducting materials to behave like conducting materials, such as doping or gating. These modifications have two outcomes: n-type and p-type. These refer to the excess or shortage of electrons, respectively. An unbalanced number of electrons would cause a current to flow through the material.

Heterojunctions

> Heterojunctions occur when two differently doped semiconducting materials are joined together. For example, a configuration could consist of p-doped and n-doped germanium. This results in an exchange of electrons and holes between the differently doped semiconducting materials. The n-doped germanium would have an excess of electrons, and the p-doped germanium would have an excess of holes. The transfer occurs until equilibrium is reached by a process called recombination, which causes the migrating electrons from the n-type to come in contact with the migrating holes from the p-type. A product of this process is charged ions, which result in an electric field.

Excited electrons

A difference in electric potential on a semiconducting material would cause it to leave thermal equilibrium and create a non-equilibrium situation. This introduces electrons and holes to the system, which interact via a process called ambipolar diffusion. Whenever thermal equilibrium is disturbed in a semiconducting material, the number of holes and electrons changes. Such disruptions can occur as a result of a temperature difference or photons, which can enter the system and create electrons and holes. The process that creates and annihilates electrons and holes are called generation and recombination.

Light emission

In certain semiconductors, excited electrons can relax by emitting light instead of producing heat. These semiconductors are used in the construction of light-emitting diodes and fluorescent quantum dots.

Thermal energy conversion

Semiconductors have large thermoelectric power factors making them useful in thermoelectric generators, as well as high thermoelectric figures of merit making them useful in thermoelectric coolers.

Materials

Silicon crystals are the most common semiconducting materials used in microelectronics and photovoltaics.

A large number of elements and compounds have semiconducting properties, including:

- Certain pure elements are found in Group 14 of the periodic table; the most commercially important of these elements are silicon and germanium. Silicon and germanium are used here effectively because they have 4 valence electrons in their outermost shell which gives them the ability to gain or lose electrons equally at the same time.
- Binary compounds, particularly between elements in Groups 13 and 15, such as gallium arsenide, Groups 12 and 16, groups 14 and 16, and between different group 14 elements, e.g. silicon carbide.
- Certain ternary compounds, oxides and alloys.
- Organic semiconductors, made of organic compounds.

Most common semiconducting materials are crystalline solids, but amorphous and liquid semiconductors are also known. These include hydrogenated amorphous silicon and mixtures of arsenic, selenium and tellurium in a variety of proportions. These compounds share with better known semiconductors the properties of intermediate conductivity and a rapid variation of conductivity with temperature, as well as occasional negative resistance. Such disordered materials lack the rigid crystalline structure of conventional semiconductors such as silicon. They are generally used in thin film structures, which do not require material of higher electronic quality, being relatively insensitive to impurities and radiation damage.

Preparation of Semiconductor Materials

Almost all of today's electronic technology involves the use of semiconductors, with the most important aspect being the integrated circuit (IC), which are found in laptops, scanners, cell-phones, etc. Semiconductors for ICs are mass-produced. To create an ideal semiconducting material, chemical purity is paramount. Any small imperfection can have a drastic effect on how the semiconducting material behaves due to the scale at which the materials are used.

A high degree of crystalline perfection is also required, since faults in crystal structure (such as dislocations, twins, and stacking faults) interfere with the semiconducting properties of the material. Crystalline faults are a major cause of defective semiconductor devices. The larger the crystal, the more difficult it is to achieve the necessary perfection. Current mass production processes use crystal ingots between 100 and 300 mm (3.9 and 11.8 in) in diameter which are grown as cylinders and sliced into wafers.

There is a combination of processes that is used to prepare semiconducting materials for ICs. One process is called thermal oxidation, which forms silicon dioxide on the surface of the silicon. This is used as a gate insulator and field oxide. Other processes are called photomasks and photolithography. This process is what creates the patterns on the circuity in the integrated circuit. Ultraviolet light is used along with a photoresist layer to create a chemical change that generates the patterns for the circuit.

Etching is the next process that is required. The part of the silicon that was not covered by the photoresist layer from the previous step can now be etched. The main process typically used today is called plasma etching. Plasma etching usually involves an etch gas pumped in a low-pressure chamber to create plasma. A common etch gas is chlorofluorocarbon, or more commonly known Freon. A high radio-frequency voltage between the cathode and anode is what creates the plasma in the chamber. The silicon wafer is located on the cathode, which causes it to be hit by the positively charged ions that are released from the plasma. The end result is silicon that is etched anisotropically.

The last process is called diffusion. This is the process that gives the semiconducting material its desired semiconducting properties. It is also known as doping. The process introduces an impure atom to the system, which creates the p-n junction. In order to get the impure atoms embedded in the silicon wafer, the wafer is first put in a 1,100 degree Celsius chamber. The atoms are injected in and eventually diffuse with the silicon. After the process is completed and the silicon has reached room temperature, the doping process is done and the semiconducting material is ready to be used in an integrated circuit.

Physics of Semiconductors

Energy Bands and Electrical Conduction

Semiconductors are defined by their unique electric conductive behavior, somewhere between that of a conductor and an insulator. The differences between these materials can be understood in terms of the quantum states for electrons, each of which may contain zero or one electron (by the Pauli exclusion principle). These states are associated with the electronic band structure of the material. Electrical conductivity arises due to the presence of electrons in states that are delocalized (extending through the material), however in order to transport electrons a state must be *partially filled*, containing an electron only part of the time. If the state is always occupied with an electron, then it is inert, blocking the passage of other electrons via that state. The energies of these quantum states are critical, since a state is partially filled only if its energy is near the Fermi level.

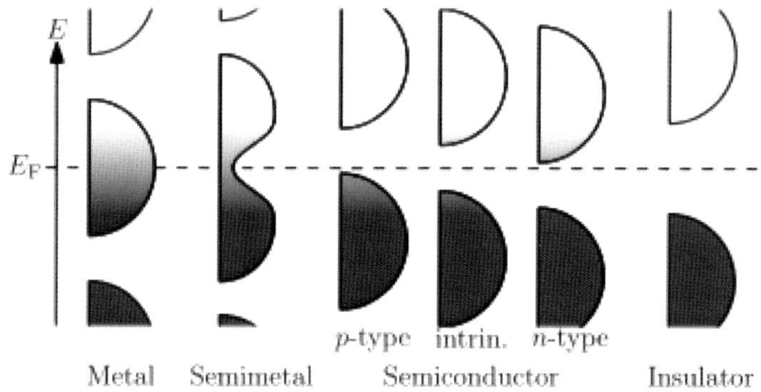

Filling of the electronic states in various types of materials at equilibrium. Here, height is energy while width is the density of available states for a certain energy in the material listed. The shade follows the Fermi–Dirac distribution (black = all states filled, white = no state filled). In metals and semimetals the Fermi level E_F lies inside at least one band. In insulators and semiconductors the Fermi level is inside a band gap; however, in semiconductors the bands are near enough to the Fermi level to be thermally populated with electrons or holes.

High conductivity in a material comes from it having many partially filled states and much state delocalization. Metals are good electrical conductors and have many partially filled states with energies near their Fermi level. Insulators, by contrast, have few partially filled states, their Fermi levels sit within band gaps with few energy states to occupy. Importantly, an insulator can be made to conduct by increasing its temperature: heating provides energy to promote some electrons across the band gap, inducing partially filled states in both the band of states beneath the band gap (valence band) and the band of states above the band gap (conduction band). An (intrinsic) semiconductor has a band gap that is smaller than that of an insulator and at room temperature significant numbers of electrons can be excited to cross the band gap.

A pure semiconductor, however, is not very useful, as it is neither a very good insulator nor a very good conductor. However, one important feature of semiconductors (and some insulators, known as *semi-insulators*) is that their conductivity can be increased and controlled by doping with impurities and gating with electric fields. Doping and gating move either the conduction or valence band much closer to the Fermi level, and greatly increase the number of partially filled states.

Some wider-band gap semiconductor materials are sometimes referred to as semi-insulators. When undoped, these have electrical conductivity nearer to that of electrical insulators, however they can be doped (making them as useful as semiconductors). Semi-insulators find niche applications in micro-electronics, such as substrates for HEMT. An example of a common semi-insulator is gallium arsenide. Some materials, such as titanium dioxide, can even be used as insulating materials for some applications, while being treated as wide-gap semiconductors for other applications.

Charge Carriers (Electrons and Holes)

The partial filling of the states at the bottom of the conduction band can be understood as adding electrons to that band. The electrons do not stay indefinitely (due to the natural thermal recombination) but they can move around for some time. The actual concentration of electrons is typically very dilute, and so (unlike in metals) it is possible to think of the electrons in the conduction band of a semiconductor as a sort of classical ideal gas, where the electrons fly around freely without being subject to the Pauli exclusion principle. In most semiconductors the conduction bands have a parabolic dispersion relation, and so these electrons respond to forces (electric field, magnetic field, etc.) much like they would in a vacuum, though with a different effective mass. Because the electrons behave like an ideal gas, one may also think about conduction in very simplistic terms such as the Drude model, and introduce concepts such as electron mobility.

For partial filling at the top of the valence band, it is helpful to introduce the concept of an electron hole. Although the electrons in the valence band are always moving around, a completely full valence band is inert, not conducting any current. If an electron is taken out of the valence band, then the trajectory that the electron would normally have taken is now missing its charge. For the purposes of electric current, this combination of the full valence band, minus the electron, can be converted into a picture of a completely empty band containing a positively charged particle that moves in the same way as the electron. Combined with the *negative* effective mass of the electrons at the top of the valence band, we arrive at a picture of a positively charged particle that responds to electric and magnetic fields just as a normal positively charged particle would do in vacuum, again with some positive effective mass. This particle is called a hole, and the collection of holes in the valence band can again be understood in simple classical terms (as with the electrons in the conduction band).

Carrier Generation and Recombination

When ionizing radiation strikes a semiconductor, it may excite an electron out of its energy level and consequently leave a hole. This process is known as *electron–hole pair generation*. Electron-hole pairs are constantly generated from thermal energy as well, in the absence of any external energy source.

Electron-hole pairs are also apt to recombine. Conservation of energy demands that these recombination events, in which an electron loses an amount of energy larger than the band gap, be accompanied by the emission of thermal energy (in the form of phonons) or radiation (in the form of photons).

In some states, the generation and recombination of electron–hole pairs are in equipoise. The number of electron-hole pairs in the steady state at a given temperature is determined by quantum statistical mechanics. The precise quantum mechanical mechanisms of generation and recombination are governed by conservation of energy and conservation of momentum.

As the probability that electrons and holes meet together is proportional to the product of their numbers, the product is in steady state nearly constant at a given temperature, providing that there is no significant electric field (which might "flush" carriers of both types, or move them from neighbour regions containing more of them to meet together) or externally driven pair generation. The product is a function of the temperature, as the probability of getting enough thermal energy to produce a pair increases with temperature, being approximately $\exp(-E_G/kT)$, where k is Boltzmann's constant, T is absolute temperature and E_G is band gap.

The probability of meeting is increased by carrier traps—impurities or dislocations which can trap an electron or hole and hold it until a pair is completed. Such carrier traps are sometimes purposely added to reduce the time needed to reach the steady state.

Doping

The conductivity of semiconductors may easily be modified by introducing impurities into their crystal lattice. The process of adding controlled impurities to a semiconductor is known as *doping*. The amount of impurity, or dopant, added to an *intrinsic* (pure) semiconductor varies its level of conductivity. Doped semiconductors are referred to as *extrinsic*. By adding impurity to the pure semiconductors, the electrical conductivity may be varied by factors of thousands or millions.

A 1 cm^3 specimen of a metal or semiconductor has of the order of 10^{22} atoms. In a metal, every atom donates at least one free electron for conduction, thus 1 cm^3 of metal contains on the order of 10^{22} free electrons, whereas a 1 cm^3 sample of pure germanium at 20 °C contains about 4.2×10^{22} atoms, but only 2.5×10^{13} free electrons and 2.5×10^{13} holes. The addition of 0.001% of arsenic (an impurity) donates an extra 10^{17} free electrons in the same volume and the electrical conductivity is increased by a factor of 10,000.

The materials chosen as suitable dopants depend on the atomic properties of both the dopant and the material to be doped. In general, dopants that produce the desired controlled changes are classified as either electron acceptors or donors. Semiconductors doped with *donor* impurities are called *n-type*, while those doped with *acceptor* impurities are known as *p-type*. The n and p type designations indicate which charge carrier acts as the material's majority carrier. The opposite carrier is called the minority carrier, which exists due to thermal excitation at a much lower concentration compared to the majority carrier.

For example, the pure semiconductor silicon has four valence electrons which bond each silicon atom to its neighbors. In silicon, the most common dopants are *group III* and *group V* elements. Group III elements all contain three valence electrons, causing them to function as acceptors when used to dope silicon. When an acceptor atom replaces a silicon atom in the crystal, a vacant state (an electron "hole") is created, which can move around the lattice and functions as a charge carrier. Group V elements have five valence electrons, which allows them to act as a donor; substitution of these atoms for silicon creates an extra free electron. Therefore, a silicon crystal doped with boron creates a p-type semiconductor whereas one doped with phosphorus results in an n-type material.

During manufacture, dopants can be diffused into the semiconductor body by contact with gaseous compounds of the desired element, or ion implantation can be used to accurately position the doped regions.

The history of the understanding of semiconductors begins with experiments on the electrical properties of materials. The properties of negative temperature coefficient of resistance, rectification, and light-sensitivity were observed starting in the early 19th century.

Thomas Johann Seebeck was the first to notice an effect due to semiconductors, in 1821. In 1833, Michael Faraday reported that the resistance of specimens of silver sulfide decreases when they are heated. This is contrary to the behavior of metallic substances such as copper. In 1839, Alexandre Edmond Becquerel reported observation of a voltage between a solid and a liquid electrolyte when struck by light, the photovoltaic effect. In 1873 Willoughby Smith observed that selenium resistors exhibit decreasing resistance when light falls on them. In 1874 Karl Ferdinand Braun observed conduction and rectification in metallic sulfides, although this effect had been discovered much earlier by Peter Munck af Rosenschold (sv) writing for the Annalen der Physik und Chemie in 1835, and Arthur Schuster found that a copper oxide layer on wires has rectification properties that ceases when the wires are cleaned. William Grylls Adams and Richard Evans Day observed the photovoltaic effect in selenium in 1876.

A unified explanation of these phenomena required a theory of solid-state physics which developed greatly in the first half of the 20th Century. In 1878 Edwin Herbert Hall demonstrated the deflection of flowing charge carriers by an applied magnetic field, the Hall effect. The discovery of the electron by J.J. Thomson in 1897 prompted theories of electron-based conduction in solids. Karl Baedeker, by observing a Hall effect with the reverse sign to that in metals, theorized that copper iodide had positive charge carriers. Johan Koenigsberger classified solid materials as metals, insulators and "variable conductors" in 1914 although his student Josef Weiss already introduced the term *Halbleiter* (semiconductor in modern meaning) in PhD thesis in 1910. Felix Bloch published a theory of the movement of electrons through atomic lattices in 1928. In 1930, B. Gudden stated that conductivity in semiconductors was due to minor concentrations of impurities. By 1931, the band theory of conduction had been established by Alan Herries Wilson and the concept of band gaps had been developed. Walter H. Schottky and Nevill Francis Mott developed models of the potential barrier and of the characteristics of a metal-semiconductor junction. By 1938, Boris Davydov had developed a theory of the copper-oxide rectifier, identifying the effect of the p–n junction and the importance of minority carriers and surface states.

Agreement between theoretical predictions (based on developing quantum mechanics) and experimental results was sometimes poor. This was later explained by John Bardeen as due to the extreme "structure sensitive" behavior of semiconductors, whose properties change dramatically based on tiny amounts of impurities. Commercially pure materials of the 1920s containing varying proportions of trace contaminants produced differing experimental results. This spurred the development of improved material refining techniques, culminating in modern semiconductor refineries producing materials with parts-per-trillion purity.

Devices using semiconductors were at first constructed based on empirical knowledge, before semiconductor theory provided a guide to construction of more capable and reliable devices.

Alexander Graham Bell used the light-sensitive property of selenium to transmit sound over a beam of light in 1880. A working solar cell, of low efficiency, was constructed by Charles Fritts in 1883 using a metal plate coated with selenium and a thin layer of gold; the device became commercially useful in photographic light meters in the 1930s. Point-contact microwave detector rectifiers made of lead sulfide were used by Jagadish Chandra Bose in 1904; the cat's-whisker detector using natural galena or other materials became a common device in the development of radio. However, it was somewhat unpredictable in operation and required manual adjustment for best performance. In 1906 H.J. Round observed light emission when electric current passed through silicon carbide crystals, the principle behind the light-emitting diode. Oleg Losev observed similar light emission in 1922 but at the time the effect had no practical use. Power rectifiers, using copper oxide and selenium, were developed in the 1920s and became commercially important as an alternative to vacuum tube rectifiers.

In the years preceding World War II, infra-red detection and communications devices prompted research into lead-sulfide and lead-selenide materials. These devices were used for detecting ships and aircraft, for infrared rangefinders, and for voice communication systems. The point-contact crystal detector became vital for microwave radio systems, since available vacuum tube devices could not serve as detectors above about 4000 MHz; advanced radar systems relied on the fast response of crystal detectors. Considerable research and development of silicon materials occurred during the war to develop detectors of consistent quality.

Detector and power rectifiers could not amplify a signal. Many efforts were made to develop a solid-state amplifier and were successful in developing a device called the point contact transistor which could amplify 20db or more. In 1922 Oleg Losev developed two-terminal, negative resistance amplifiers for radio, and he perished in the Siege of Leningrad after successful completion. In 1926 Julius Edgar Lilienfeld patented a device resembling a modern field-effect transistor, but it was not practical. R. Hilsch and R. W. Pohl in 1938 demonstrated a solid-state amplifier using a structure resembling the control grid of a vacuum tube; although the device displayed power gain, it had a cut-off frequency of one cycle per second, too low for any practical applications, but an effective application of the available theory. At Bell Labs, William Shockley and A. Holden started investigating solid-state amplifiers in 1938. The first p–n junction in silicon was observed by Russell Ohl about 1941, when a specimen was found to be light-sensitive, with a sharp boundary between p-type impurity at one end and n-type at the other. A slice cut from the specimen at the p–n boundary developed a voltage when exposed to light.

In France, during the war, Herbert Mataré had observed amplification between adjacent point contacts on a germanium base. After the war, Mataré's group announced their "Transistron" amplifier only shortly after Bell Labs announced the "transistor".

Semiconductor Materials

Semiconductor materials are nominally small band gap insulators. The defining property of a semiconductor material is that it can be doped with impurities that alter its electronic properties in a controllable way.

Because of their application in the computer and photovoltaic industry—in devices such as transistors, lasers, and solar cells—the search for new semiconductor materials and the improvement of existing materials is an important field of study in materials science.

Most commonly used semiconductor materials are crystalline inorganic solids. These materials are classified according to the periodic table groups of their constituent atoms.

Different semiconductor materials differ in their properties. Thus, in comparison with silicon, compound semiconductors have both advantages and disadvantages. For example, gallium arsenide (GaAs) has six times higher electron mobility than silicon, which allows faster operation; wider band gap, which allows operation of power devices at higher temperatures, and gives lower thermal noise to low power devices at room temperature; its direct band gap gives it more favorable optoelectronic properties than the indirect band gap of silicon; it can be alloyed to ternary and quaternary compositions, with adjustable band gap width, allowing light emission at chosen wavelengths, and allowing e.g. matching to wavelengths with lowest losses in optical fibers. GaAs can be also grown in a semi-insulating form, which is suitable as a lattice-matching insulating substrate for GaAs devices. Conversely, silicon is robust, cheap, and easy to process, whereas GaAs is brittle and expensive, and insulation layers can not be created by just growing an oxide layer; GaAs is therefore used only where silicon is not sufficient.

By alloying multiple compounds, some semiconductor materials are tunable, e.g., in band gap or lattice constant. The result is ternary, quaternary, or even quinary compositions. Ternary compositions allow adjusting the band gap within the range of the involved binary compounds; however, in case of combination of direct and indirect band gap materials there is a ratio where indirect band gap prevails, limiting the range usable for optoelectronics; e.g. AlGaAs LEDs are limited to 660 nm by this. Lattice constants of the compounds also tend to be different, and the lattice mismatch against the substrate, dependent on the mixing ratio, causes defects in amounts dependent on the mismatch magnitude; this influences the ratio of achievable radiative/nonradiative recombinations and determines the luminous efficiency of the device. Quaternary and higher compositions allow adjusting simultaneously the band gap and the lattice constant, allowing increasing radiant efficiency at wider range of wavelengths; for example AlGaInP is used for LEDs. Materials transparent to the generated wavelength of light are advantageous, as this allows more efficient extraction of photons from the bulk of the material. That is, in such transparent materials, light production is not limited to just the surface. Index of refraction is also composition-dependent and influences the extraction efficiency of photons from the material.

Solid-state materials are commonly grouped into three classes: insulators, semiconductors, and conductors. (At low temperatures some conductors, semiconductors, and insulators may become superconductors.) The figure shows the conductivities σ (and the corresponding resistivities $\rho = 1/\sigma$) that are associated with some important materials in each of the three classes. Insulators, such as fused quartz and glass, have very low conductivities, on the order of 10^{-18} to 10^{-10} siemens per centimetre; and conductors, such as aluminum, have high conductivities, typically from 10^4 to 10^6 siemens per centimetre. The conductivities of semiconductors are between these extremes and are generally sensitive to temperature, illumination, magnetic fields, and minute amounts of impurity atoms. For example, the addition of about 10 atoms of boron (known as a dopant) per million atoms of silicon can increase its electrical conductivity a thousandfold (partially accounting for the wide variability shown in the preceding figure).

The study of semiconductor materials began in the early 19th century. The elemental semiconductors are those composed of single species of atoms, such as silicon (Si), germanium (Ge), and tin (Sn) in column IV and selenium (Se) and tellurium (Te) in column VI of the periodic table. There are, however, numerous compound semiconductors, which are composed of two or more elements. Gallium arsenide (GaAs), for example, is a binary III-V compound, which is a combination of gallium (Ga) from column III and arsenic (As) from column V. Ternary compounds can be formed by elements from three different columns—for instance, mercury indium telluride ($HgIn_2Te_4$), a II-III-VI compound. They also can be formed by elements from two columns, such as aluminum gallium arsenide ($Al_xGa_{1-x}As$), which is a ternary III-V compound, where both Al and Ga are from column III and the subscript x is related to the composition of the two elements from 100 percent Al ($x = 1$) to 100 percent Ga ($x = 0$). Pure silicon is the most important material for integrated circuit applications, and III-V binary and ternary compounds are most significant for light emission.

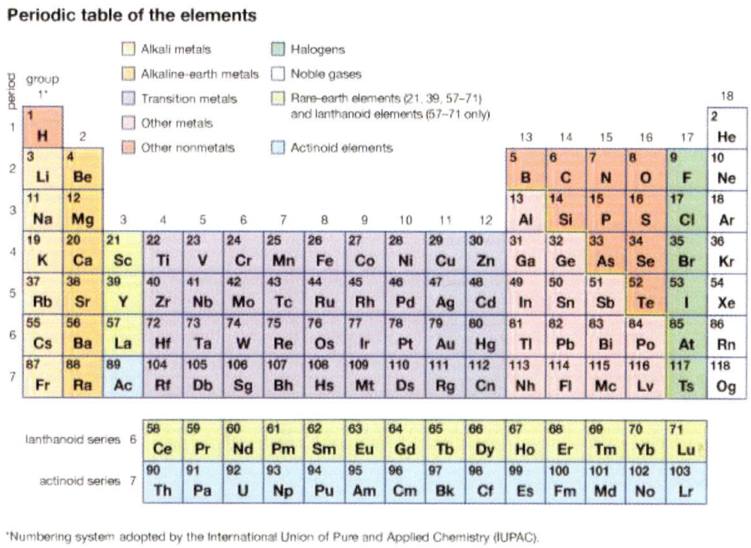

periodic table : Modern version of the periodic table of the elements

Prior to the invention of the bipolar transistor in 1947, semiconductors were used only as two-terminal devices, such as rectifiers and photodiodes. During the early 1950s germanium was the major semiconductor material. However, it proved unsuitable for many applications, because devices made of the material exhibited high leakage currents at only moderately elevated temperatures. Since the early 1960s silicon has become by far the most widely used semiconductor, virtually supplanting germanium as a material for device fabrication. The main reasons for this are twofold: (1) silicon devices exhibit much lower leakage currents, and (2) silicon dioxide (SiO_2), which is a high-quality insulator, is easy to incorporate as part of a silicon-based device. Thus, silicon technology has become very advanced and pervasive, with silicon devices constituting more than 95 percent of all semiconductor products sold worldwide.

Many of the compound semiconductors have some specific electrical and optical properties that are superior to their counterparts in silicon. These semiconductors, especially gallium arsenide, are used mainly for optoelectronic and certain radio frequency (RF) applications.

Elemental Semiconductors

Single-element semiconductors are silicon (Si), germanium (Ge), arsenic (As), astatine (At), boron (B), polonium (Po), tellurium (Te), and antimony (Sb). Out of the elemental semiconductors Si and Ge (column IV of periodic table) are the most common and are composed of single species of atoms.

Germanium and SiGe Alloys

Silicon and germanium have similar proprieties such as belonging to the same group of the periodic table and being both indirect band gap semiconductors. Germanium has high bulk mobility values for electrons and holes, being this values two and four times higher than for silicon. These higher values imply a higher current with a smaller applied electric field leading to faster switching speeds. Also, the closer values for the mobility give rise to a more symmetrical CMOS. In the table below (reproduced from Dobbie) are stated some proprieties for germanium and silicon.

Property	Silicon	Germanium
Electron bulk mobility (cm^2/Vs)	1450	3900
Hole bulk mobility (cm^2/Vs)	450	1900
Electron effective mass, m^*/m_0	$m_l^* = 0.98$ $m_t^* = 0.19$	$m_l^* = 1.64$ $m_t^* = 0.082$
Hole effective mass, m^*/m_0	$m_{lh}^* = 0.16$ $m_{hh}^* = 0.49$	$m_{lh}^* = 0.044$ $m_{hh}^* = 0.28$
Melting point (°C)	1415	937
Band gap (eV)	1.12	0.66

Germanium has an indirect band gap of 0.66 eV. Its band strucure is plotted below (reproduced from Seeger) aginst the wave number, k. As we can observe the band gap is located in the L minimum od the {111} direction. The conducion band is 8-fold degenerate at this point.

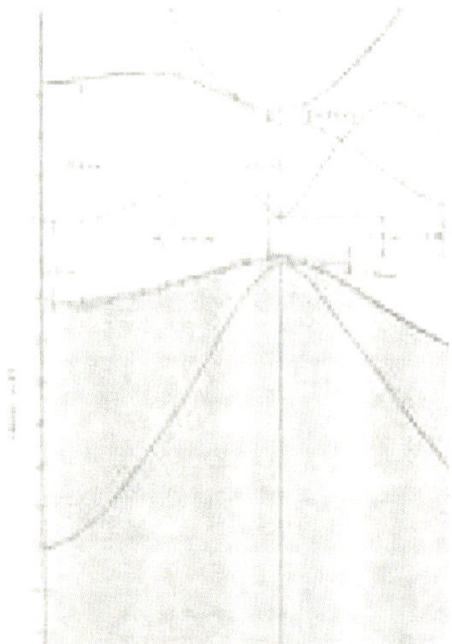

A way to lift the degeneracy is to is to grow germanium in silicon creating the alloy SiGe, usually designated by $Si_{1-x}Ge_x$, this is possible because, as already mentioned, silicon and germanium have the same crystal structure, diamond lattice. Strain in the alloy SiGe has a strong effect in the band structure, as x increases the indirect band gap decreases, also it is possible to observe an linear increase in the lattice parameter with x, as we can see it in the graph below (reproduced from Schaffer):

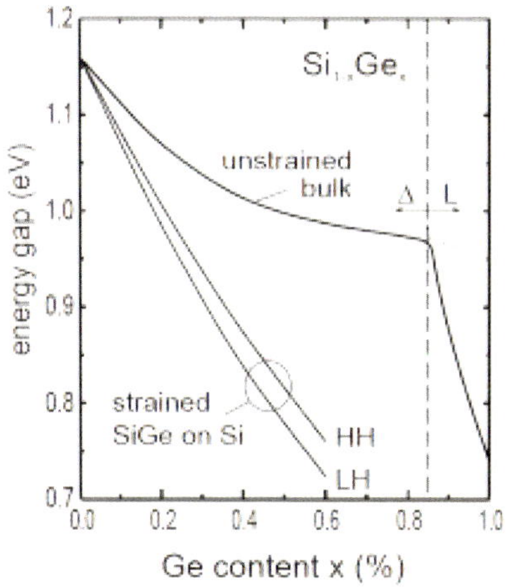

Compound Semiconductors

Compound semiconductors are combinations of atoms of column III and column V and some atoms from column II and VI (combination of two atoms results in binary compounds). Gallium arsenide (GaAs), indium phosphide (InP), gallium nitride (GaN), silicon carbide (SiC) and silicon germanium (SiGe) are also commonly used compound semiconductors. There are also three-element (ternary) compounds (GaAsP) and four-elements (quaternary) compounds such as InGaAsP.

Since electrons in compound semiconductors move much faster than electrons in silicon, the most popular single-element semiconductor, compound semiconductors are ideal for high-speed signal processing. They operate at a lower voltage, respond to light, and generate microwaves.

Owing to their superior light receiving/emitting function, magnetic sensitivity and heat resistance, compound semiconductors are also used in various electronic systems/devices familiar to us. Some examples of such systems/devices include lasers and light-receiving elements for optical fiber communication, power amplifiers for mobile phones and other wireless communication devices, light sources for DVDs and Blu-ray discs, white LEDs for illumination, and solar batteries.

Impressive as the impact of silicon has been on our lives, it has a very basic and limited set of properties that restricts its application in many new and emerging technology areas that demand ultra-high performance levels along with sensing and other capabilities.

By atomically engineering crystal structures that combine elements either side of those in group IV of the periodic table (eg groups III and V), a set of new semiconductor materials has emerged whose enhanced properties offer significant performance improvements over those of silicon alone.

These compound semiconductors enable high speed processing in excess of 100 times that of silicon, as well as an array of other properties including the ability to emit and sense light, all the way from the infrared, through the visible and into the ultra-violet part of the spectrum.

Compound semiconductors have already complimented silicon in areas such as wireless communications, where chips made from material combinations such as gallium and arsenic (gallium arsenide, or GaAs) are found in virtually every smartphone where they enable high speed, high efficiency wireless communications in cellular and WiFi networks.

Other properties offered by compound semiconductor materials include the ability to emit and sense light in the form of general lighting (LEDs) and communications (lasers and receivers for fibre-optics).

The photonic and power efficiency properties offered by compound semiconductors that could not be achieved with silicon alone, will enable technologies essential in areas such as safety and security systems, healthcare technologies, aerospace and automotive applications including electrically powered and autonomous vehicles.

It is our ability to harness the advanced properties of the full range of semiconducting materials that will drive the digital revolution for generations to come.

Difference between Elemental and Compound Semiconductors

Elements are pure chemical substances that have one or one type of atom. Compounds are chemical substances that are formed from one or more elements, held together by chemical bonds.

Elements and Compounds are most commonly heard in chemistry and are considered as basics of the subject. In order to understand chemistry, a person needs to know the difference between elements and compounds.

An element is a pure chemical substance that has one or one type of atom, distinguished by its atomic number. The atomic number is derived from the number of protons present in the element's nucleus. There are a total of 118 elements that have been identified, divided between metal, metalloids and non-metals. Most elements are available on the earth, while a few have been developed artificially through nuclear reactions. An element is already in its rawest form and cannot be further broken down. All of the elements can be found in the Periodic Table, listed by atomic number.

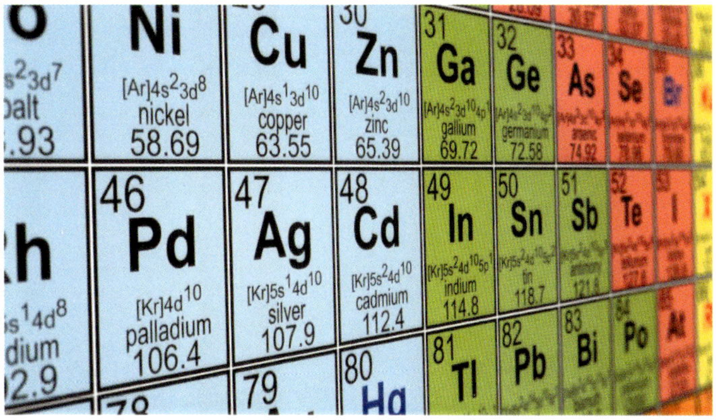

1. Elemental Semiconductors are made from single element also known as mixed element. Compound Semiconductors as the name say are made from compound.

2. Elemental Semiconductors are made up from IV group and VI group elements. Compound Semiconductors are made from III and V group elements or II and VI group elements.

3. Elemental Semiconductors are called as indirect band gap semiconductor in which electron-hole recombination takes place through traps. Compound Semiconductors are called as direct band gap semiconductor in which electron-hole recombination takes place directly.

4. For Elemental semiconductors, Heat is produced in the recombination while for Compound semiconductors, Photons are emitted during recombination.

5. For Elemental semiconductors, Life time of charge carriers is more due to indirect recombination. For Compound semiconductors, Life time of charge carriers is less due to direct recombination.

6. In elemental, current amplification is more as compaired to compond semiconductors.

7. Elemental Semiconductors are used for making diodes, transistor, etc. Compound Semiconductors are used for making LED, laser diodes, etc.

8. Example of Elemental semiconductors are Ge, Si. Example of Compound semiconductors are GaAs, GaP, CdS, MgO.

Doping Semiconductor

In a piece of pure germanium or silicon, no free charge carriers are available at 0^oK. The conductivity of a semiconductor may be increased by adding a controlled amounts of an impurity (e.g. an impurity concentration of one part per million can change a sample of Si from a poor conductor to a good conductor of electric current). This process is called doping.

Semiconductor may be Classified as:

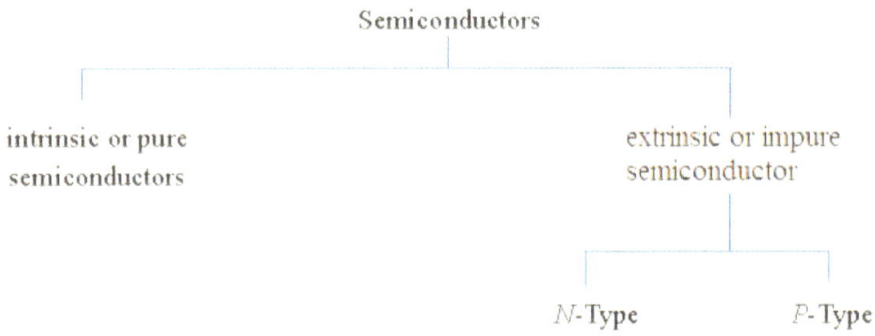

Carrier Concentration

The concentration of dopant affects many electrical properties. Most important is the material's charge carrier concentration. In an intrinsic semiconductor under thermal equilibrium, the concentrations of electrons and holes are equivalent. That is,

$$n = p = n_i.$$

In a non-intrinsic semiconductor under thermal equilibrium, the relation becomes (for low doping):

$$n_0 \cdot p_0 = n_i^2$$

where n_o is the concentration of conducting electrons, p_o is the electron hole concentration, and n_i is the material's intrinsic carrier concentration. The intrinsic carrier concentration varies between materials and is dependent on temperature. Silicon's n_i, for example, is roughly 1.08×10^{10} cm^{-3} at 300 kelvins, about room temperature.

In general, increased doping leads to increased conductivity due to the higher concentration of carriers. Degenerate (very highly doped) semiconductors have conductivity levels comparable to metals and are often used in integrated circuits as a replacement for metal. Often superscript plus and minus symbols are used to denote relative doping concentration in semiconductors. For example, n^+ denotes an n-type semiconductor with a high, often degenerate, doping concentration. Similarly, p^- would indicate a very lightly doped p-type material. Even degenerate levels of doping imply low concentrations of impurities with respect to the base semiconductor. In intrinsic crystalline silicon, there are approximately 5×10^{22} atoms/cm^3. Doping concentration for silicon semiconductors may range anywhere from 10^{13} cm^{-3} to 10^{18} cm^{-3}. Doping concentration above about 10^{18} cm^{-3} is considered degenerate at room temperature. Degenerately doped silicon contains a proportion of impurity to silicon on the order of parts per thousand. This proportion may be reduced to parts per billion in very lightly doped silicon. Typical concentration values fall somewhere in this range and are tailored to produce the desired properties in the device that the semiconductor is intended for.

Effect on Band Structure

Doping a semiconductor in a good crystal introduces allowed energy states within the band gap, but very close to the energy band that corresponds to the dopant type. In other words, electron donor impurities create states near the conduction band while electron acceptor impurities create states near the valence band. The gap between these energy states and the nearest energy band is usually referred to as dopant-site bonding energy or E_B and is relatively small. For example, the E_B for boron in silicon bulk is 0.045 eV, compared with silicon's band gap of about 1.12 eV. Because E_B is so small, room temperature is hot enough to thermally ionize practically all of the dopant atoms and create free charge carriers in the conduction or valence bands.

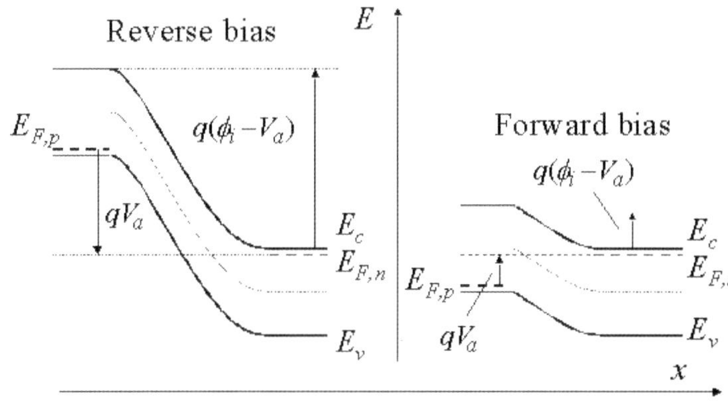

Band diagram of PN junction operation in forward bias mode showing reducing depletion width. Both p and n junctions are doped at a $1\times10^{15}/cm^3$ doping level, leading to built-in potential of ~0.59 V. Reducing depletion width can be inferred from the shrinking charge profile, as fewer dopants are exposed with increasing forward bias.

Dopants also have the important effect of shifting the energy bands relative to the Fermi level. The energy band that corresponds with the dopant with the greatest concentration ends up closer to the Fermi level. Since the Fermi level must remain constant in a system in thermodynamic equilibrium, stacking layers of materials with different properties leads to many useful electrical properties induced by band bending, if the interfaces can be made cleanly enough. For example, the p-n junction's properties are due to the band bending that happens as a result of the necessity to line up the bands in contacting regions of p-type and n-type material. This effect is shown in a band diagram. The band diagram typically indicates the variation in the valence band and conduction band edges versus some spatial dimension, often denoted *x*. The Fermi level is also usually indicated in the diagram. Sometimes the *intrinsic Fermi level*, E_i, which is the Fermi level in the absence of doping, is shown. These diagrams are useful in explaining the operation of many kinds of semiconductor devices.

Relationship to Carrier Concentration (Low Doping)

For low levels of doping, the relevant energy states are populated sparsely by electrons (conduction band) or holes (valence band). It is possible to write simple expressions for the electron and hole carrier concentrations, by ignoring Pauli exclusion (via Maxwell–Boltzmann statistics):

$$n_e = N_C(T)\exp((E_F - E_C)/kT), \quad n_h = N_V(T)\exp((E_V - E_F)/kT),$$

where E_F is the Fermi level, E_C is the minimum energy of the conduction band, and E_V is the maximum energy of the valence band. These are related to the value of the intrinsic concentration via

$$n_i^2 = n_h n_e = N_V(T)N_C(T)\exp((E_V - E_C)/kT),$$

an expression which is independent of the doping level, since $E_C - E_V$ (the band gap) does not change with doping.

The concentration factors $N_C(T)$ and $N_V(T)$ are given by

$$N_C(T) = 2(2\pi m_e^* kT / h^2)^{3/2} \quad N_V(T) = 2(2\pi m_h^* kT / h^2)^{3/2}.$$

where m_e^* and m_h^* are the density of states effective masses of electrons and holes, respectively, quantities that are roughly constant over temperature.

Techniques of Doping and Synthesis

The synthesis of n-type semiconductors may involve the use of vapor-phase epitaxy. In vapor-phase epitaxy, a gas containing the negative dopant is passed over the substrate wafer. In the case of n-type GaAs doping, hydrogen sulfide is passed over the gallium arsenide, and sulfur is incorporated into the structure. This process is characterized by a constant concentration of sulfur on the surface. In the case of semiconductors in general, only a very thin layer of the wafer needs to be doped in order to obtain the desired electronic properties. The reaction conditions typically range from 600 to 800 °C for the n-doping with group VI elements, and the time is typically 6–12 hours depending on the temperature.

Process

Some dopants are added as the (usually silicon) boule is grown, giving each wafer an almost uniform initial doping. To define circuit elements, selected areas — typically controlled by photolithography — are further doped by such processes as diffusion and ion implantation, the latter method being more popular in large production runs because of increased controllability.

Small numbers of dopant atoms can change the ability of a semiconductor to conduct electricity. When on the order of one dopant atom is added per 100 million atoms, the doping is said to be *low* or *light*. When many more dopant atoms are added, on the order of one per ten thousand atoms, the doping is referred to as *high* or *heavy*. This is often shown as *n+* for n-type doping or *p+* for p-type doping.

Dopant Elements

Group IV Semiconductors

(Note: When discussing periodic table groups, semiconductor physicists always use an older notation, not the current IUPAC group notation. For example, the carbon group is called “Group IV”, not “Group 4”.)

For the Group IV semiconductors such as diamond, silicon, germanium, silicon carbide, and silicon germanium, the most common dopants are acceptors from Group III or donors from Group V elements. Boron, arsenic, phosphorus, and occasionally gallium are used to dope silicon. Boron is the p-type dopant of choice for silicon integrated circuit production because it diffuses at a rate that makes junction depths easily controllable. Phosphorus is typically used for bulk-doping of silicon wafers, while arsenic is used to diffuse junctions, because it diffuses more slowly than phosphorus and is thus more controllable.

By doping pure silicon with Group V elements such as phosphorus, extra valence electrons are added that become unbonded from individual atoms and allow the compound to be an electrically conductive n-type semiconductor. Doping with Group III elements, which are missing the fourth valence electron, creates "broken bonds" (holes) in the silicon lattice that are free to move. The result is an electrically conductive p-type semiconductor. In this context, a Group V element is said to behave as an electron donor, and a group III element as an acceptor. This is a key concept in the physics of a diode.

A very heavily doped semiconductor behaves more like a good conductor (metal) and thus exhibits more linear positive thermal coefficient. Such effect is used for instance in sensistors. Lower dosage of doping is used in other types (NTC or PTC) thermistors.

Silicon Dopants

- Acceptors, p-type
  - Boron is a p-type dopant. Its diffusion rate allows easy control of junction depths. Common in CMOS technology. Can be added by diffusion of diborane gas. The only acceptor with sufficient solubility for efficient emitters in transistors and other applications requiring extremely high dopant concentrations. Boron diffuses about as fast as phosphorus.
  - Aluminium, used for deep p-diffusions. Not popular in VLSI and ULSI. Also a common unintentional impurity.
  - Gallium is a dopant used for long-wavelength infrared photoconduction silicon detectors in the 8–14 μm atmospheric window. Gallium-doped silicon is also promising for solar cells, due to its long minority carrier lifetime with no lifetime degradation; as such it is gaining importance as a replacement of boron doped substrates for solar cell applications.
  - Indium is a dopant used for long-wavelength infrared photoconduction silicon detectors in the 3–5 μm atmospheric window.
- Donors, n-type
  - Phosphorus is a n-type dopant. It diffuses fast, so is usually used for bulk doping, or for well formation. Used in solar cells. Can be added by diffusion of phosphine gas. Bulk doping can be achieved by nuclear transmutation, by irradiation of pure silicon with neutrons in a nuclear reactor. Phosphorus also traps gold atoms, which otherwise quickly diffuse through silicon and act as recombination centers.
  - Arsenic is a n-type dopant. Its slower diffusion allows using it for diffused junctions. Used for buried layers. Has similar atomic radius to silicon, high concentrations can be achieved. Its diffusivity is about a tenth of phosphorus or boron, so is used where the dopant should stay in place during subsequent thermal processing. Useful for shallow diffusions where well-controlled abrupt boundary is desired. Preferred dopant in VLSI circuits. Preferred dopant in low resistivity ranges.
  - Antimony is a n-type dopant. It has a small diffusion coefficient. Used for buried layers. Has diffusivity similar to arsenic, is used as its alternative. Its diffusion is virtually purely substitutional, with no interstitials, so it is free of anomalous effects. For this superior property, it is sometimes used in VLSI instead of arsenic. Heavy doping with antimony is important for power devices. Heavily antimony-doped silicon has lower

concentration of oxygen impurities; minimal autodoping effects make it suitable for epitaxial substrates.

- Bismuth is a promising dopant for long-wavelength infrared photoconduction silicon detectors, a viable n-type alternative to the p-type gallium-doped material.
- Lithium is used for doping silicon for radiation hardened solar cells. The lithium presence anneals defects in the lattice produced by protons and neutrons. Lithium can be introduced to boron-doped p+ silicon, in amounts low enough to maintain the p character of the material, or in large enough amount to counterdope it to low-resistivity n type.

- Other
    - Germanium can be used for band gap engineering. Germanium layer also inhibits diffusion of boron during the annealing steps, allowing ultrashallow p-MOSFET junctions. Germanium bulk doping suppresses large void defects, increases internal gettering, and improves wafer mechanical strength.
    - Silicon, germanium and xenon can be used as ion beams for pre-amorphization of silicon wafer surfaces. Formation of an amorphous layer beneath the surface allows forming ultrashallow junctions for p-MOSFETs.
    - Nitrogen is important for growing defect-free silicon crystal. Improves mechanical strength of the lattice, increases bulk microdefect generation, suppresses vacancy agglomeration.
    - Gold and platinum are used for minority carrier lifetime control. They are used in some infrared detection applications. Gold introduces a donor level 0.35 eV above the valence band and an acceptor level 0.54 eV below the conduction band. Platinum introduces a donor level also at 0.35 eV above the valence band, but its acceptor level is only 0.26 eV below conduction band; as the acceptor level in n-type silicon is shallower, the space charge generation rate is lower and therefore the leakage current is also lower than for gold doping. At high injection levels platinum performs better for lifetime reduction. Reverse recovery of bipolar devices is more dependent on the low-level lifetime, and its reduction is better performed by gold. Gold provides a good tradeoff between forward voltage drop and reverse recovery time for fast switching bipolar devices, where charge stored in base and collector regions must be minimized. Conversely, in many power transistors a long minority carrier lifetime is required to achieve good gain, and the gold/platinum impurities must be kept low.

Other Semiconductors

- Gallium arsenide
    - n-type: tellurium, sulphur (substituting As), tin, silicon, germanium (substituting Ga)
    - p-type: beryllium, zinc, chromium (substituting Ga), silicon, germanium (substituting As)
- Gallium phosphide
    - n-type: tellurium, selenium, sulphur (substituting phosphorus)

  - p-type: zinc, magnesium (substituting Ga), tin (substituting P)
- Gallium nitride, Indium gallium nitride, Aluminium gallium nitride
  - n-type: silicon (substituting Ga), germanium (substituting Ga, better lattice match), carbon (substituting Ga, naturally embedding into MOVPE-grown layers in low concentration)
  - p-type: magnesium (substituting Ga) - challenging due to relatively high ionisation energy above the valence band edge, strong diffusion of interstitial Mg, hydrogen complexes passivating of Mg acceptors and by Mg self-compensation at higher concentrations)
- Cadmium telluride
  - n-type: indium, aluminium (substituting Cd), chlorine (substituting Te)
  - p-type: phosphorus (substituting Te), lithium, sodium (substituting Cd)
- Cadmium sulfide
  - n-type: gallium (substituting Cd), iodine, fluorine (substituting S)
  - p-type: lithium, sodium (substituting Cd)

Compensation

In most cases many types of impurities will be present in the resultant doped semiconductor. If an equal number of donors and acceptors are present in the semiconductor, the extra core electrons provided by the former will be used to satisfy the broken bonds due to the latter, so that doping produces no free carriers of either type. This phenomenon is known as *compensation*, and occurs at the p-n junction in the vast majority of semiconductor devices. Partial compensation, where donors outnumber acceptors or vice versa, allows device makers to repeatedly reverse (invert) the type of a given portion of the material by applying successively higher doses of dopants, so-called counterdoping. Most modern semiconductors are made by successive selective counterdoping steps to create the necessary P and N type areas.

Although compensation can be used to increase or decrease the number of donors or acceptors, the electron and hole mobility is always decreased by compensation because mobility is affected by the sum of the donor and acceptor ions.

Doping in Conductive Polymers

Conductive polymers can be doped by adding chemical reactants to oxidize, or sometimes reduce, the system so that electrons are pushed into the conducting orbitals within the already potentially conducting system. There are two primary methods of doping a conductive polymer, both of which use an oxidation-reduction (i.e., redox) process.

1. Chemical doping involves exposing a polymer such as melanin, typically a thin film, to an oxidant such as iodine or bromine. Alternatively, the polymer can be exposed to a reductant; this method is far less common, and typically involves alkali metals.

2. Electrochemical doping involves suspending a polymer-coated, working electrode in an electrolyte solution in which the polymer is insoluble along with separate counter and reference electrodes. An electric potential difference is created between the electrodes that causes a charge and the appropriate counter ion from the electrolyte to enter the polymer in the form of electron addition (i.e., n-doping) or removal (i.e., p-doping).

N-doping is much less common because the Earth's atmosphere is oxygen-rich, thus creating an oxidizing environment. An electron-rich, n-doped polymer will react immediately with elemental oxygen to *de-dope* (i.e., reoxidize to the neutral state) the polymer. Thus, chemical n-doping must be performed in an environment of inert gas (e.g., argon). Electrochemical n-doping is far more common in research, because it is easier to exclude oxygen from a solvent in a sealed flask. However, it is unlikely that n-doped conductive polymers are available commercially.

Doping in Organic Molecular Semiconductors

Molecular dopants are preferred in doping molecular semiconductors due to their compatibilities of processing with the host, that is, similar evaporation temperatures or controllable solubility. Besides, molecular dopants are also benefitted from their relatively large size comparing metal ion dopants (such as Li^+ and Mo^{6+}), which makes them have excellent spatial confinement in order to be used in multilayer structure, such as OLEDs and Organic solar cells. The typical p-type dopants are F4-TCNQ and $Mo(tfd)_3$. However, as the similar problem encountered in doping conductive polymers, air-stable n-dopants suitable for materials with low electron affinity (EA) are still elusive. Recently, photo-activation with the combination of cleavable dimeric dopants, such as $[RuCp^*Mes]_2$, suggest a new path to realize effective n-doping in those low-EA materials.

Magnetic Doping

Research on magnetic doping has shown that considerable alteration of certain properties such as specific heat may be affected by small concentrations of an impurity; for example, dopant impurities in semiconducting ferromagnetic alloys can generate different properties as first predicted by White, Hogan, Suhl and Nakamura. The inclusion of dopant elements to impart dilute magnetism is of growing significance in the field of Magnetic semiconductors. The presence of disperse ferromagnetic species is key to the functionality of emerging Spintronics, a class of systems that utilise electron spin in addition to charge. Using Density functional theory(DFT) the temperature dependent magnetic behaviour of dopants within a given lattice can be modeled to identify candidate semiconductor systems.

Single Dopants in Semiconductors

The sensitive dependence of a semiconductor's properties on dopants has provided an extensive range of tunable phenomena to explore and apply to devices. It is possible to identify the effects of a solitary dopant on commercial device performance as well as on the fundamental properties of a semiconductor material. New applications have become available that require the discrete character of a single dopant, such as single-spin devices in the area of quantum information or single-dopant transistors. Dramatic advances in the past decade towards observing, controllably creating and manipulating single dopants, as well as their application in novel devices have allowed opening the new field of solotronics (solitary dopant optoelectronics).

Neutron Transmutation Doping

Neutron transmutation doping (NTD) is an unusual doping method for special applications. Most commonly, it is used to dope silicon n-type in high-power electronics. It is based on the conversion of the Si-30 isotope into phosphorus atom by neutron absorption as follows:

$$^{30}\mathrm{Si}(n,\gamma)^{31}\mathrm{Si} \rightarrow {}^{31}\mathrm{P} + \beta^- \ (\mathrm{T}_{1/2} = 2.62h).$$

In practice, the silicon is typically placed near a nuclear reactor to receive the neutrons. As neutrons continue to pass through the silicon, more and more phosphorus atoms are produced by transmutation, and therefore the doping becomes more and more strongly n-type. NTD is a far less common doping method than diffusion or ion implantation, but it has the advantage of creating an extremely uniform dopant distribution.

Modulation Doping

Modulation doping is a synthesis technique in which the dopants are spatially separated from the carriers. In this way, carrier-donor scattering is suppressed, allowing very high mobility to be attained.

Intrinsic Semiconductor

An intrinsic (pure) semiconductor, also called an undoped semiconductor or i-type semiconductor, is a pure semiconductor without any significant dopant species present. The number of charge carriers is therefore determined by the properties of the material itself instead of the amount of impurities. In intrinsic semiconductors the number of excited electrons and the number of holes are equal: n = p. This may even be the case after doping the semiconductor, though only if it is doped with both donors and acceptors equally. In this case, n = p still holds, and the semiconductor remains intrinsic, though doped.

The electrical conductivity of intrinsic semiconductors can be due to crystallographic defects or electron excitation. In an intrinsic semiconductor the number of electrons in the conduction band is equal to the number of holes in the valence band. An example is $Hg_{0.8}Cd_{0.2}Te$ at room temperature.

An indirect band gap intrinsic semiconductor is one in which the maximum energy of the valence band occurs at a different k (k-space wave vector) than the minimum energy of the conduction band. Examples include silicon and germanium. A direct band gap intrinsic semiconductor is one where the maximum energy of the valence band occurs at the same k as the minimum energy of the conduction band. Examples include gallium arsenide.

A silicon crystal is different from an insulator because at any temperature above absolute zero, there is a non-zero probability that an electron in the lattice will be knocked loose from its position, leaving behind an electron deficiency called a "hole". If a voltage is applied, then both the electron and the hole can contribute to a small current flow.

The conductivity of a semiconductor can be modeled in terms of the band theory of solids. The band model of a semiconductor suggests that at ordinary temperatures there is a finite possibility that electrons can reach the conduction band and contribute to electrical conduction.

The term intrinsic here distinguishes between the properties of pure "intrinsic" silicon and the dramatically different properties of doped n-type or p-type semiconductors.

Semiconductor Current

The current which will flow in an intrinsic semiconductor consists of both electron and hole current. That is, the electrons which have been freed from their lattice positions into the conduction band can move through the material. In addition, other electrons can hop between lattice positions to fill the vacancies left by the freed electrons. This additional mechanism is called hole conduction because it is as if the holes are migrating across the material in the direction opposite to the free electron movement. The current flow in an intrinsic semiconductor is influenced by the density of energy states which in turn influences the electron density in the conduction band. This current is highly temperature dependent.

Extrinsic Semiconductor

An extrinsic semiconductor is one that has been doped, that is, into which a doping agent has been introduced, giving it different electrical properties than the intrinsic (pure) semiconductor. This doping involves adding dopant atoms to an intrinsic semiconductor, which changes the electron and hole carrier concentrations of the semiconductor at thermal equilibrium, the temperature at which two adjacent substances exchange no heat energy. Dominant carrier concentrations in an extrinsic semiconductor classify it as either an n-type or p-type semiconductor. The electrical properties of extrinsic semiconductors make them essential components of many electronic devices.

Semiconductor Doping

Semiconductor doping is the process that changes an intrinsic semiconductor to an extrinsic semiconductor. During doping, impurity atoms are introduced to an intrinsic semiconductor. Impurity atoms are atoms of a different element than the atoms of the intrinsic semiconductor. Impurity atoms act as either donors or acceptors to the intrinsic semiconductor, changing the electron and hole concentrations of the semiconductor. Impurity atoms are classified as either donor or acceptor atoms based on the effect they have on the intrinsic semiconductor.

Donor impurity atoms have more valence electrons than the atoms they replace in the intrinsic semiconductor lattice. Donor impurities "donate" their extra valence electrons to a semiconductor's conduction band, providing excess electrons to the intrinsic semiconductor. Excess electrons increase the electron carrier concentration (n_0) of the semiconductor, making it n-type.

Acceptor impurity atoms have fewer valence electrons than the atoms they replace in the intrinsic semiconductor lattice. They "accept" electrons from the semiconductor's valence band. This provides excess holes to the intrinsic semiconductor. Excess holes increase the hole carrier concentration (p_0) of the semiconductor, creating a p-type semiconductor.

Semiconductors and dopant atoms are defined by the column of the periodic table in which they fall. The column definition of the semiconductor determines how many valence electrons its atoms have and whether dopant atoms act as the semiconductor's donors or acceptors.

Group IV semiconductors use group V atoms as donors and group III atoms as acceptors.

Group III-V semiconductors, the compound semiconductors, use group VI atoms as donors and group II atoms as acceptors. Group III-V semiconductors can also use group IV atoms as either donors or acceptors. When a group IV atom replaces the group III element in the semiconductor lattice, the group IV atom acts as a donor. Conversely, when a group IV atom replaces the group V element, the group IV atom acts as an acceptor. Group IV atoms can act as both donors and acceptors; therefore, they are known as amphoteric impurities.

	Intrinsic semiconductor	Donor atoms	Acceptor atoms
Group IV semiconductors	Silicon, Germanium	Phosphorus, Arsenic, Antimony	Boron, Aluminium, Gallium
Group III-V semiconductors	Aluminum phosphide, Aluminum arsenide, Gallium arsenide, Gallium nitride	Selenium, Tellurium, Silicon, Germanium	Beryllium, Zinc, Cadmium, Silicon, Germanium

Use of Extrinsic Semiconductors

Extrinsic semiconductors are components of many common electrical devices. A semiconductor diode (devices that allow current in only one direction) consists of p-type and n-type semiconductors placed in junction with one another. Currently, most semiconductor diodes use doped silicon or germanium.

Transistors (devices that enable current switching) also make use of extrinsic semiconductors. Bipolar junction transistors (BJT), which amplify current, are one type of transistor. The most common BJTs are NPN and PNP type. NPN transistors have two layers of n-type semiconductors sandwiching a p-type semiconductor. PNP transistors have two layers of p-type semiconductors sandwiching an n-type semiconductor.

Field-effect transistors (FET) are another type of transistor which amplify current implementing extrinsic semiconductors. As opposed to BJTs, they are called unipolar because they involve single carrier type operation – either N-channel or P-channel. FETs are broken into two families, junction gate FET (JFET), which are three terminal semiconductors, and insulated gate FET (IGFET), which are four terminal semiconductors.

Other devices implementing the extrinsic semiconductor:

- Lasers
- Solar cells
- Photodetectors
- Light-emitting diodes
- Thyristors

Intrinsic semiconductors, in which some suitable impurity or doping has been added in extremely small amounts (about 1 part in 108) are called extrinsic semiconductors. By doping, a semiconductor can be altered to have a predominance of either electrons or holes.

Depending on the type of doping material used, extrinsic semiconductors can be sub-divided into two classes:

i. N-type semiconductors and

ii. P-type semiconductors.

There are two types of doped semiconductors, n-type (mostly electrons) and p-type (mostly holes). When a crystal is doped such that the equilibrium carrier concentrations no and po are different from the intrinsic carrier concentration ni, the material is said to be extrinsic.

N-type Extrinsic Semiconductor

By adding a pentavalent material (element from fifth group of periodic table) like antimony (Sb) to pure semiconductor e.g Si, n type semiconductor is obtained. Each antimony atom forms covalent bonds with the surrounding four Si atoms with the help of four of its five electrons. The fifth electron is extra and is loosely bound to the antimony atom. This extra electron can be easily excited from the valence band to the conduction band by the application of electric field or increase in thermal energy. In a N-type semiconductors, electrons are the majority carriers while holes constitute the minority carriers.

When pentavalent impurity is added to an intrinsic or pure semiconductor (silicon or germanium), then it is said to be an n-type semiconductor. Pentavalent impurities such as phosphorus, arsenic, antimony etc are called donor impurity.

Let us consider, pentavalent impurity phosphorus is added to silicon as shown in below figure. Phosphorus atom has 5 valence electrons and silicon has 4 valence electrons. Phosphorus atom has one excess valence electron than silicon. The four valence electrons of each phosphorus atom form 4 covalent bonds with the 4 neighboring silicon atoms. The fifth valence electron of the phosphorus atom cannot able to form the covalent bond with the silicon atom because silicon atom does not have the fifth valence electron to form the covalent bond.

Thus, fifth valence electron of phosphorus atom does not involve in the formation of covalent bonds. Hence, it is free to move and not attached to the parent atom.

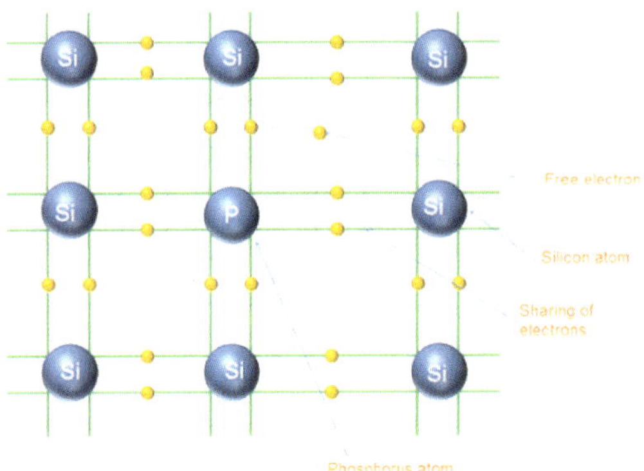

This shows that each phosphorus atom donates one free electron. Therefore, all the pentavalent impurities are called donors. The number of free electrons are depends on the amount of impurity (phosphorus) added to the silicon. A small addition of impurity (phosphorus) generates millions of free electrons.

Charge on n-type Semicondctor

So many people think that n-type semiconductor has large number of free electrons. So, the total electric charge of n-type semiconductor is negative. But this assumption is wrong. Even though n-type semiconductor has large number of free electrons, but these free electrons is given by the pentavalent atoms that are electrically neutral. Therefore, the total electric charge of n-type semiconductor is also neutral.

Conduction in n-type Semiconductor

Let us consider an n-type semiconductor as shown in below figure. When voltage is applied to n-type semiconductor; the free electrons moves towards positive terminal of applied voltage. Similarly holes moves towards negative terminal of applied voltage.

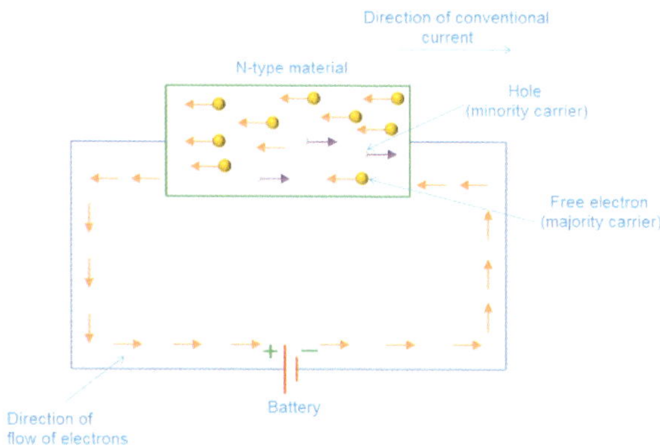

In n-type semiconductor, the population of free electrons is more whereas the population of holes is less. Hence in n-type semiconductor free electrons are called majority carriers and holes are called minority carriers. Therefore, in a n-type semiconductor conduction is mainly because of motion of free electrons.

P-type Extrinsic Semiconductor

When a trivalent like boron (B) is added to a pure semiconductor, p type semiconductor is formed. Three valence electrons of boron atom form covalent bonds with four surrounding germanium atoms but one bond is left incomplete and gives rise to a hole. Thus, boron which is called an acceptor impurity causes as many positive holes in a germanium crystal as there are boron atoms thereby producing a P-type (P for positive) extrinsic semiconductor. In this type of semiconductor, conduction is by the movement of holes in the valence band.

When the trivalent impurity is added to an intrinsic or pure semiconductor (silicon or germanium), then it is said to be an p-type semiconductor. Trivalent impurities such as Boron (B), Gallium (G), Indium(In), Aluminium(Al) etc are called acceptor impurity.

Let us consider, trivalent impurity boron is added to silicon as shown in below figure. Boron atom has three valence electrons and silicon has four valence electrons. The three valence electrons of each boron atom form 3 covalent bonds with the 3 neighboring silicon atoms.

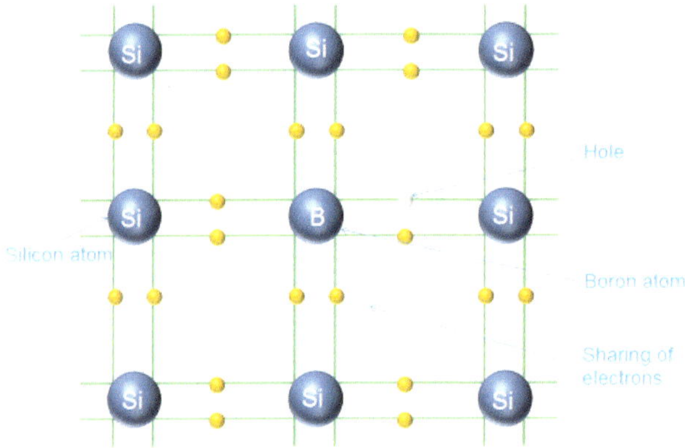

In the fourth covalent bond, only silicon atom contributes one valence electron, while the boron atom has no valence electron to contribute. Thus, the fourth covalent bond is incomplete with shortage of one electron. This missing electron is called hole.

This shows each boron atom accept one electron to fill the hole. Therefore, all the trivalent impurities are called acceptors. A small addition of impurity (boron) provides millions of holes.

Charge on p-type Semiconductor

So many people think that p-type semiconductor has large number of holes and current conduction is mainly due to these holes. So, the total electric charge of p-type semiconductor is positive. But this assumption is wrong. Even though p-type semiconductor has large number of holes, but these holes is provided by the trivalent atoms that are electrically neutral. Therefore, the total electric charge of p-type semiconductor is also neutral.

Conduction in p-type Semiconductor

Let us consider a p-type semiconductor as shown in below figure. When voltage is applied to p-type semiconductor; the holes in valence band moves towards negative terminal of applied voltage. Similarly free electrons move towards positive terminal of applied voltage.

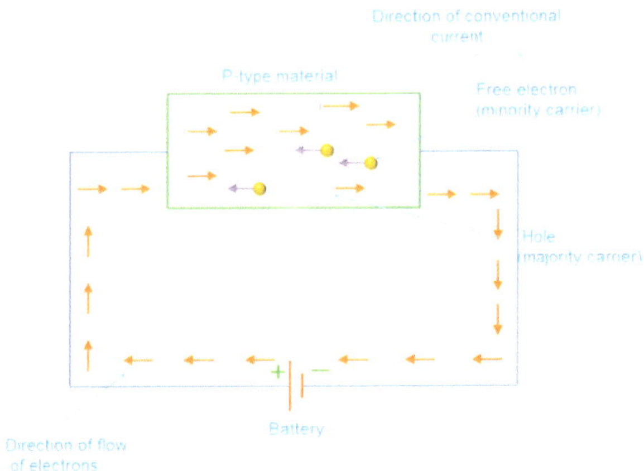

In p-type semiconductor, the population of holes in valence band is more, whereas the population of free electrons in conduction band is less. So, current conduction is mainly because of holes in valence band. Free electrons in conduction band constitute little current. Hence in p-type semiconductor, holes are called majority carriers and free electrons are called minority carriers.

Band Gap Semiconductor

A band gap is the distance between the valence band of electrons and the conduction band. Essentially, the band gap represents the minimum energy that is required to excite an electron up to a state in the conduction band where it can participate in conduction. The lower energy level is the valence band, and thus if a gap exists between this level and the higher energy conduction band, energy must be input for electrons to become free. The size and existence of this band gap allows one to visualize the difference between conductors, semiconductors, and insulators. These distances can be seen in diagrams known as *band diagrams*.

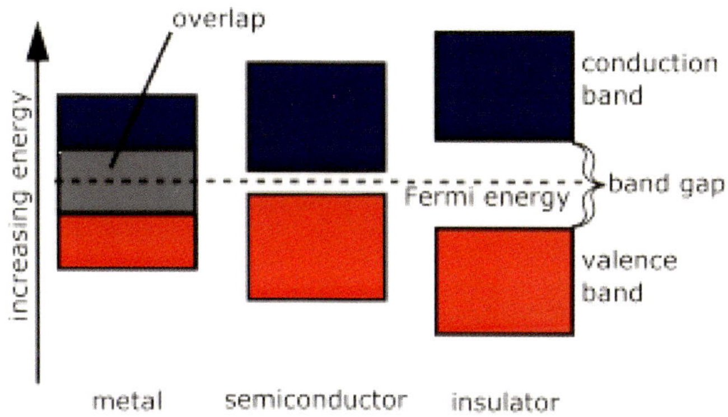

A band gap diagram showing the different sizes of band gaps for conductors, semiconductors, and insulators.

Band Gap Sizes

Figure above illustrates the difference in size of the band gap for insulators, conductors, and semiconductors. The size of this band gap gives the materials some of their distinct properties. In insulators, the electrons in the valence band are separated by a large band gap from the conduction band. This means that there is a large "forbidden" gap in energies preventing electrons from the valence band from jumping up into the conduction band and participating in conduction. This is provides an explanation for why insulators do not conduct electricity well.

In conductors, the valence band overlaps with the conduction band. This overlap causes the valence electrons to be essentially free to move into the conduction band and participate in conduction. Since it is not a full overlap, only a fraction of the valence electrons can move through the material, but this is still enough to make conductors conductive.

In semiconductors, the gap is small enough that it can be bridged by some sort of excitation - perhaps from the Sun in the case of photovoltaic cells. The gap is essentially some size "in-between" that of a conductor or insulator. In this model, a finite number of electrons are able to reach the conduction band and conduct small amounts of electricity. The excitation of this electron also allows additional conduction processes to occur as a result of the electron hole left behind. An electron from an atom close by can occupy this space, creating a chain reaction of holes and electron movement that creates current. A small amount of doping material can drastically increase the conductivity of this material.

Until now, we saw how electrons behave in a crystal and visualized, at least to some extent, the energy band diagrams. However, a complete energy band diagram of a semiconductor is much more complex than what we have seen. For example Si band diagram. Can you identify those symmetry point indications?.

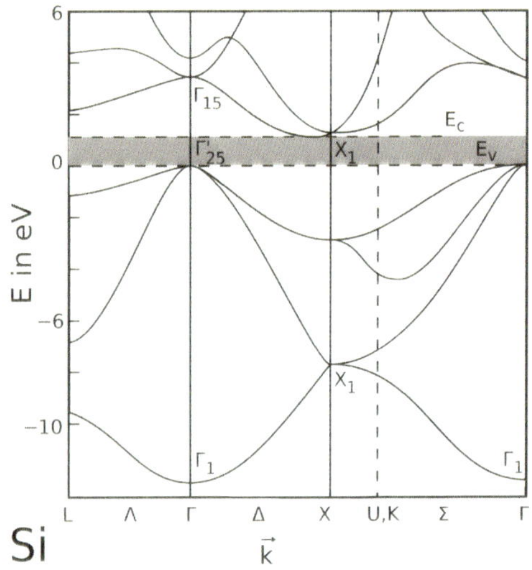

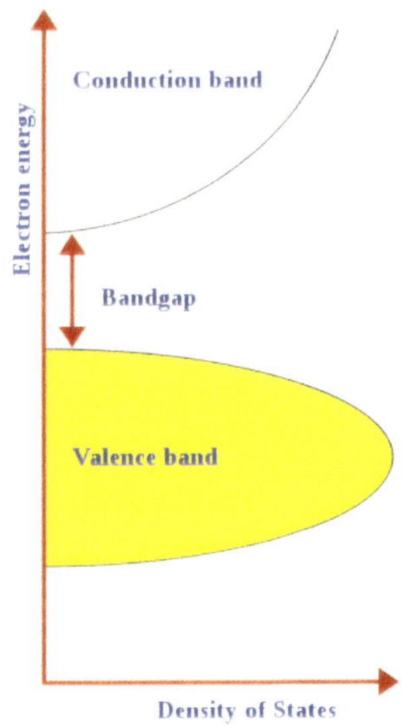

Semiconductor band structure.

Every solid has its own characteristic energy-band structure. This variation in band structure is responsible for the wide range of electrical characteristics observed in various materials. In semiconductors and insulators, electrons are confined to a number of bands of energy, and forbidden from other regions. The term "band gap" refers to the energy difference between the top of the valence band and the bottom of the conduction band. Electrons are able to jump from one band to another. However, in order for an electron to jump from a valence band to a conduction band, it requires a specific minimum amount of energy for the transition. The required energy differs with different materials. Electrons can gain enough energy to jump to the conduction band by absorbing either a phonon (heat) or a photon (light).

A semiconductor is a material with a small but non-zero band gap that behaves as an insulator at absolute zero but allows thermal excitation of electrons into its conduction band at temperatures that are below its melting point. In contrast, a material with a large band gap is an insulator. In conductors, the valence and conduction bands may overlap, so they may not have a band gap.

The conductivity of intrinsic semiconductors is strongly dependent on the band gap. The only available charge carriers for conduction are the electrons that have enough thermal energy to be excited across the band gap and the electron holes that are left off when such an excitation occurs.

Band-gap engineering is the process of controlling or altering the band gap of a material by controlling the composition of certain semiconductor alloys, such as GaAlAs, InGaAs, and InAlAs. It is also possible to construct layered materials with alternating compositions by techniques like molecular-beam epitaxy. These methods are exploited in the design of heterojunction bipolar transistors (HBTs), laser diodes and solar cells.

The distinction between semiconductors and insulators is a matter of convention. One approach is to think of semiconductors as a type of insulator with a narrow band gap. Insulators with a larger band gap, usually greater than 4 eV, are not considered semiconductors and generally do not exhibit semiconductive behaviour under practical conditions. Electron mobility also plays a role in determining a material's informal classification.

The band-gap energy of semiconductors tends to decrease with increasing temperature. When temperature increases, the amplitude of atomic vibrations increase, leading to larger interatomic spacing. The interaction between the lattice phonons and the free electrons and holes will also affect the band gap to a smaller extent. The relationship between band gap energy and temperature can be described by Varshni's empirical expression (named after Y. P. Varshni),

$$E_g(T) = E_g(0) - \frac{\alpha T^2}{T + \beta},$$ where Eg(0), α and β are material constants.

In a regular semiconductor crystal, the band gap is fixed owing to continuous energy states. In a quantum dot crystal, the band gap is size dependent and can be altered to produce a range of energies between the valence band and conduction band. It is also known as quantum confinement effect.

Band gaps also depend on pressure. Band gaps can be either direct or indirect, depending on the electronic band structure.

Photovoltaic Cells

The optical band gap determines what portion of the solar spectrum a photovoltaic cell absorbs. A semiconductor will not absorb photons of energy less than the band gap; and the energy of the electron-hole pair produced by a photon is equal to the bandgap energy. A luminescent solar converter uses a luminescent medium to downconvert photons with energies above the band gap to photon energies closer to the band gap of the semiconductor comprising the solar cell.

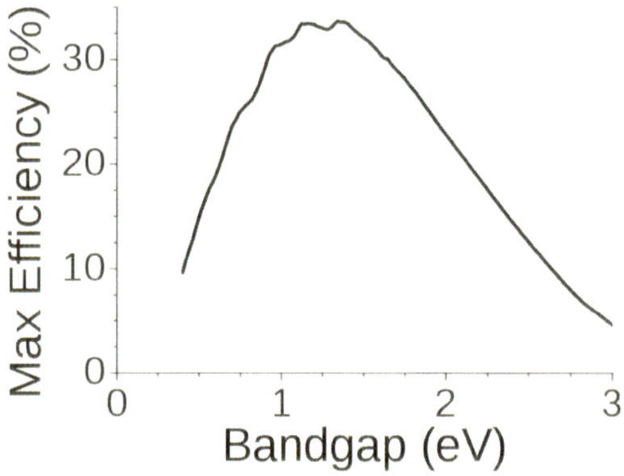

The Shockley–Queisser limit gives the maximum possible efficiency of a single-junction solar cell under un-concentrated sunlight, as a function of the semiconductor band gap. If the band gap is too high, most daylight photons cannot be absorbed; if it is too low, then most photons have much more energy than necessary to excite electrons across the band gap, and the rest is wasted. The semiconductors commonly used in commercial solar cells have band gaps near the peak of this curve, for example silicon (1.1eV) or CdTe (1.5eV). The Shockley–Queisser limit has been exceeded experimentally by combining materials with different band gap energies to make tandem solar cells.

List of Band Gaps

Below are band gap values for some selected materials.

Group	Material	Symbol	Band gap (eV) @ 302K
III-V	Aluminium nitride	AlN	6.0
IV	Diamond	C	5.5
IV	Silicon	Si	1.14
IV	Germanium	Ge	0.67
III–V	Gallium nitride	GaN	3.4
III–V	Gallium phosphide	GaP	2.26
III–V	Gallium arsenide	GaAs	1.43
IV–V	Silicon nitride	Si_3N_4	5
IV–VI	Lead sulfide	PbS	0.37
IV–VI	Silicon dioxide	SiO_2	9
	Copper oxide	Cu_2O	2.1

Optical versus Electronic Bandgap

In materials with a large exciton binding energy, it is possible for a photon to have just barely enough energy to create an exciton (bound electron–hole pair), but not enough energy to separate the electron and hole (which are electrically attracted to each other). In this situation, there is a distinction between "optical bandgap" and "electrical band gap" (or "transport gap"). The optical bandgap is the threshold for photons to be absorbed, while the transport gap is the threshold for creating an electron–hole pair that is *not* bound together. The optical bandgap is at a lower energy than the transport gap.

In almost all inorganic semiconductors, such as silicon, gallium arsenide, etc., there is very little interaction between electrons and holes (very small exciton binding energy), and therefore the optical and electronic bandgap are essentially identical, and the distinction between them is ignored. However, in some systems, including organic semiconductors and single-walled carbon nanotubes, the distinction may be significant.

In Photonics and Phononics

In photonics, band gaps or stop bands are ranges of photon frequencies where, if tunneling effects are neglected, no photons can be transmitted through a material. A material exhibiting this behaviour is known as a photonic crystal. The concept of hyperuniformity has broadened the range of photonic band gap materials, beyond photonic crystals. By applying the technique in supersymmetric quantum mechanics, a new class of optical disordered materials have been suggested, which support band gaps perfectly equivalent to those of crystals or quasicrystals. Similar physics applies to phonons in a phononic crystal.

Wide Band Gap Semiconductors

Room temperature band gap of typical semiconductors and insulators are shown in table below. Materials with band gap less than 3 eV are semiconductors while greater than 3 eV are insulators. ZnO is usually considered a wide band gap semiconductor rather than an insulator.

Material	Band gap *(eV)*
Si	1.11
Ge	0.67
GaAs	1.43
CdS	2.42
ZnO	3.37
Al_2O_3	7.0
SiO_2	9.0

Wide-bandgap semiconductors (WBG or WBGS) are semiconductor materials which have a relatively large band gap compared to typical semiconductors. Typical semiconductors like silicon have a bandgap in the range of 1 - 1.5 electronvolt (eV), whereas wide-bandgap materials have bandgaps in the range of 2 - 4 eV. Generally, wide-bandgap semiconductors have electronic properties which fall in between those of typical semiconductors and insulators.

Compared to circuits built with usual semiconductors, wide-bandgap semiconductors permit devices to operate at much higher voltages, frequencies and temperatures than conventional semiconductor materials like silicon and gallium arsenide. They are the key component used to make green and blue LEDs and lasers, and are also used in certain radio frequency applications, notably military radars. Their inherent qualities make them suitable for a wide range of roles, and they are one of the leading contenders for next-generation devices for general semiconductor use.

The wider bandgap is particularly important for allowing wide bandgap devices to operate at much higher temperatures, on the order of 300 °C. This makes them highly attractive for military applications, where they have seen a fair amount of use. The high temperature tolerance also means that these devices can be operated at much higher power levels under normal conditions. Additionally, most wide bandgap materials also have a much higher critical electrical field density, on the order of ten times that of conventional semiconductors. Combined, these properties allow them to operate at much higher voltages and currents, which makes them highly valuable in military, radio and energy conversion settings. The US Department of Energy believes they will be a foundational technology in new electrical grid and alternative energy devices, as well as the robust and efficient power components used in high energy vehicles from electric trains to plug-in electric vehicles. Most wide-bandgap materials also have high free-electron velocities, which allows them to work at higher switching speeds, which adds to their value in radio applications. A single WBG device can be used to make a complete radio system, eliminating the need for separate signal and radio frequency components, while operating at higher frequencies and power levels.

Currently, WBG materials are far behind the development level of conventional semiconductors, which has seen massive investment since the 1970s. However, their clear inherent advantages in many roles, combined with some unique features of the materials themselves, has led to increasing interest as an everyday material, replacing silicon in even more commonplace roles. Their higher energy density handling is particularly interesting in attempts to extend Moore's law, where conventional technologies appear to be reaching a density plateau.

Use in Devices

Wide bandgap materials have several characteristics that make them useful compared to lower bandgap materials. The higher energy gap gives devices the ability to operate at higher temperatures, and for some applications, allows devices to switch larger voltages. The wide bandgap also brings the electronic transition energy into the range of the energy of visible light, and hence light-emitting devices such as light-emitting diodes (LEDs) and semiconductor lasers can be made that emit in the visible spectrum, or even produce ultraviolet emission.

Solid-state lighting could reduce the amount of energy required to provide lighting as compared with incandescent lights, which are associated with a luminous efficacy of less than 20 lumens per watt. The efficacy of LEDs is on the order of 160 lumens per watt.

Wide bandgap semiconductors can also be used in RF signal processing. Silicon-based power transistors are reaching limits of operating frequency, breakdown voltage, and power density. Wide bandgap materials can be used in high-temperature and power switching applications.

Materials

There are many III–V and II–VI compound semiconductors with high bandgaps. The only high bandgap materials in group IV are diamond and silicon carbide (SiC).

Aluminum nitride (AlN) can be used to fabricate ultraviolet LEDs with wavelengths down to 200–250 nm.

Gallium nitride (GaN) is used to make blue LEDs and lasers.

Boron nitride (BN) is used in cubic boron nitride.

Material Properties

Wide bandgap materials are semiconductors with bandgaps greater than 3 eV.

Bandgap

Quantum mechanics gives rise to a series of distinct electron energy levels, or *bands*, that vary from material to material. Each band can hold a certain number of electrons, if the atom has more electrons then they are forced into higher energy bands. In the presence of external energy some of the electrons will gain energy and move back up the energy bands, before releasing this and falling back down the bands again. With the constant application of external energy, like the thermal energy present at room temperature, an equilibrium is reached where the population of electrons moving up and down the bands is equal.

Depending on the distribution of the energy bands, and the "band gap" between them, the materials will have very different electrical properties. For instance, at room temperature most metals have a series of partially filled bands that allow electrons to be added or removed with little applied energy. When tightly packed together, electrons can easily move from atom to atom, making them excellent conductors. In comparison, most plastic materials have widely spaced energy levels that requires considerable energy to move electrons between their atoms, making them natural insulators. Semiconductors are those materials that have both types of bands, and at normal operational temperatures, some electrons are in both bands.

In semiconductors, adding a small amount of energy pushes more electrons into the *conduction band*, making them more conductive and allowing current to flow like a conductor. Reversing the polarity of this applied energy pushes the electrons into the more widely separated bands, making them insulators and stopping the flow. Since the amount of energy needed to push the electrons between these two levels is very small, semiconductors allow switching with very little energy input. However, this switching process depends on the electrons being naturally distributed between the two states, so small inputs cause the population statistics to change rapidly. As the external temperature changes, due to the Maxwell–Boltzmann distribution more and more electrons will normally find themselves in one state or the other, causing the switching action to occur on its own, or stop entirely.

The size of the atoms and the number of protons in the atom are the primary predictors of the strength and layout of the bandgaps. Materials with small atoms and strong, electronegative atomic bonds are associated with wide bandgaps. Elements high on the periodic table are more likely to be wide bandgap materials. With regard to III-V compounds, nitrides are associated with the largest bandgaps, and, in the II-VI family, oxides are generally considered to be insulators. Bandgaps can often be engineered by alloying, and Vegard's Law states that there is a linear relation between lattice constant and composition of a solid solution at constant temperature. The position of the conduction band minima versus maxima in the band structure determine whether a bandgap is direct or indirect. Most wide bandgap materials are associated with a direct bandgap, with SiC and GaP as exceptions.

Optical Properties

The bandgap determines the wavelength at which LEDs can emit light and the wavelength at which photovoltaics operate most efficiently. Wide-bandgap devices therefore are useful at shorter wavelengths than other semiconductor devices. The bandgap for GaAs of 1.4 eV, for example, corresponds to a wavelength of approximately 890 nm, which is invisible infrared light (the equivalent wavelength for light energy can be determined by dividing the constant 1240 nm-eV by the energy in eV, so 1240 nm-eV/1.4 eV=886 nm). Therefore, GaAs photovoltaics are not ideal for converting shorter-wavelength visible light into electricity. Silicon at 1.1 eV (1100 nm) is even worse. For solar-energy conversion using a single junction photovoltaic cell, the ideal bandgap has been variously estimated from around 1.0 eV up to around 1.5 eV (depending on various assumptions) because that low wavelength threshold covers nearly the entire solar spectrum that reaches the Earth's surface, but a lower-bandgap single-junction cell wastes a large portion of that power by inefficiently converting the shorter-wavelength parts of the solar spectrum. Because of this, a major area in solar energy research is developing multi-junction solar cells that collect separate parts of the spectrum with more efficiency, and wide bandgap photovoltaics are a key component for collecting the part of the spectrum beyond the infrared.

The use of LEDs in lighting applications depended particularly on the development of wide-bandgap nitride semiconductors.

The connection between the wavelength and the bandgap is that the energy of the bandgap is the minimum energy that is needed to excite an electron into the conduction band. In order for an unassisted photon to cause this excitation, it must have at least that much energy. In the opposite process, when excited electron-hole pairs undergo recombination, photons are generated with energies that correspond to the magnitude of the bandgap.

A phonon is required in the process of absorption or emission in the case of an indirect bandgap semiconductor, so indirect bandgap semiconductors are usually very inefficient emitters, although they work reasonably well as absorbers also (as with silicon photovoltaics).

Breakdown Field

Impact ionization is often attributed to be the cause of breakdown. At the point of breakdown, electrons in a semiconductor are associated with sufficient kinetic energy to produce carriers when they collide with lattice atoms.

Wide bandgap semiconductors are associated with a high breakdown voltage. This is due to a larger electric field required to generate carriers through impact mechanism.

At high electric fields, drift velocity saturates due to scattering from optical phonons. A higher optical phonon energy results in fewer optical phonons at a particular temperature, and there are therefore fewer scattering centers, and electrons in wide bandgap semiconductors can achieve high peak velocities.

The drift velocity reaches a peak at an intermediate electric field and undergoes a small drop at higher fields. Intervalley scattering is an additional scattering mechanism at large electric fields, and it is due to a shift of carriers from the lowest valley of the conduction band to the upper valleys, where the lower band curvature raises the effective mass of the electrons and lowers electron mobility. The drop in drift velocity at high electric fields due to intervalley scattering is small in comparison to high saturation velocity that results from low optical phonon scattering. There is therefore an overall higher saturation velocity.

Saturation Velocity

High effective masses of charge carriers are a result of low band curvatures, which correspond to low mobility. Fast response times of devices with wide bandgap semiconductors is due to the high carrier drift velocity at large electric fields, or saturation velocity.

Bandgap Discontinuity

When wide bandgap semiconductors are used in heterojunctions, band discontinuities formed at equilibrium can be a design feature, although the discontinuity can result in complications when creating ohmic contacts.

Polarization

Wurtzite and zincblende structures characterize most wide bandgap semiconductors. Wurtzite phases allow spontaneous polarization in the (0001) direction. A result of the spontaneous polarization and piezoelectricity is that the polar surfaces of the materials are associated with higher sheet carrier density than the bulk. The polar face produces a strong electric field, which creates high interface charge densities.

Thermal Properties

Silicon and other common materials have a bandgap on the order of 1 to 1.5 electronvolt (eV), which implies that such semiconductor devices can be controlled by relatively low voltages. However, it also implies that they are more readily activated by thermal energy, which interferes with their proper operation. This limits silicon based devices to operational temperatures below about 100 °C, beyond which the uncontrolled thermal activation of the devices makes it difficult for them to operate correctly. Wide-bandgap materials typically have bandgaps on the order of 2 to 4 eV, allowing them to operate at much higher temperatures on the order of 300 °C. This makes them highly attractive in military applications, where they have seen a fair amount of use.

Melting temperatures, thermal expansion coefficients, and thermal conductivity can be considered to be secondary properties that are essential in processing, and these properties are related to the bonding in wide bandgap materials. Strong bonds result in higher melting temperatures and lower thermal expansion coefficients. A high Debye temperature results in a high thermal conductivity. With such thermal properties, heat is easily removed.

Applications

High Power Applications

The high breakdown voltage of wide bandgap semiconductors is a useful property in high-power applications that require large electric fields.

Devices for high power and high temperature applications have been developed. Both gallium nitride and silicon carbide are robust materials well suited for such applications. Due to its robustness and ease of manufacture, semiconductors using silicon carbide are expected to be used widely, create simpler and higher efficiency charging for hybrid and all-electric vehicles, reduced energy loss and longer life solar and wind energy power converters, and elimination of bulky grid substation transformers. Cubic boron nitride is used as well. Most of these are for specialist applications in space programmes and military systems. They have not begun to displace silicon from its leading place in the general power semiconductor market.

Light-emitting Diodes

In the near future, white LEDs with the features of more brightness and longer life may replace incandescent bulbs in many situations. The next generation of DVD players (The Blu-ray and HD DVD formats) uses GaN based blue lasers.

Transducers

Large piezoelectric effects allow wide bandgap materials to be used as transducers.

HEMT

Very high speed GaN uses the phenomenon of high interface-charge densities. Due to its cost, aluminum nitride is so far used mostly in military applications.

Direct and Indirect Band Gaps

In semiconductor physics, the band gap of a semiconductor is of two types, a direct band gap or an indirect band gap. The minimal-energy state in the conduction band and the maximal-energy state in the valence band are each characterized by a certain crystal momentum (k-vector) in the Brillouin zone. If the k-vectors are the same, it is called a "direct gap". If they are different, it is called an "indirect gap". The band gap is called "direct" if the crystal momentum of electrons and holes is the same in both the conduction band and the valence band; an electron can directly emit a photon. In an "indirect" gap, a photon cannot be emitted because the electron must pass through an intermediate state and transfer momentum to the crystal lattice.

Examples of direct bandgap material includes some III-V materials such as InAs, GaAs. Indirect bandgap materials include Si, Ge. Some III-V materials are indirect bandgap as well, for example AlSb.

Semiconductors have a filled valence band and an empty conduction band. Electrons from the valence band can be excited to the conduction band by either thermal excitation or by optical absorption. When the electron returns to the valence band the energy is released either as heat or as photons. Based on this, one another way of classifying semiconductors is:

i. Direct band gap semiconductor

ii. Indirect band gap semiconductor

In direct band gap semiconductors the dominant mechanism by which the energy is released (in the absence of defects) is by photons (electromagnetic radiation) while in indirect band gap semiconductors the energy is released by phonons (heat).

Direct band gap semiconductors are useful for opto electronic devices like LEDs, and lasers.

The difference between direct and indirect band gap semiconductors is related to their band structure. In the case of Si or Ge, the valence band maximum and the conduction band minimum are not on top of each other. This makes Si and Ge an indirect band gap semiconductor. In a direct band gap semiconductor, on the other hand, the top of the valence band and bottom of the conduction band lie on top of each other e.g. GaAs, (Fig a & b below)

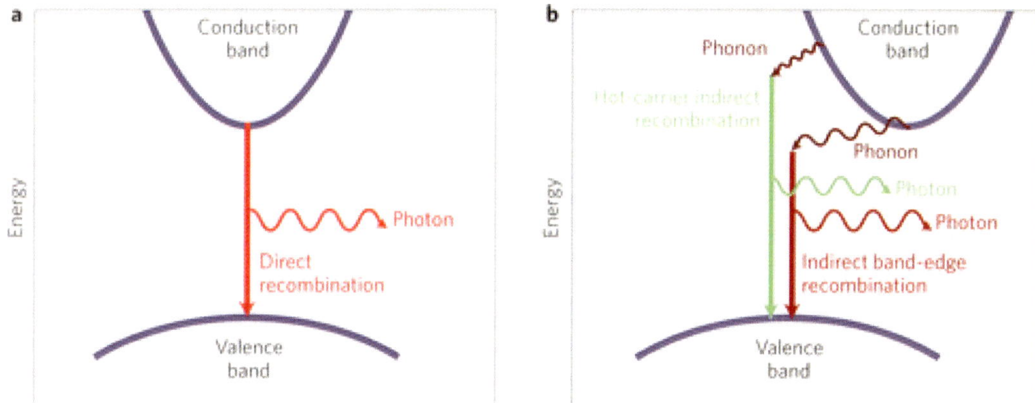

Implications for radiative recombination

Interactions among electrons, holes, phonons, photons, and other particles are required to satisfy conservation of energy and crystal momentum (i.e., conservation of total k-vector). A photon with an energy near a semiconductor band gap has almost zero momentum. One important process is called radiative recombination, where an electron in the conduction band annihilates a hole in the valence band, releasing the excess energy as a photon. This is possible in a direct band gap semiconductor if the electron has a k-vector near the conduction band minima (the hole will share the same k-vector), but not possible in an indirect band gap semiconductor, as photons cannot carry crystal momentum, and thus conservation of crystal momentum would be violated. For radiative recombination to occur in an indirect band gap material, the process must also involve the absorption or emission of a phonon, where the phonon momentum equals the difference between the electron and hole momentum. (It can also, instead, involve a crystallographic defect, which performs essentially the same role.) The involvement of the phonon makes this process much less likely to occur in a given span of time, which is why *radiative recombination is far slower in indirect band gap materials than direct band gap ones*. This is why light-emitting and laser diodes are almost always made of direct band gap materials, and not indirect band gap ones like silicon.

The fact that radiative recombination is slow in indirect band gap materials also means that, under most circumstances, radiative recombinations will be a small proportion of total recombinations, with most recombinations being non-radiative, taking place at point defects or at grain boundaries. However, if the excited electrons are prevented from reaching these recombination places, they have no choice but to eventually fall back into the valence band by radiative recombination. This can be done by creating a dislocation loop in the material. At the edge of the loop, the planes above and beneath the "dislocation disk" are pulled apart, creating a negative pressure, which raises the energy of the conduction band substantially, with the result that the electrons cannot pass this edge. Provided that the area directly above the dislocation loop is defect-free (no non-radiative recombination possible), the electrons will fall back into the valence shell by radiative recombination, thus emitting light. This is the principle on which "DELEDs" (Dislocation Engineered LEDs) are based.

Implications for Light Absorption

The exact reverse of radiative recombination is light absorption. For the same reason as above, light with a photon energy close to the band gap can penetrate much farther before being absorbed in an indirect band gap material than a direct band gap one (at least insofar as the light absorption is due to exciting electrons across the band gap).

This fact is very important for photovoltaics (solar cells). Silicon is the most common solar-cell material, despite the fact that it is indirect-gap and therefore does not absorb light very well. Silicon solar cells are typically hundreds of microns thick; if it was much thinner, much of the light (particularly in the infrared) would simply pass through. By comparison, thin-film solar cells are made of direct band gap materials (such as CdTe, CIGS or CZTS), which absorb the light in a much thinner region, and consequently can be made with a very thin active layer (often less than 1 micron thick).

The absorption spectrum of an indirect band gap material usually depends more on temperature than that of a direct material, because at low temperatures there are fewer phonons, and therefore it is less likely that a photon and phonon can be simultaneously absorbed to create an indirect transition. For example, silicon is opaque to visible light at room temperature, but transparent to red light at liquid helium temperatures, because red photons can only be absorbed in an indirect transition.

Formulae for Absorption

A common and simple method for determining whether a band gap is direct or indirect uses absorption spectroscopy. By plotting certain powers of the absorption coefficient against photon energy, one can normally tell both what value the band gap has, and whether or not it is direct.

For a direct band gap, the absorption coefficient α is related to light frequency according to the following formula:

$$\alpha \approx A^* \sqrt{h_\nu - E_g} \text{, with } A^* = \frac{q^2 x_{vc}^2 (2m_r)^{3/2}}{\lambda_0 \epsilon_0 \hbar^3 n}$$

where:

- α is the absorption coefficient, a function of light frequency
- ν is light frequency
- h is Planck's constant ($h\nu$ is the energy of a photon with frequency ν)
- $\hbar$ is reduced Planck's constant ($\hbar = h/2\pi$)

- E_g is the band gap energy
- A^* is a certain frequency-independent constant, with formula above
- $m_r = \frac{m_h^* m_e^*}{m_h^* + m_e^*}$, where m_e^* and m_h^* are the effective masses of the electron and hole, respectively (m_r is called a "reduced mass")
- q is the elementary charge
- n is the (real) index of refraction
- ϵ_0 is the vacuum permittivity
- x_{vc} is a "matrix element", with units of length and typical value the same order of magnitude as the lattice constant.

This formula is valid only for light with photon energy larger, but not too much larger, than the band gap (more specifically, this formula assumes the bands are approximately parabolic), and ignores all other sources of absorption other than the band-to-band absorption in question, as well as the electrical attraction between the newly created electron and hole. It is also invalid in the case that the direct transition is forbidden, or in the case that many of the valence band states are empty or conduction band states are full.

On the other hand, for an indirect band gap, the formula is:

$$\alpha \propto \frac{(h\nu - E_g + E_p)^2}{\exp(\frac{E_p}{kT}) - 1} + \frac{(h\nu - E_g - E_p)^2}{1 - \exp(-\frac{E_p}{kT})}$$

where:

- E_p is the energy of the phonon that assists in the transition
- k is Boltzmann's constant
- T is the thermodynamic temperature

(This formula involves the same approximations mentioned above.)

Therefore, if a plot of $h\nu$ versus α^2 forms a straight line, it can normally be inferred that there is a direct band gap, measurable by extrapolating the straight line to the $\alpha = 0$ axis. On the other hand, if a plot of $h\nu$ versus $\alpha^{1/2}$ forms a straight line, it can normally be inferred that there is an indirect band gap, measurable by extrapolating the straight line to the $\alpha = 0$ axis (assuming $E_p \approx 0$).

Other Aspects

In some materials with an indirect gap, the value of the gap is negative. The top of the valence band is higher than the bottom of the conduction band in energy. Such materials are known as semimetals.

References

- J. W. Allen (1960). "Gallium Arsenide as a semi-insulator". Nature. 187 (4735): 403–405. Bibcode:1960Natur.187..403A. doi:10.1038/187403b0
- Shockley, William (1950). Electrons and holes in semiconductors : with applications to transistor electronics. R. E. Krieger Pub. Co. ISBN 0-88275-382-7
- Abass, A. K.; Ahmad, N. H. (1986). "Indirect band gap investigation of orthorhombic single crystals of sulfur". Journal of Physics and Chemistry of Solids. 47 (2): 143. Bibcode:1986JPCS...47..143A. doi:10.1016/0022-3697(86)90123-X
- Evans, D A; McGlynn, A G; Towlson, B M; Gunn, M; Jones, D; Jenkins, T E; Winter, R; Poolton, N R J (2008). "Determination of the optical band-gap energy of cubic and hexagonal boron nitride using luminescence excitation spectroscopy". Journal of Physics: Condensed Matter. 20 (7): 075233. Bibcode:2008JPCM...20g5233E. doi:10.1088/0953-8984/20/7/075233
- Middleman, S. (1993). Process Engineering Analysis in Semiconductor Device Fabrication. pp. 29, 330–337. ISBN 0-07-041853-5
- Levy, Roland Albert (1989). Microelectronic Materials and Processes. Dordrecht: Kluwer Academic. pp. 6–7. ISBN 0-7923-0154-4. Retrieved 2008-02-23
- Christopher M. Parry (1981). "Bismuth-Doped Silicon: An Extrinsic Detector for Long-Wavelength Infrared (LWIR) Applications". Bismuth-Doped Silicon: An Extrinsic Detector For Long-Wavelength Infrared (LWIR) Applications. Mosaic Focal Plane Methodologies I. 0244. p. 2. doi:10.1117/12.959299
- Neamen, Donald A. (2003). Semiconductor Physics and Devices: Basic Principles (3rd ed.). McGraw-Hill Higher Education. ISBN 0-07-232107-5
- H. Unlu (1992). "A Thermodynamic Model for Determining Pressure and Temperature Effects on the Bandgap Energies and other Properties of some Semiconductors". Solid State Electronics. 35 (9): 1343–1352. Bibcode:1992SSEle..35.1343U. doi:10.1016/0038-1101(92)90170-H
- Shen, Shyh-Chiang. "Wide-bandgap device research and development at SRL". Georgia Institute of Technology Semiconductor Research Laboratory. Retrieved 2014-09-03

Semiconductor Device and Junction

Semiconductor devices exhibit properties of easy conductance of electric current, variable resistance and sensitivity to heat and light. These are therefore utilized for the design of logic gates and digital circuits. Science and technology have undergone rapid developments in the past decade, which has resulted in the design of innovative semiconductor devices and junctions, in the field of semiconductor technology, which have been extensively detailed in this chapter.

Semiconductor Device

The Semiconductor device is made up of a material that is neither a good conductor nor a good insulator, it is called a semiconductor. Such devices have established wide applications because of their reliability, compactness, and low cost. These are discrete components which are used in power devices, compactness optical sensors, and light emitters, including solid-state lasers. They have a wide range of current and voltage handling capabilities, with current ratings more than 5,000 amperes and voltage ratings more than 100,000 volts. More importantly, semiconductor devices lend themselves to integration into complex but readily build-up microelectronic circuits. They are having probable future, the key elements of the majority of electronic systems including communications with data-processing, consumer, and industrial-control equipment.

Semiconductor devices are nothing but electronic components that exploit the electronic properties of semiconductor materials, like as silicon, germanium, and gallium arsenide, as well as organic semiconductors. Semiconductor devices have replaced vacuum tubes in many applications. They use electronic conduction in the solid state as opposed to the thermionic emission in a high vacuum. Semiconductor devices are manufactured for both discrete devices and integrated circuits, which consist of from a few to billions of devices manufactured and interconnected on a single semiconductor substrate or wafer.

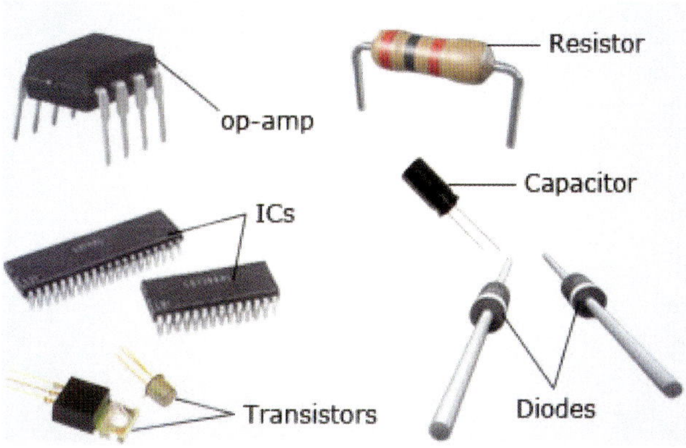

Semiconductor Devices

Semiconductor materials are useful by their behavior which can be easily manipulated by the addition of impurities is known as doping. Semiconductor conductivity can be controlled by the electric or magnetic field, by exposure to light or heat, or by the mechanical deformation of a doped mono crystalline grid; thus, semiconductors can make excellent sensors. Current conduction in a semiconductor occurs free of electrons and holes, collectively known as charge carriers. Doping of silicon is done by adding a small amount of impurity atoms and also for phosphorus or boron, significantly increases the number of electrons or holes within the semiconductor.

When a doped semiconductor contains excess holes it is called "p-type"(positive for holes)semiconductor, and when it contains some excess of free electrons, it is known as "n-type"(negative for electrons) semiconductor, is the sign of charge of the majority mobile charge carriers. The junctions which formed where n-type and p-type semiconductors are joined together is called p–n junction.

Diode

A semiconductor diode is a device typically made up of a single p-n junction. The junction of a p-type and n-type semiconductor forms a depletion region where current conduction is reserved by the lack of mobile charge carriers. When the device is forward biased, this depletion region is reduced, allowing for significant conduction, when the diode is reverse biased, the only less current can be achieved and the depletion region can be extended. Exposing a semiconductor to light can produce electron hole pairs, which increases the number of free carriers and thereby the conductivity. Diodes optimized to take advantage of this phenomenon is known as photodiodes. Compound semiconductor diodes are also being used to generate light, light-emitting diodes and laser diodes.

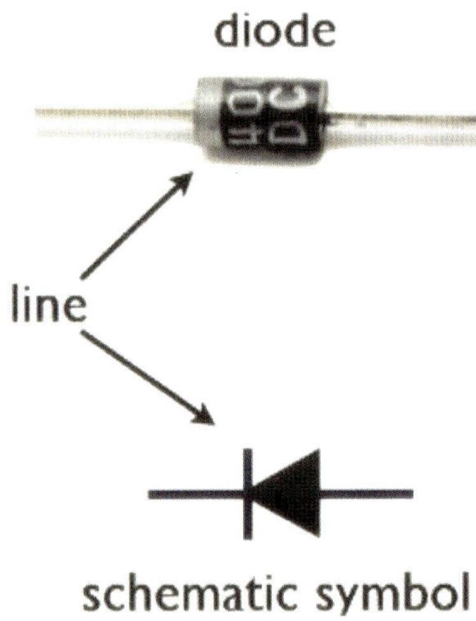

Diode

Transistor

Bipolar junction transistors are formed by two p-n junctions, in either p-n-p or n-p-n configuration. The middle or base, the region between the junctions is typically very narrow. The other regions, and their related terminals, are known as the emitter and collector. A small current injected through the junction between the base and emitter change the properties of the base collector junction so it can be conduct current even though it is reverse biased. This creates a larger current between the collector and emitter, and controlled by the base-emitter current.

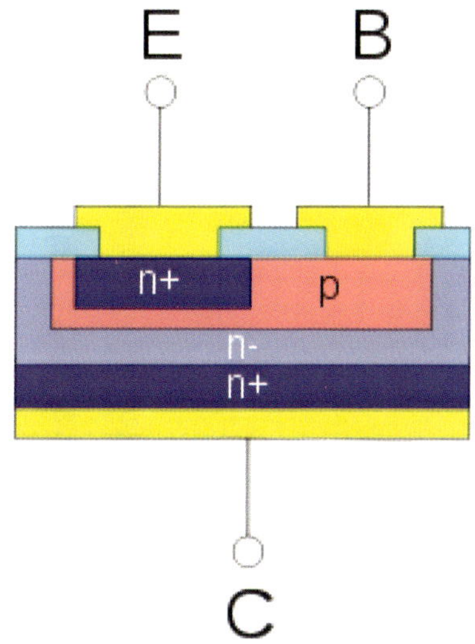

Transistor

Another type of transistor named as field-effect transistor, it operates on the principle that semiconductor conductivity can increased or decreased by the presence of an electric field. An electric field can increase the number of electrons and holes in a semiconductor, thus changing its conductivity. The electric field may be applied by a reverse-biased p-n junction, and it's forms a junction field-effect transistor (JFET) or by an electrode insulated from the bulk material by an oxide layer, and it forms a metal-oxide semiconductor field-effect transistor (MOSFET).

Now a day's most used in the MOSFET, a solid-state device, and semiconductor devices. The gate electrode is charged to produce an electric field that can control the conductivity of a "channel" between two terminals, is called the source and drain. Depending upon the type of carrier in the channel, the device may be n-channel (for electrons) or p-channel (for holes) MOSFET.

Semiconductor Device Materials

The silicon (Si) is most widely used material in semiconductor devices. It's having lower raw material cost and relatively simple process. Its useful temperature range makes it currently the best compromise among the various competing materials. Silicon used in semiconductor device manufacturing is presently fabricated into bowls that are large enough in diameter to allow the manufacture of 300 mm (12 in.) wafers.

Germanium (Ge) was a widely used in early semiconductor material, but its thermal sensitivity makes less useful than silicon. Nowadays, germanium is often alloyed with (Si) silicon for use in very-high-speed SiGe devices; IBM is a main producer of such devices.

Gallium arsenide (GaAs) is also widely used with high-speed devices, but so far, it has been difficult to form large-diameter bowls of this material, limiting the wafer diameter sizes significantly smaller than silicon wafers thus making mass production of Gallium arsenide (GaAs) devices significantly more expensive than silicon.

List of Common Semiconductor Devices

The list of common semiconductor devices mainly includes two terminals, three terminals and four terminal devices.

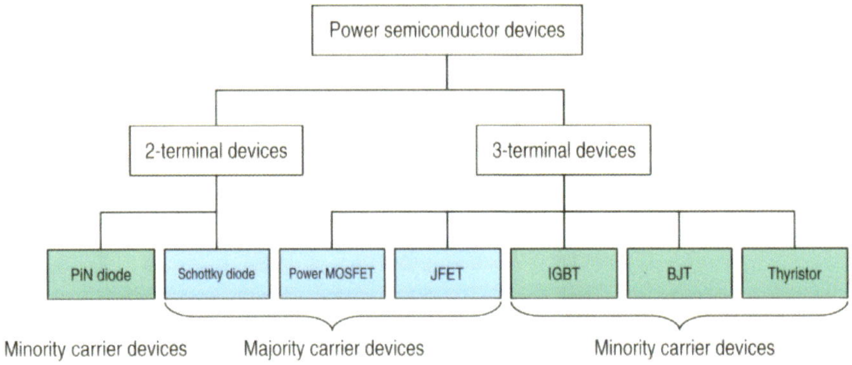

Common Semiconductor Devices

The two-terminal devices are:

- Diode (rectifier diode)
- Gunn diode
- IMPATT diode
- Laser diode
- Zener diode
- Schottky diode
- PIN diode
- Tunnel diode
- Light-emitting diode (LED)
- Photo transistor
- Photocell
- Solar cell
- Transient-voltage-suppression diode
- VCSEL

Three-terminal devices are:

- Bipolar transistor
- Field-effect transistor

- Darlington transistor
- Insulated-gate bipolar transistor (IGBT)
- Unijunction transistor
- Silicon-controlled rectifier
- Thyristor
- TRIAC

Four-terminal devices are:

- Photo coupler (Optocoupler)
- Hall effect sensor (magnetic field sensor)

Semiconductor Device Applications

All types of transistor can be used as the building blocks of logic gates, which is useful to design of digital circuits. In digital circuits like as microprocessors, transistors so which is acting as a switch (on-off); in the MOSFET, for example, the voltage applied to the gate determines whether the switch is on or off.

The transistors are used for analog circuits do not act as switches (on-off); relatively, they respond to a continuous range of input with a continuous range of output. Common analog circuits include oscillators and amplifiers. The circuits that interface or translate between analog circuits and digital circuits are known as the mixed-signal circuits.

Advantages of Semiconductor Devices

- As semiconductor devices have no filaments, hence no power is needed to heat them to cause the emission of electrons.
- Since no heating is required, semiconductor devices are set into operation as soon as the circuit is switched on.
- During operation, semiconductor devices do not produce any humming noise.
- Semiconductor devices require low voltage operation as compared to vacuum tubes.
- Owing to their small sizes, the circuits involving semiconductor devices are very compact.
- Semiconductor devices are shock proof.
- Semiconductor devices are cheaper as compared to vacuum tubes.
- Semiconductor devices have an almost unlimited life.
- As no vacuum has to be created in semiconductor devices, they have no vacuum deterioration trouble.

Disadvantages of Semiconductor Devices

- The noise level is higher in semiconductor devices as compared to that in the vacuum tubes.
- Ordinary semiconductor devices cannot handle as more power as ordinary vacuum tubes can do.
- In high frequency range, they have poor responder.

Thus, this is all about different types of semiconductor devices include two terminals, three terminals and four terminal devices.

P-n Junction

If a block of P-type semiconductor is placed in contact with a block of N-type semiconductor in Figure below(a), the result is of no value. We have two conductive blocks in contact with each other, showing no unique properties. The problem is two separate and distinct crystal bodies. The number of electrons is balanced by the number of protons in both blocks. Thus, neither block has any net charge.

However, a single semiconductor crystal manufactured with P-type material at one end and N-type material at the other in Figure below (b) has some unique properties. The P-type material has positive majority charge carriers, holes, which are free to move about the crystal lattice. The N-type material has mobile negative majority carriers, electrons. Near the junction, the N-type material electrons diffuse across the junction, combining with holes in P-type material. The region of the P-type material near the junction takes on a net negative charge because of the electrons attracted. Since electrons departed the N-type region, it takes on a localized positive charge. The thin layer of the crystal lattice between these charges has been depleted of majority carriers, thus, is known as the *depletion region*. It becomes nonconductive intrinsic semiconductor material. In effect, we have nearly an insulator separating the conductive P and N doped regions.

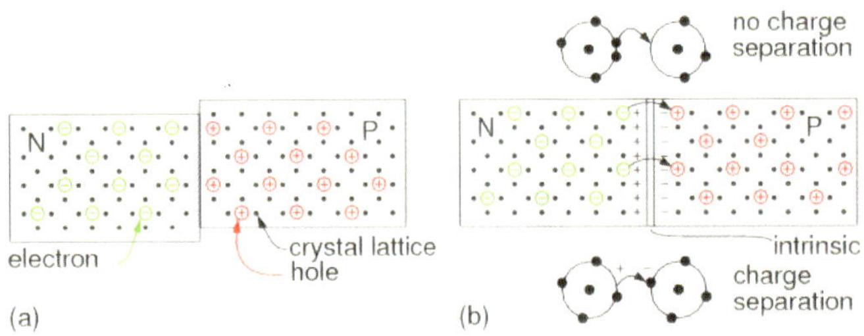

(a) Blocks of P and N semiconductor in contact have no exploitable properties.
(b) Single crystal doped with P and N type impurities develops a potential barrier.

This separation of charges at the PN junction constitutes a potential barrier. This potential barrier must be overcome by an external voltage source to make the junction conduct. The formation of the junction and potential barrier happens during the manufacturing process. The magnitude of the potential barrier is a function of the materials used in manufacturing. Silicon PN junctions have a higher potential barrier than germanium junctions.

In Figure below(a) the battery is arranged so that the negative terminal supplies electrons to the N-type material. These electrons diffuse toward the junction. The positive terminal removes electrons from the P-type semiconductor, creating holes that diffuse toward the junction. If the battery voltage is great enough to overcome the junction potential (0.6V in Si), the N-type electrons and P-holes combine annihilating each other. This frees up space within the lattice for more carriers to flow toward the junction. Thus, currents of N-type and P-type majority carriers flow toward the junction. The recombination at the junction allows a battery current to flow through the PN junction diode. Such a junction is said to be *forward biased*.

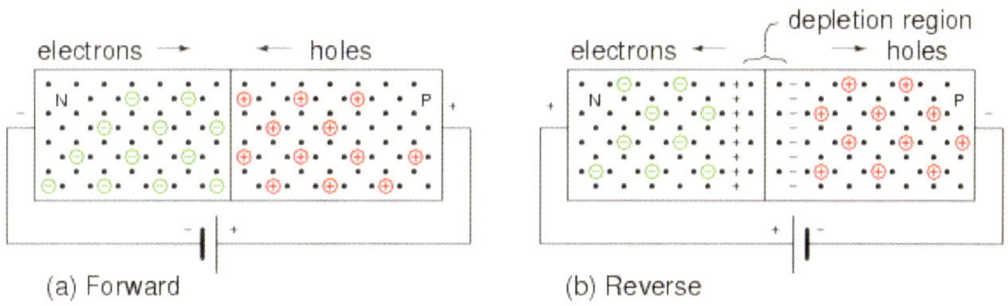

(a) Forward battery bias repels carriers toward junction, where recombination results in battery current. (b) Reverse battery bias attracts carriers toward battery terminals, away from junction. Depletion region thickness increases. No sustained battery current flows.

If the battery polarity is reversed as in Figure above(b) majority carriers are attracted away from the junction toward the battery terminals. The positive battery terminal attracts N-type majority carriers, electrons, away from the junction. The negative terminal attracts P-type majority carriers, holes, away from the junction. This increases the thickness of the nonconducting depletion region. There is no recombination of majority carriers; thus, no conduction. This arrangement of battery polarity is called *reverse bias*.

The diode schematic symbol is illustrated in Figure below(b) corresponding to the doped semiconductor bar at (a). The diode is a *unidirectional* device. Electron current only flows in one direction, against the arrow, corresponding to forward bias. The cathode, bar, of the diode symbol corresponds to N-type semiconductor. The anode, arrow, corresponds to the P-type semiconductor. To remember this relationship, Not-pointing (bar) on the symbol corresponds to N-type semiconductor. Pointing (arrow) corresponds to P-type.

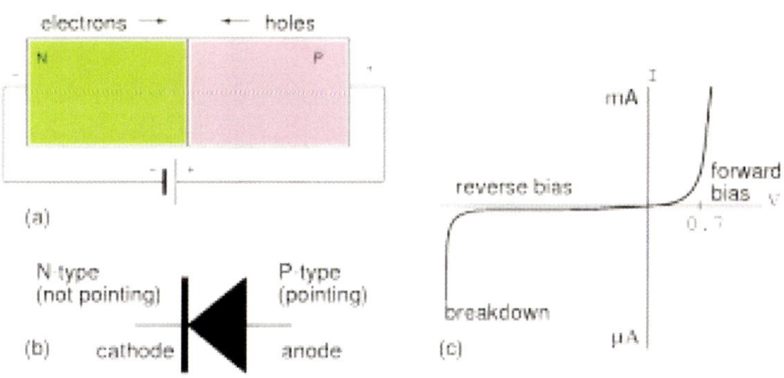

(a) Forward biased PN junction, (b) Corresponding diode schematic symbol (c) Silicon Diode I vs V characteristic curve.

If a diode is forward biased as in Figure above(a), current will increase slightly as voltage is increased from 0 V. In the case of a silicon diode a measurable current flows when the voltage approaches 0.6 V in Figure above(c). As the voltage increases past 0.6 V, current increases considerably after the knee. Increasing the voltage well beyond 0.7 V may result in high enough current to destroy the diode. The forward voltage, V_F, is a characteristic of the semiconductor: 0.6 to 0.7 V for silicon, 0.2 V for germanium, a few volts for Light Emitting Diodes (LED). The forward current ranges from a few mA for point contact diodes to 100 mA for small signal diodes to tens or thousands of amperes for power diodes.

If the diode is reverse biased, only the leakage current of the intrinsic semiconductor flows. This is plotted to the left of the origin in Figure above(c). This current will only be as high as 1 µA for the most extreme conditions for silicon small signal diodes. This current does not increase appreciably with increasing reverse bias until the diode breaks down. At breakdown, the current increases so greatly that the diode will be destroyed unless a high series resistance limits current. We normally select a diode with a higher reverse voltage rating than any applied voltage to prevent this. Silicon diodes are typically available with reverse break down ratings of 50, 100, 200, 400, 800 V and higher. It is possible to fabricate diodes with a lower rating of a few volts for use as voltage standards.

The reverse leakage current of under a µA for silicon diodes is due to conduction of the intrinsic semiconductor. This is the leakage that can be explained by theory. Thermal energy produces few electron hole pairs, which conduct leakage current until recombination. In actual practice this predictable current is only part of the leakage current. Much of the leakage current is due to surface conduction, related to the lack of cleanliness of the semiconductor surface. Both leakage currents increase with increasing temperature, approaching a µA for small silicon diodes.

For germanium, the leakage current is orders of magnitude higher. Since germanium semiconductors are rarely used today, this is not a problem in practice.

Properties

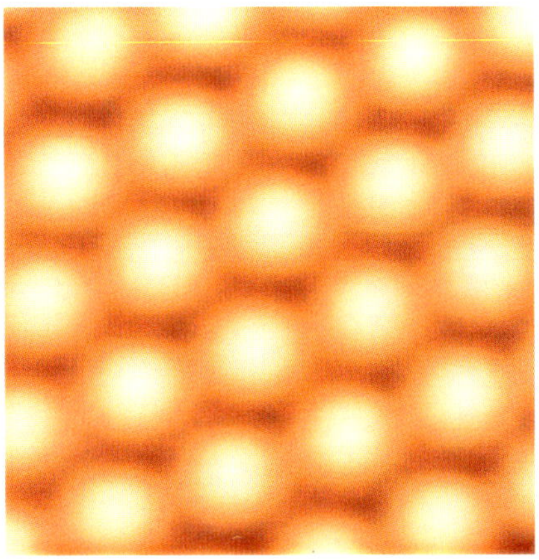

Image silicon atoms (Si) enlarged about 45,000,000x.

The p–n junction possesses essential properties for modern electronics. A p-doped semiconductor is relatively conductive. The same is true of an n-doped semiconductor, but the junction between them can become depleted of charge carriers, and hence non-conductive, depending on the relative voltages of the two semiconductor regions. By manipulating this non-conductive layer, p–n junctions are commonly used as diodes: circuit elements that allow a flow of electricity in one direction but not in the other (opposite) direction. *Bias* is the application of a voltage across a p–n junction; *forward bias* is in the direction of easy current flow, and *reverse bias* is in the direction of little or no current flow.

The forward-bias and the reverse-bias properties of the p–n junction imply that it can be used as a diode. A p–n junction diode allows electric charges to flow in one direction, but not in the opposite direction; negative charges (electrons) can easily flow through the junction from n to p but not from p to n, and the reverse is true for holes. When the p–n junction is forward-biased, electric charge flows freely due to reduced resistance of the p–n junction. When the p–n junction is reverse-biased, however, the junction barrier (and therefore resistance) becomes greater and charge flow is minimal.

Equilibrium (Zero Bias)

In a p–n junction, without an external applied voltage, an equilibrium condition is reached in which a potential difference is formed across the junction. This potential difference is called built-in potential V_{bi}.

After joining p-type and n-type semiconductors, electrons from the n region near the p–n interface tend to diffuse into the p region leaving behind positively charged ions in the n region and being recombined with holes, forming negatively charged ions in the p region. Likewise, holes from the p-type region near the p–n interface begin to diffuse into the n-type region, leaving behind negatively charged ions in the p region and recombining with electrons, forming positive ions in the n region. The regions near the p–n interface lose their neutrality and most of their mobile carriers, forming the space charge region or depletion layer.

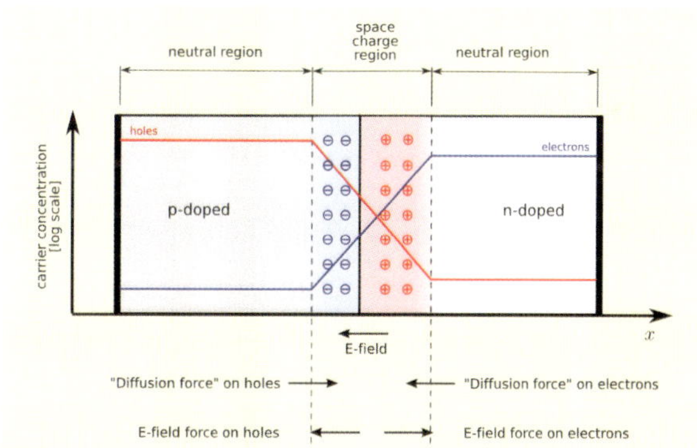

A p–n junction in thermal equilibrium with zero-bias voltage applied. Electron and hole concentration are reported with blue and red lines, respectively.

Gray regions are charge-neutral. Light-red zone is positively charged. Light-blue zone is negatively charged. The electric field is shown on the bottom, the electrostatic force on electrons and holes and the direction in which the diffusion tends to move electrons and holes. (The log concentration curves should actually be smoother with slope varying with field strength.)

The electric field created by the space charge region opposes the diffusion process for both electrons and holes. There are two concurrent phenomena: the diffusion process that tends to generate more space charge, and the electric field generated by the space charge that tends to counteract the diffusion. The carrier concentration profile at equilibrium is shown in figure A with blue and red lines. Also shown are the two counterbalancing phenomena that establish equilibrium.

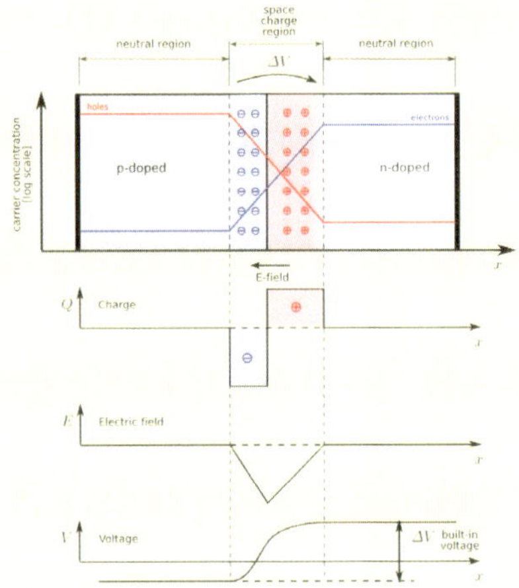

A p–n junction in thermal equilibrium with zero-bias voltage applied. Under the junction, plots for the charge density, the electric field, and the voltage are reported. (The log concentration curves should actually be smoother, like the voltage.)

The space charge region is a zone with a net charge provided by the fixed ions (donors or acceptors) that have been left *uncovered* by majority carrier diffusion. When equilibrium is reached, the charge density is approximated by the displayed step function. In fact, since the y-axis of figure A is log-scale, the region is almost completely depleted of majority carriers (leaving a charge density equal to the net doping level), and the edge between the space charge region and the neutral region is quite sharp. The space charge region has the same magnitude of charge on both sides of the p–n interfaces, thus it extends farther on the less doped side in this example (the n side in figures A and B).

Forward Bias

In forward bias, the p-type is connected with the positive terminal and the n-type is connected with the negative terminal.

With a battery connected this way, the holes in the p-type region and the electrons in the n-type region are pushed toward the junction and start to neutralize the depletion zone, reducing its width. The positive potential applied to the p-type material repels the holes, while the negative potential applied to the n-type material repels the electrons. The change in potential between the p side and the n side decreases or switches sign. With increasing forward-bias voltage, the depletion zone eventually becomes thin enough that the zone's electric field cannot counteract charge carrier motion across the p–n junction, which as a consequence reduces electrical resistance. The electrons that cross the p–n junction into the p-type material (or holes that cross into the n-type material) will diffuse into the nearby neutral region. The amount of minority diffusion in the near-neutral zones determines the amount of current that may flow through the diode.

Only majority carriers (electrons in n-type material or holes in p-type) can flow through a semiconductor for a macroscopic length. With this in mind, consider the flow of electrons across the junction. The forward bias causes a force on the electrons pushing them from the N side toward the P side. With forward bias, the depletion region is narrow enough that electrons can cross the junction and *inject* into the p-type material. However, they do not continue to flow through the p-type material indefinitely, because it is energetically favorable for them to recombine with holes.

The average length an electron travels through the p-type material before recombining is called the *diffusion length*, and it is typically on the order of micrometers.

Although the electrons penetrate only a short distance into the p-type material, the electric current continues uninterrupted, because holes (the majority carriers) begin to flow in the opposite direction. The total current (the sum of the electron and hole currents) is constant in space, because any variation would cause charge buildup over time (this is Kirchhoff's current law). The flow of holes from the p-type region into the n-type region is exactly analogous to the flow of electrons from N to P (electrons and holes swap roles and the signs of all currents and voltages are reversed).

Therefore, the macroscopic picture of the current flow through the diode involves electrons flowing through the n-type region toward the junction, holes flowing through the p-type region in the opposite direction toward the junction, and the two species of carriers constantly recombining in the vicinity of the junction. The electrons and holes travel in opposite directions, but they also have opposite charges, so the overall current is in the same direction on both sides of the diode, as required.

The Shockley diode equation models the forward-bias operational characteristics of a p–n junction outside the avalanche (reverse-biased conducting) region.

Reverse Bias

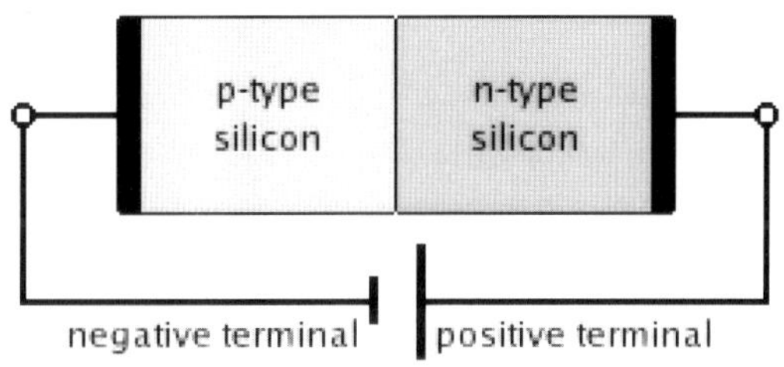

A silicon p–n junction in reverse bias.

Connecting the *p-type* region to the *negative* terminal of the battery and the *n-type* region to the *positive* terminal corresponds to reverse bias. If a diode is reverse-biased, the voltage at the cathode is comparatively higher than at the anode. Therefore, very little current will flow until the diode breaks down. The connections are illustrated in the adjacent diagram.

Because the p-type material is now connected to the negative terminal of the power supply, the 'holes' in the p-type material are pulled away from the junction, leaving behind charged ions and causing the width of the depletion region to increase. Likewise, because the n-type region is connected to the positive terminal, the electrons will also be pulled away from the junction, with similar effect. This increases the voltage barrier causing a high resistance to the flow of charge carriers, thus allowing minimal electric current to cross the p–n junction. The increase in resistance of the p–n junction results in the junction behaving as an insulator.

The strength of the depletion zone electric field increases as the reverse-bias voltage increases. Once the electric field intensity increases beyond a critical level, the p–n junction depletion zone breaks down and current begins to flow, usually by either the Zener or the avalanche breakdown processes. Both of these breakdown processes are non-destructive and are reversible, as long as the amount of current flowing does not reach levels that cause the semiconductor material to overheat and cause thermal damage.

This effect is used to advantage in Zener diode regulator circuits. Zener diodes have a low breakdown voltage. A standard value for breakdown voltage is for instance 5.6 V. This means that the voltage at the cathode cannot be more than about 5.6 V higher than the voltage at the anode (although there is a slight rise with current), because the diode will break down – and therefore conduct – if the voltage gets any higher. This in effect limits the voltage over the diode.

Another application of reverse biasing is Varicap diodes, where the width of the depletion zone (controlled with the reverse bias voltage) changes the capacitance of the diode.

Governing Equations

Size of Depletion Region

For a p–n junction, letting $C_A(x)$ and $C_D(x)$ be the concentrations of acceptor and donor atoms respectively, and letting $N_0(x)$ and $P_0(x)$ be the equilibrium concentrations of electrons and holes respectively, yields, by Poisson's equation:

$$-\frac{\mathrm{d}^2 V}{\mathrm{d}x^2} = \frac{\rho}{\varepsilon} = \frac{q}{\varepsilon}\left[(N_0 - P_0) + (C_D - C_A)\right]$$

where V is the electric potential, ρ is the charge density, ε is permittivity and q is the magnitude of the electron charge. Letting d_p be the width of the depletion region within the p-side, and letting d_n be the width of the depletion region within the n-side, it must be that

$$d_p C_A = d_n C_D$$

because the total charge on either side of the depletion region must cancel out. Therefore, letting D and ΔV represent the entire depletion region and the potential difference across it,

$$\Delta V = \int_D \int \frac{q}{\varepsilon}\left[(N_0 - P_0) + (C_D - C_A)\right]\mathrm{d}x\mathrm{d}x$$

$$= \frac{C_A C_D}{C_A + C_D}\frac{q}{2\varepsilon}(d_p + d_n)^2$$

where $P_0 = N_0 = 0$, because we are in the depletion region. And thus, letting d be the total width of the depletion region, we get

$$d = \sqrt{\frac{2\varepsilon}{q}\frac{C_A + C_D}{C_A C_D}\Delta V}$$

ΔV can be written as $\Delta V_0 + \Delta V_{\text{ext}}$, where we have broken up the voltage difference into the equilibrium plus external components. The equilibrium potential results from diffusion forces, and thus we can calculate ΔV_0 by implementing the Einstein relation and assuming the semiconductor is nondegenerate (i.e. the product P_0N_0 is independent of the Fermi energy):

$$\Delta V_0 = \frac{kT}{q}\ln\left(\frac{C_AC_D}{P_0N_0}\right)$$

where T is the temperature of the semiconductor and k is Boltzmann constant.

Current Across Depletion Region

The *Shockley ideal diode equation* characterizes the current across a p–n junction as a function of external voltage and ambient conditions (temperature, choice of semiconductor, etc.). To see how it can be derived, we must examine the various reasons for current. The convention is that the forward (+) direction be pointed against the diode's built-in potential gradient at equilibrium.

- Forward Current ($\mathbf{J}_F$)
    - ◇ Diffusion Current: current due to local imbalances in carrier concentration $\mathbf{J}_D \propto -q\nabla n$, via the equation
- Reverse Current ($\mathbf{J}_R$)
    - ◇ Field Current
    - ◇ Generation Current

Non-rectifying Junctions

In the above diagrams, contact between the metal wires and the semiconductor material also creates metal–semiconductor junctions called Schottky diodes. In a simplified ideal situation a semiconductor diode would never function, since it would be composed of several diodes connected back-to-front in series. But, in practice, surface impurities within the part of the semiconductor that touches the metal terminals will greatly reduce the width of those depletion layers to such an extent that the metal-semiconductor junctions do not act as diodes. These *non-rectifying junctions* behave as ohmic contacts regardless of applied voltage polarity.

Semiconductor Device Fabrication: Methods and Techniques

Semiconductor device fabrication is the process that is used to build integrated circuits. The core concepts of crystal growth techniques and thin film growth techniques have been elucidated in great detail in this chapter. It also discusses in detail the theories and methodologies related to semiconductor device fabrication and the growth techniques of fabricating a pn junction.

Semiconductor device fabrication is the process used to create chips, the integrated circuits that are present in everyday electrical and electronic devices. It is a multiple-step sequence of photographic and chemical processing steps during which electronic circuits are gradually created on a wafer made of pure semiconducting material. Silicon is the most commonly used semiconductor material today, along with various compound semiconductors.

The entire manufacturing process, from start to packaged chips ready for shipment, takes six to eight weeks and is performed in highly specialized facilities referred to as fabs. In more advanced semiconductor devices, such as modern 14/10/7 nm nodes, fabrication can take up to 15 weeks with 11–13 weeks being the industry average.

Wafers

A typical wafer is made out of extremely pure silicon that is grown into mono-crystalline cylindrical ingots (boules) up to 300 mm (slightly less than 12 inches) in diameter using the Czochralski process. These ingots are then sliced into wafers about 0.75 mm thick and polished to obtain a very regular and flat surface.

Processing

In semiconductor device fabrication, the various processing steps fall into four general categories: deposition, removal, patterning, and modification of electrical properties.

- *Deposition* is any process that grows, coats, or otherwise transfers a material onto the wafer. Available technologies include physical vapor deposition (PVD), chemical vapor deposition (CVD), electrochemical deposition (ECD), molecular beam epitaxy (MBE) and more recently, atomic layer deposition (ALD) among others.

- *Removal* is any process that removes material from the wafer; examples include etch processes (either wet or dry) and chemical-mechanical planarization (CMP).

- *Patterning* is the shaping or altering of deposited materials, and is generally referred to as lithography. For example, in conventional lithography, the wafer is coated with a chemical called a *photoresist*; then, a machine called a *stepper* focuses, aligns, and moves a mask, exposing select portions of the wafer below to short wavelength light; the exposed regions are washed away by a developer solution. After etching or other processing, the remaining photoresist is removed by plasma ashing.
- *Modification of electrical properties* has historically entailed doping transistor *sources* and *drains* (originally by diffusion furnaces and later by ion implantation). These doping processes are followed by furnace annealing or, in advanced devices, by rapid thermal annealing (RTA); annealing serves to activate the implanted dopants. Modification of electrical properties now also extends to the reduction of a material's dielectric constant in low-k insulators via exposure to ultraviolet light in UV processing (UVP). Modification is frequently achieved by oxidation, which can be carried out to create semiconductor-insulator junctions, such as in the local oxidation of silicon (LOCOS) to fabricate Field Effect Transistors.

Modern chips have up to eleven metal levels produced in over 300 sequenced processing steps.

Front-end-of-line (FEOL) Processing

FEOL processing refers to the formation of the transistors directly in the silicon. The raw wafer is engineered by the growth of an ultrapure, virtually defect-free silicon layer through epitaxy. In the most advanced logic devices, *prior* to the silicon epitaxy step, tricks are performed to improve the performance of the transistors to be built. One method involves introducing a *straining step* wherein a silicon variant such as silicon-germanium (SiGe) is deposited. Once the epitaxial silicon is deposited, the crystal lattice becomes stretched somewhat, resulting in improved electronic mobility. Another method, called *silicon on insulator* technology involves the insertion of an insulating layer between the raw silicon wafer and the thin layer of subsequent silicon epitaxy. This method results in the creation of transistors with reduced parasitic effects.

Gate Oxide and Implants

Front-end surface engineering is followed by growth of the gate dielectric (traditionally silicon dioxide), patterning of the gate, patterning of the source and drain regions, and subsequent implantation or diffusion of dopants to obtain the desired complementary electrical properties. In dynamic random-access memory (DRAM) devices, storage capacitors are also fabricated at this time, typically stacked above the access transistor (the now defunct DRAM manufacturer Qimonda implemented these capacitors with trenches etched deep into the silicon surface).

Back-end-of-line (BEOL) Processing

Metal Layers

Once the various semiconductor devices have been created, they must be interconnected to form the desired electrical circuits. This occurs in a series of wafer processing steps collectively referred to as BEOL. BEOL processing involves creating metal interconnecting wires that are isolated by dielectric layers. The insulating material has traditionally been a form of SiO_2 or a silicate glass, but recently new low dielectric constant materials are being used (such as silicon oxycarbide), typically providing dielectric constants around 2.7 (compared to 3.82 for SiO_2), although materials with constants as low as 2.2 are being offered to chipmakers.

Interconnect

Synthetic detail of a standard cell through four layers of planarized copper interconnect, down to the polysilicon (pink), wells (greyish) and substrate (green).

Historically, the metal wires have been composed of aluminum. In this approach to wiring (often called *subtractive aluminum*), blanket films of aluminum are deposited first, patterned, and then etched, leaving isolated wires. Dielectric material is then deposited over the exposed wires. The various metal layers are interconnected by etching holes (called *"vias")* in the insulating material and then depositing tungsten in them with a CVD technique; this approach is still used in the fabrication of many memory chips such as dynamic random-access memory (DRAM), because the number of interconnect levels is small (currently no more than four).

More recently, as the number of interconnect levels for logic has substantially increased due to the large number of transistors that are now interconnected in a modern microprocessor, the timing delay in the wiring has become so significant as to prompt a change in wiring material (from aluminum to copper interconnect layer) and a change in dielectric material (from silicon dioxides to newer low-K insulators). This performance enhancement also comes at a reduced cost via damascene processing, which eliminates processing steps. As the number of interconnect levels increases, planarization of the previous layers is required to ensure a flat surface prior to subsequent lithography. Without it, the levels would become increasingly crooked, extending outside the depth of focus of available lithography, and thus interfering with the ability to pattern. CMP (chemical-mechanical planarization) is the primary processing method to achieve such planarization, although dry *etch back* is still sometimes employed when the number of interconnect levels is no more than three.

Wafer Test

The highly serialized nature of wafer processing has increased the demand for metrology in between the various processing steps. For example, thin film metrology based on ellipsometry or reflectometry is used to tightly control the thickness of gate oxide, as well as the thickness, refractive index and extinction coefficient of photoresist and other coatings. Wafer test metrology equipment is used to verify that the wafers haven't been damaged by previous processing steps up until testing; if too many dies on one wafer have failed, the entire wafer is scrapped to avoid the costs of further processing. Virtual metrology has been used to predict wafer properties based on statistical methods without performing the physical measurement itself.

Device Test

Once the front-end process has been completed, the semiconductor devices are subjected to a variety of electrical tests to determine if they function properly. The proportion of devices on the wafer found to perform properly is referred to as the yield. Manufacturers are typically secretive about their yields, but it can be as low as 30%. Process variation is one among many reasons for low yield.

The fab tests the chips on the wafer with an electronic tester that presses tiny probes against the chip. The machine marks each bad chip with a drop of dye. Currently, electronic dye marking is possible if wafer test data is logged into a central computer database and chips are "binned" (i.e. sorted into virtual bins) according to the predetermined test limits. The resulting binning data can be graphed, or logged, on a wafer map to trace manufacturing defects and mark bad chips. This map can also be used during wafer assembly and packaging.

Chips are also tested again after packaging, as the bond wires may be missing, or analog performance may be altered by the package. This is referred to as the "final test".

Usually, the fab charges for testing time, with prices in the order of cents per second. Testing times vary from a few milliseconds to a couple of seconds, and the test software is optimized for reduced testing time. Multiple chip (multi-site) testing is also possible, because many testers have the resources to perform most or all of the tests in parallel.

Chips are often designed with "testability features" such as scan chains or a "built-in self-test" to speed testing, and reduce testing costs. In certain designs that use specialized analog fab processes, wafers are also laser-trimmed during the testing, in order to achieve tightly-distributed resistance values as specified by the design.

Good designs try to test and statistically manage *corners* (extremes of silicon behavior caused by a high operating temperature combined with the extremes of fab processing steps). Most designs cope with at least 64 corners.

Die Preparation

Once tested, a wafer is typically reduced in thickness before the wafer is scored and then broken into individual dice, a process known as wafer dicing. Only the good, unmarked chips are packaged.

Packaging

Plastic or ceramic packaging involves mounting the die, connecting the die pads to the pins on the package, and sealing the die. Tiny wires are used to connect the pads to the pins. In the old days, wires were attached by hand, but now specialized machines perform the task. Traditionally, these wires have been composed of gold, leading to a lead frame (pronounced "leed frame") of solder-plated copper; lead is poisonous, so lead-free "lead frames" are now mandated by RoHS.

Chip scale package (CSP) is another packaging technology. A plastic dual in-line package, like most packages, is many times larger than the actual die hidden inside, whereas CSP chips are nearly the size of the die; a CSP can be constructed for each die *before* the wafer is diced.

The packaged chips are retested to ensure that they were not damaged during packaging and that the die-to-pin interconnect operation was performed correctly. A laser then etches the chip's name and numbers on the package.

List of Steps

This is a list of processing techniques that are employed numerous times throughout the construction of a modern electronic device; this list does not necessarily imply a specific order.

- Wafer processing
    - Wet cleans
        - Cleaning by solvents such as acetone, trichloroethylene
        - Piranha solution
        - RCA clean
    - Photolithography
    - Ion implantation (in which dopants are embedded in the wafer creating regions of increased (or decreased) conductivity)
    - Dry etching
    - Wet etching
    - Plasma ashing
    - Thermal treatments
        - Rapid thermal anneal
        - Furnace anneals
        - Thermal oxidation
    - Chemical vapor deposition (CVD)
    - Physical vapor deposition (PVD)
    - Molecular beam epitaxy (MBE)
    - Electrochemical deposition (ECD).
    - Chemical-mechanical planarization (CMP)
    - Wafer testing (where the electrical performance is verified)

    - Wafer backgrinding (to reduce the thickness of the wafer so the resulting chip can be put into a thin device like a smartcard or PCMCIA card.)
- Die preparation
    - Wafer mounting
    - Die cutting
- IC packaging
    - Die attachment
    - IC bonding
        - Wire bonding
        - Thermosonic bonding
        - Flip chip
        - Wafer bonding
        - Tape Automated Bonding (TAB)
    - IC encapsulation
        - Baking
        - Plating
        - Lasermarking
        - Trim and form
- IC testing

Hazardous Materials

Many toxic materials are used in the fabrication process. These include:

- poisonous elemental dopants, such as arsenic, antimony, and phosphorus.
- poisonous compounds, such as arsine, phosphine, and silane.
- highly reactive liquids, such as hydrogen peroxide, fuming nitric acid, sulfuric acid, and hydrofluoric acid.

It is vital that workers should not be directly exposed to these dangerous substances. The high degree of automation common in the IC fabrication industry helps to reduce the risks of exposure. Most fabrication facilities employ exhaust management systems, such as wet scrubbers, combustors, heated absorber cartridges, etc., to control the risk to workers and to the environment.

Silicon Dioxide

Front end surface engineering is followed by: growth of the gate dielectric, traditionally silicon dioxide (SiO_2), patterning of the gate, patterning of the source and drain regions, and subsequent implantation or diffusion of dopants to obtain the desired complementary electrical properties. In memory devices, storage cells, conventionally capacitors, are also fabricated at this time, either into the silicon surface or stacked above the transistor.

Once the various semiconductor devices have been created they must be interconnected to form the desired electrical circuits. This "Back End Of Line" involves creating metal interconnecting wires that are isolated by insulating dielectrics. The insulating material was traditionally a form of SiO_2 or a silicate glass, but recently new low dielectric constant materials are being used. These dielectrics presently take the form of SiOC and have dielectric constants around 2.7 (compared to 3.9 for SiO_2), although materials with constants as low as 2.2 are being offered to chipmakers.

Historically, the metal wires consisted of aluminum. In this approach to wiring often called "subtractive aluminum", blanket films of aluminum are deposited first, patterned, and then etched, leaving isolated wires. Dielectric material is then deposited over the exposed wires. The various metal layers are interconnected by etching holes, called "vias," in the insulating material and depositing tungsten in them with a CVD technique. This approach is still used in the fabrication of many memory chips such as dynamic random access memory (DRAM) as the number of interconnect levels is small, currently no more than four.

More recently, as the number of interconnect levels for logic has substantially increased due to the large number of transistors that are now interconnected in a modern microprocessor, the timing delay in the wiring has become significant prompting a change in wiring material from aluminum to copper and from the aforementioned silicon dioxides to newer low-K material. This performance enhancement also comes at a reduced cost via damascene processing that eliminates processing steps. In damascene processing, in contrast to subtractive aluminum technology, the dielectric material is deposited first as a blanket film and is patterned and etched leaving holes or trenches. In "single damascene" processing, copper is then deposited in the holes or trenches surrounded by a thin barrier film resulting in filled vias or wire "lines" respectively. In "dual damascene" technology, both the trench and via are fabricated before the deposition of copper resulting in formation of both the via and line simultaneously, further reducing the number of processing steps. The thin barrier film, called Copper Barrier Seed (CBS), is necessary to prevent copper diffusion into the dielectric. The ideal barrier film is effective, but is barely there. As the presence of excessive barrier film competes with the available copper wire cross section, formation of the thinnest yet continuous barrier represents one of the greatest ongoing challenges in copper processing today.

As the number of interconnect levels increases, planarization of the previous layers is required to ensure a flat surface prior to subsequent lithography. Without it, the levels would become increasingly crooked and extend outside the depth of focus of available lithography, interfering with the ability to pattern. CMP is the primary processing method to achieve such planarization although dry "etch back" is still sometimes employed if the number of interconnect levels is no more than three.

Semiconductor Growth Techniques

There is one reason why semiconductors have become the choice material for the electronic industry is the existence of highly growth technquies.one of the method is Czochralski Method. It is the most important method for the growth of bulk crystals of semiconductors. Czochralski method involves melting the raw materials in a crucible. A technique for achieving high purity includes growing the crystal using Czochralski process. In a Bridgman method a seed crystal is usually used for the contact with the metal like Czochralski Method. A temperature gradient is created along the along the length of the crucible so that the temperature around the seed crystal is below the melting point. As the seed crystal grows, the temperature profile is translated along the crucible by controlling the heaters along the furnace or by slowly moving the ampoule containing the seed crystal within the furnace. In Chemical Vapour Deposition gases containing the required chemical elements are made to react in the vicinity of the substrate.

The semiconductor produced as a result of the reaction is deposited as a thin film on a substrate inside the reactor. In CVD the gases are let into the reactor at relatively high pressure. As a result, the reactor may contain high concentration of contaminants in the form of residual gases. Typically several molecular beams containing the necessary elements for forming the semiconductor and for doping the sample are aimed at the substrate, where the film grows epitaxial. Hence these growth techniques are named as Molecular Beam Epitaxy. This Liquid Phase Epitaxy (LPE) growth technique has been very successful in growing GAAs laser diodes. Usually a group III metal, such as GA or In, is utilized as the solvent for As. When the solvent is cooled in contact with a GaAs substrate it becomes supersaturated with As and nucleation of GaAs starts on the substrate.

- Integrated circuit design and fabrication
- Deposition of Thin films, Photolithography, Etching, Implantation and annealing
- CMOS digital gates
- MOSFET amplifiers

Crystal Growth Techniques

Crystal growth is an interdisciplinary subject covering physics, chemistry, material science, chemical engineering, metallurgy, crystallography, mineralogy etc. In the past few decades, there has been a growing interest on crystal growth processes, particularly in view of the increasing demand of materials for technological applications (Binay Kumar et al., 2007). Natural crystals have often been formed at relatively low temperatures by crystallisation from solutions, sometimes in the course of hundreds and thousands of years. Nowadays, crystals are produced artificially to satisfy the needs of science and technology. Crystal growth is rather an art than a science (Brian R. Pamplin 1980). Many attempts have been made for a long time to produce good crystals of desired material. Presently, crystal growth specialists have been moved from the periphery to the center of the materials-based technology (Alan Holden et al., 1982). This chapter reviews the different methods of crystal growth and various experimental techniques which are employed to obtain good quality crystals.

Semiconductors can be in bulk as well as thin film form. The basic techniques for growth of semiconductor single crystals are as follows:

Czochralski Method

The Czochralski process is a method of crystal growth used to obtain single crystals of semiconductors (e.g. silicon, germanium and gallium arsenide), metals (e.g. palladium, platinum, silver, gold), salts and synthetic gemstones. The process is named after Polish scientist Jan Czochralski, who invented the method in 1915 while investigating the crystallization rates of metals. He made this discovery by accident, while studying the crystallization rate of metals when, instead of dipping his pen into the ink, he did so in molten tin and drew a tin filament, that later proved to be a single crystal.

The most important application may be the growth of large cylindrical ingots, or boules, of single crystal silicon used in the electronics industry to make semiconductor devices like integrated circuits. Other semiconductors, such as gallium arsenide, can also be grown by this method, although lower defect densities in this case can be obtained using variants of the Bridgman-Stockbarger technique.

This method utilizes a seed crystal of the material to be grown. A crucible containing the raw material in polycrystalline form is heated so as to melt the material. The seed crystal is lowered into the crucible and withdrawn with continuous rotation. The material solidifies into the same orientation as that of the seed crystal. This technique uses an apparatus called a crystal puller whose schematic is shown in Schematic of crystal puller to grow Si crystals using Czochralski Method.

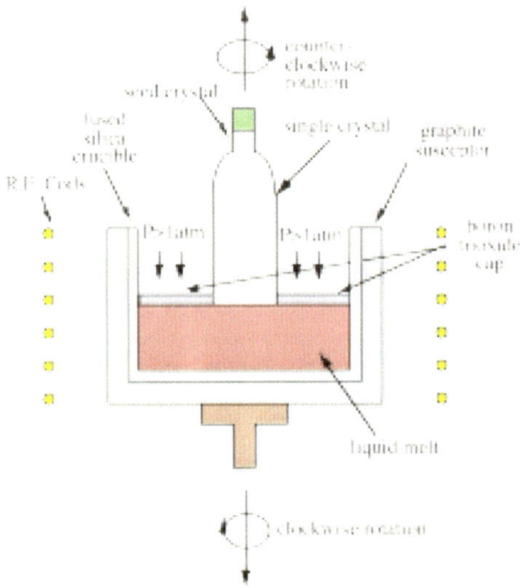

The puller has three main components:

- A furnace with power supply which includes a crucible made of fused-silicon(SiO_2), arrangement for rotation (clockwise), a graphite suspector, a rotation mechanism (clockwise) and a heating element.
- An arrangement for pulling crystal including a seed holder and rotation mechanism (counter clockwise);

- An ambient control which includes a gas source (such as argon), a flow control and exhaust system. The puller microcontrolled control system for controlling various process parameter such as temperature, crystal diameter, pull rate and rotation speeds. In crystal growing process let us say of silicon, polycrystalline silicon (EGS) is placed in the crucible and the furnace is heated above the melting temperature of silicon. For large diameters silicon ingots an external field is applied to the basic Czochralski puller. The purpose of external magnetic field is to control the concentration of defects, impurities and oxygen contents. The boron trioxide cap is essential for capping the silicon melt and also in GaAs melt it prevents the arsenic loss when the pressure on the surface is above atmospheric pressure. In order to produce doped semiconductors, the dopant can be introduced into the melt along with the raw material. However, the limitation of the method is that the crucible material (usually graphite) and ambient gas can lead to incorporation of impurities.

Application

Monocrystalline silicon (mono-Si) grown by the *Czochralski process* is often referred to as *monocrystalline Czochralski silicon* (Cz-Si). It is the basic material in the production of integrated circuits used in computers, TVs, mobile phones and all types of electronic equipment and semiconductor devices. Monocrystalline silicon is also used in large quantities by the photovoltaic industry for the production of conventional mono-Si solar cells. The almost perfect crystal structure yields the highest light-to-electricity conversion efficiency for silicon.

Production of Czochralski Silicon

Crystal of Czochralski grown silicon.

High-purity, semiconductor-grade silicon (only a few parts per million of impurities) is melted in a crucible at 1,425 °C (2,597 °F; 1,698 K), usually made of quartz. Dopant impurity atoms such as boron or phosphorus can be added to the molten silicon in precise amounts to dope the silicon, thus changing it into p-type or n-type silicon, with different electronic properties. A precisely oriented rod-mounted seed crystal is dipped into the molten silicon. The seed crystal's rod is slowly pulled upwards and rotated simultaneously. By precisely controlling the temperature gradients, rate of pulling and speed of rotation, it is possible to extract a large, single-crystal, cylindrical ingot from the melt. Occurrence of unwanted instabilities in the melt can be avoided by investigating and visualizing the temperature and velocity fields during the crystal growth process. This process is normally performed in an inert atmosphere, such as argon, in an inert chamber, such as quartz.

Crystal Sizes

Silicon crystal being grown by Czochralski process at Raytheon, 1956. The induction heating coil is visible, and the end of the crystal just emerging from the melt. The technician is measuring the temperature with an optical pyrometer. One of the earliest Si plants, the crystals produced by this early apparatus were only one inch in diameter.

Due to the efficiencies of common wafer specifications, the semiconductor industry has used wafers with standardized dimensions. In the early days, the boules were smaller, only a few inches wide. With advanced technology, high-end device manufacturers use 200 mm and 300 mm diameter wafers. The width is controlled by precise control of the temperature, the speeds of rotation and the speed the seed holder is withdrawn. The crystal ingots from which these wafers are sliced can be up to 2 metres in length, weighing several hundred kilograms. Larger wafers allow improvements in manufacturing efficiency, as more chips can be fabricated on each wafer, so there has been a steady drive to increase silicon wafer sizes. The next step up, 450 mm, is currently scheduled for introduction in 2018. Silicon wafers are typically about 0.2–0.75 mm thick, and can be polished to great flatness for making integrated circuits or textured for making solar cells.

The process begins when the chamber is heated to approximately 1500 degrees Celsius, melting the silicon. When the silicon is fully melted, a small seed crystal mounted on the end of a rotating shaft is slowly lowered until it just dips below the surface of the molten silicon. The shaft rotates counterclockwise and the crucible rotates clockwise. The rotating rod is then drawn upwards very slowly—about 25 mm per hour when making a crystal of ruby—allowing a roughly cylindrical boule to be formed. The boule can be from one to two metres, depending on the amount of silicon in the crucible.

The electrical characteristics of the silicon are controlled by adding material like phosphorus or boron to the silicon before it is melted. The added material is called dopant and the process is called doping. This method is also used with semiconductor materials other than silicon, such as gallium arsenide.

Incorporating Impurities

A puller rod with seed crystal for growing single-crystal silicon by the Czochralski process.

When silicon is grown by the Czochralski method, the melt is contained in a silica (quartz) crucible. During growth, the walls of the crucible dissolve into the melt and Czochralski silicon therefore contains oxygen at a typical concentration of 10^{18} cm^{-3}. Oxygen impurities can have beneficial or detrimental effects. Carefully chosen annealing conditions can give rise to the formation of oxygen precipitates. These have the effect of trapping unwanted transition metal impurities in a process known as gettering, improving the purity of surrounding silicon. However, formation of oxygen precipitates at unintended locations can also destroy electrical structures. Additionally, oxygen impurities can improve the mechanical strength of silicon wafers by immobilising any dislocations which may be introduced during device processing. It was experimentally shown in the 1990s that the high oxygen concentration is also beneficial for the radiation hardness of silicon particle detectors used in harsh radiation environment (such as CERN's LHC/HL-LHC projects). Therefore, radiation detectors made of Czochralski- and Magnetic Czochralski-silicon are considered to be promising candidates for many future high-energy physics experiments. It has also been shown that the presence of oxygen in silicon increases impurity trapping during post-implantation annealing processes.

However, oxygen impurities can react with boron in an illuminated environment, such as that experienced by solar cells. This results in the formation of an electrically active boron–oxygen complex that detracts from cell performance. Module output drops by approximately 3% during the first few hours of light exposure.

Mathematical Expression of Impurity Incorporation from Melt

The impurity concentration in the solid crystal that results from freezing an incremental amount of volume can be obtained from consideration of the segregation coefficient.

k_O : Segregation coefficient

V_0 : Initial volume

I_0 : Number of impurities

C_0 : Impurity concentration in the melt

V_L : Volume of the melt

I_L : Number of impurities in the melt

C_L : Concentration of impurities in the melt

V_S : Volume of solid

C_S : Concentration of impurities in the solid

During the growth process, volume of melt dV freezes, and there are impurities from the melt that are removed.

$$dI = -k_O C_L dV$$

$$dI = -k_O \frac{I_L}{V_O - V_S} dV$$

$$\int_{I_O}^{I_L} \frac{dI}{I_L} = -k_O \int_0^{V_S} \frac{dV}{V_O - V_S}$$

$$\ln\left(\frac{I_L}{I_O}\right) = \ln\left(1 - \frac{V_S}{V_O}\right)^{k_O}$$

$$I_L = I_O \left(1 - \frac{V_S}{V_O}\right)^{k_O}$$

$$C_S = -\frac{dI_L}{dV_S}$$

$$C_S = C_O k_O (1 - f)^{k_o - 1}$$

$$f = V_S / V_O$$

Float Zone Process

In order to grow semiconductor crystals with lower contaminations, the float zone process is preferred over Czochialski technique. The seed crystal placed on a high purity polycrystalline rod at the bottom is held in a vertical position and rotated. This whole arrangement is enclosed in quartz envelope having maintained inert atmosphere (argon). Now, a small zone of the crystal is melted using a radio-frequency heater, which traverses the length of the rod while moving upward from the seed. The molten silicon is retained by surface tension between the melting and growing silicon faces. As the floating zone move upwards, a single crystal silicon freezes at the zone's retreating end and grows as an extension of the seed crystal. Float zone process generally produces materials having higher resistivities as compared to Czochralski process because it can be used to purify the crystal more easily.

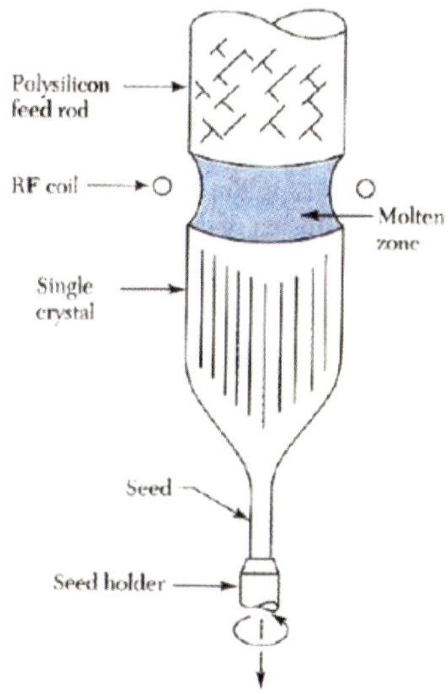

Schematic of float zone process

Process Details

The principle is that the segregation coefficient *k* (the ratio of an impurity in the solid phase to that in the liquid phase) is usually less than one. Therefore, at the solid/liquid boundary, the impurity atoms will diffuse to the liquid region. Thus, by passing a crystal boule through a thin section of furnace very slowly, such that only a small region of the boule is molten at any time, the impurities will be segregated at the end of the crystal. Because of the lack of impurities in the leftover regions which solidify, the boule can grow as a perfect single crystal if a seed crystal is placed at the base to initiate a chosen direction of crystal growth. When high purity is required, such as in semiconductor industry, the impure end of the boule is cut off, and the refining is repeated.

In zone refining, solutes are segregated at one end of the ingot in order to purify the remainder, or to concentrate the impurities. In zone leveling, the objective is to distribute solute evenly throughout the purified material, which may be sought in the form of a single crystal. For example, in the preparation of a transistor or diode semiconductor, an ingot of germanium is first purified by zone refining. Then a small amount of antimony is placed in the molten zone, which is passed through the pure germanium. With the proper choice of rate of heating and other variables, the antimony can be spread evenly through the germanium. This technique is also used for the preparation of silicon for use in computer chips.

Heaters

A variety of heaters can be used for zone melting, with their most important characteristic being the ability to form short molten zones that move slowly and uniformly through the ingot. Induction coils, ring-wound resistance heaters, or gas flames are common methods. Another method is to pass an electric current directly through the ingot while it is in a magnetic field, with the resulting magnetomotive force carefully set to be just equal to the weight in order to hold the liquid suspended. Optical heaters using high powered halogen or xenon lamps are used extensively in research facilities particularly for the production of insulators, but their use in industry is limited by the relatively low power of the lamps, which limits the size of crystals produced by this method.

Zone melting can be done as a batch process, or it can be done continuously, with fresh impure material being continually added at one end and purer material being removed from the other, with impure zone melt being removed at whatever rate is dictated by the impurity of the feed stock.

Indirect-heating floating zone methods use an induction-heated tungsten ring to heat the ingot radiatively, and are useful when the ingot is of a high-resistivity semiconductor on which classical induction heating is ineffective.

Mathematical Expression of Impurity Concentration

When the liquid zone moves by a distance dx, the number of impurities in the liquid change. Impurities are incorporated in the melting liquid and freezing solid.

k_O : Segregation coefficient

L : Zone length

C_O : Initial uniform impurity concentration of the rod

C_L : Concentration of impurities in the liquid

I : Number of impurities in the liquid

I_O : Number of impurities in zone when first formed at bottom

The number of impurities in the liquid changes in accordance with the expression below during the movement dx of the molten zone

$$dI = (C_O - k_O C_L)dx$$

$$C_L = I / L$$

$$\int_0^x dx = \int_{I_O}^{I} \frac{dI}{C_O - \frac{k_O I}{L}}$$

$$I_O = C_O L$$

$$C_S = k_O I / L$$

$$C_S(x) = C_O\left(1 - (1 - k_O)e^{-\frac{k_O x}{L}}\right)$$

Applications

Solar Cells

In solar cells float zone processing is particularly useful because the single crystal silicon grown has desirable properties. The bulk charge carrier lifetime in float-zone silicon is the highest among various manufacturing processes. Float-zone carrier lifetimes are around 1000 microseconds compared to 20-200 microseconds with Czochralski process, and 1–30 microseconds with cast multi-crystalline silicon. A longer bulk lifetime increases the efficiency of solar cells significantly.

Related Processes

Zone Remelting

Another related process is zone remelting, in which two solutes are distributed through a pure metal. This is important in the manufacture of semiconductors, where two solutes of opposite conductivity type are used. For example, in germanium, pentavalent elements of group V such as antimony and arsenic produce negative (n-type) conduction and the trivalent elements of group III such as aluminum and boron produce positive (p-type) conduction. By melting a portion of such an ingot and slowly refreezing it, solutes in the molten region become distributed to form the desired n-p and p-n junctions.

Bridgman Method

This method is similar to Czochralski method except that in this case the temperature in the vicinity of the seed crystal is kept below its melting point. The seed crystal is also placed in the crucible along with the polycrystalline material. However, a temperature gradient is maintained across the furnace in which crucible is placed. The schematic of the process is shown in figure.

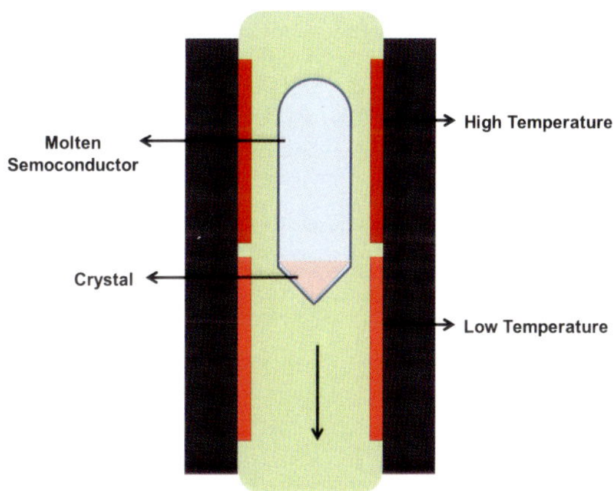

Bridgman process

After a crystal is grown, it is separated from the seed and the ingot left is shaped and sliced into wafers. This wafer can then be used as the preliminary material for fabrication of semiconductors devices and ICs.

The methods involve heating polycrystalline material above its melting point and slowly cooling it from one end of its container, where a seed crystal is located. A single crystal of the same crystallographic orientation as the seed material is grown on the seed and is progressively formed along the length of the container. The process can be carried out in a horizontal or vertical orientation, and usually involves a rotating crucible/ampoule to stir the melt.

The Bridgman method is a popular way of producing certain semiconductor crystals such as gallium arsenide, for which the Czochralski process is more difficult. The process can reliably produce single crystal ingots, but does not necessarily result in uniform properties through the crystal.

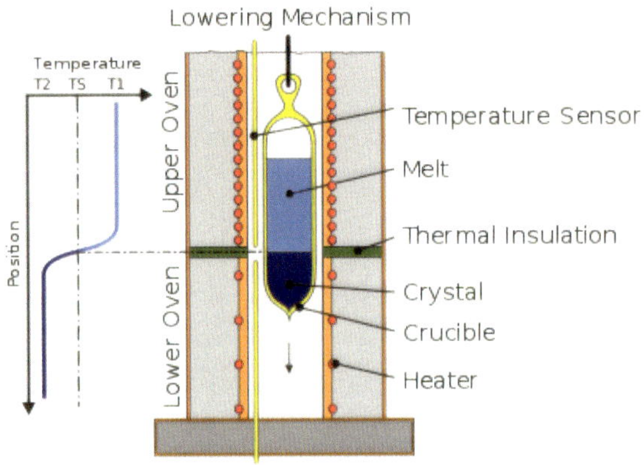

The difference between the Bridgman technique and Stockbarger technique is subtle: While both methods utilize a temperature gradient and a moving crucible, the Bridgman technique utilizes the relatively uncontrolled gradient produced at the exit of the furnace; the Stockbarger technique introduces a baffle, or shelf, separating two coupled furnaces with temperatures above and below the freezing point. Stockbarger's modification of the Bridgman technique allows for better control over the temperature gradient at the melt/crystal interface.

When seed crystals are not employed as described above, polycrystalline ingots can be produced from a feedstock consisting of rods, chunks, or any irregularly shaped pieces once they are melted and allowed to re-solidify. The resultant microstructure of the ingots so obtained are characteristic of directionally solidified metals and alloys with their aligned grains.

A variant of the technique known as the horizontal directional solidification method or HDSM developed by Khachik Bagdasarov starting in the 1960s in the Soviet Union uses a flat-bottomed crucible with short sidewalls rather than an enclosed ampoule, and has been used to grow various large oxide crystals including Yb:YAG (a laser host crystal), and sapphire crystals 45 cm wide and over 1 meter long.

Thin Film Growth Techniques

Apart from bulk single crystals, semiconductors can also be grown in the form of thin films over a substrate. They provide higher surface to volume ratio and are extensively used in device and IC fabrication. There are numerous techniques to produce semiconductor thin films. However, to ensure good quality films with minimum strain, the structure of the film and the substrate should be similar, which is also known as epitaxial growth. The single crystalline substrate in this case, acts as the seed. When the same material is deposited over the substrate, it is known as homoepitaxy whereas when a different material (but of similar structure) is deposited, it is known as heteroepitaxy.

The properties and versatility of the thin films can be obtained by selecting proper technique of film deposition. Thin film deposition methods can be broadly classified as either chemical or physical methods. The difference between the chemical and physical thin film deposition methods depends upon the method of depositing thin film material on the substrate.

In chemical deposition technique, fluid precursor is used which chemically react with the substrate. Since the thin film material is conducted through the fluid precursor, chemical deposition is conformal approaching the substrate without preference to a particular direction. A conformal is an uneven interface with the body and has a constant thickness on horizontal and vertical surfaces.

Chemical Vapor Deposition (CVD)

Various gases containing the material to be deposited are introduced into a reactor. Upon reacting, the material is deposited in the form of a thin film on a substrate placed inside the reactor. In a typical CVD system, the reactant gases, known as precursor gases, are inserted into the chamber. Upon passing through this chamber, they encounter the substrate which is placed at some temperature where they react and form solid films shown in the figure below.

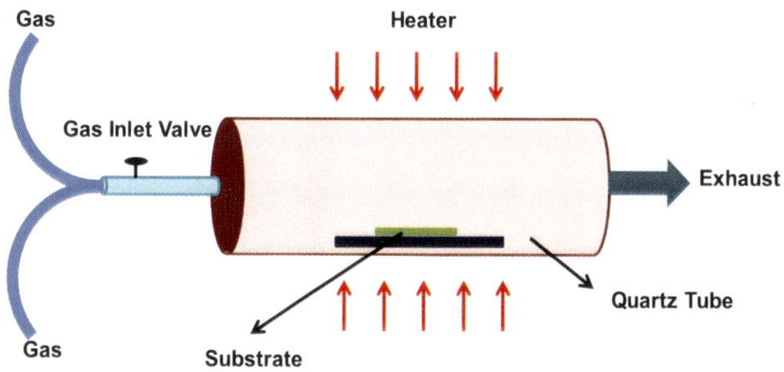

Schematic of the CVD process

The gas flow and the temperature of the substrate are the two parameters that should be controlled in order to produce good quality films.CVD has number of advantages such as flexibility of wide range of precursors, uniform thickness, conformality etc. Moreover, it does not require ultra high vacuum. But it has certain drawbacks in terms of safety and health hazards of the precursors.

Types

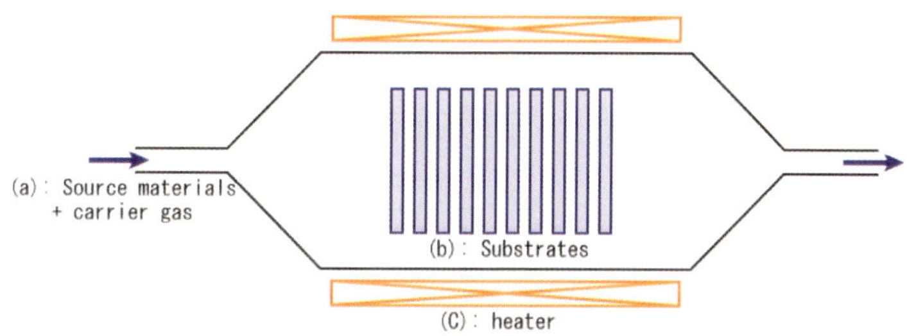

Hot-wall thermal CVD (batch operation type)

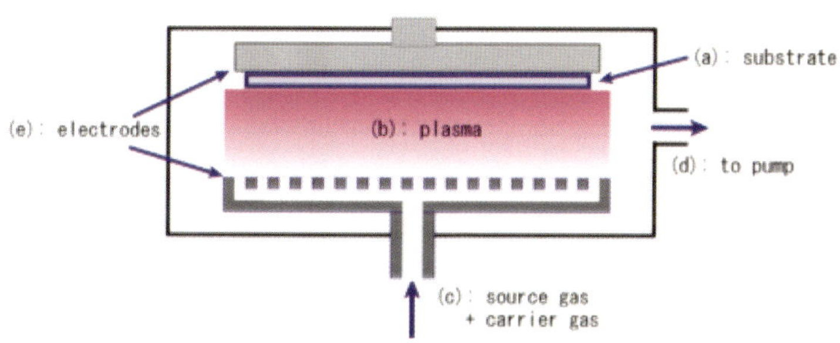

Plasma assisted CVD

CVD is practiced in a variety of formats. These processes generally differ in the means by which chemical reactions are initiated.

- Classified by operating conditions:
  - *Atmospheric pressure CVD* (APCVD) – CVD at atmospheric pressure.
  - *Low-pressure CVD* (LPCVD) – CVD at sub-atmospheric pressures. Reduced pressures tend to reduce unwanted gas-phase reactions and improve film uniformity across the wafer.
  - *Ultrahigh vacuum CVD* (UHVCVD) – CVD at very low pressure, typically below 10^{-6} Pa ($\sim10^{-8}$ torr). Note that in other fields, a lower division between high and ultra-high vacuum is common, often 10^{-7} Pa.

Most modern CVD is either LPCVD or UHVCVD.

- Classified by physical characteristics of vapor:
  - *Aerosol assisted CVD* (AACVD) – CVD in which the precursors are transported to the substrate by means of a liquid/gas aerosol, which can be generated ultrasonically. This technique is suitable for use with non-volatile precursors.
  - *Direct liquid injection CVD* (DLICVD) – CVD in which the precursors are in liquid form (liquid or solid dissolved in a convenient solvent). Liquid solutions are injected in a vaporization chamber towards injectors (typically car injectors). The precursor vapors are then transported to the substrate as in classical CVD. This technique is suitable for use on liquid or solid precursors. High growth rates can be reached using this technique.
- Classified by type of substrate heating:
  - Hot wall CVD – CVD in which the chamber is heated by an external power source and the substrate is heated by radiation from the heated chamber walls.
  - Cold wall CVD – CVD in which only the substrate is directly heated either by induction or by passing current through the substrate itself or a heater in contact with the substrate. The chamber walls are at room temperature.
- Plasma methods:
  - *Microwave plasma-assisted CVD* (MPCVD)
  - *Plasma-Enhanced CVD* (PECVD) – CVD that utilizes plasma to enhance chemical reaction rates of the precursors. PECVD processing allows deposition at lower temperatures, which is often critical in the manufacture of semiconductors. The lower temperatures also allow for the deposition of organic coatings, such as plasma polymers, that have been used for nanoparticle surface functionalization.
  - *Remote plasma-enhanced CVD* (RPECVD) – Similar to PECVD except that the wafer substrate is not directly in the plasma discharge region. Removing the wafer from the plasma region allows processing temperatures down to room temperature.

- *Atomic-layer CVD* (ALCVD) – Deposits successive layers of different substances to produce layered, crystalline films.
- *Combustion Chemical Vapor Deposition* (CCVD) – Combustion Chemical Vapor Deposition or flame pyrolysis is an open-atmosphere, flame-based technique for depositing high-quality thin films and nanomaterials.
- *Hot filament CVD* (HFCVD) – also known as catalytic CVD (Cat-CVD) or more commonly, initiated CVD (iCVD), this process uses a hot filament to chemically decompose the source gases. The filament temperature and substrate temperature thus are independently controlled, allowing colder temperatures for better adsorption rates at the substrate and higher temperatures necessary for decomposition of precursors to free radicals at the filament.
- *Hybrid Physical-Chemical Vapor Deposition* (HPCVD) – This process involves both chemical decomposition of precursor gas and vaporization of a solid source.
- *Metalorganic chemical vapor deposition* (MOCVD) – This CVD process is based on metalorganic precursors.
- *Rapid thermal CVD* (RTCVD) – This CVD process uses heating lamps or other methods to rapidly heat the wafer substrate. Heating only the substrate rather than the gas or chamber walls helps reduce unwanted gas-phase reactions that can lead to particle formation.
- *Vapor-phase epitaxy* (VPE)
- *Photo-initiated CVD* (PICVD) – This process uses UV light to stimulate chemical reactions. It is similar to plasma processing, given that plasmas are strong emitters of UV radiation. Under certain conditions, PICVD can be operated at or near atmospheric pressure.
- *Laser Chemical vapor deposition (LCVD)* - This CVD process uses lasers to heat spots or lines on a substrate in semiconductor applications. In MEMS and in fiber production the lasers are used rapidly to break down the precursor gas-process temperature can exceed 2000oC-to build up a solid structure in much the same way as laser sintering based 3-D printers build up solids from powders.

Uses

CVD is commonly used to deposit conformal films and augment substrate surfaces in ways that more traditional surface modification techniques are not capable of. CVD is extremely useful in the process of atomic layer deposition at depositing extremely thin layers of material. A variety of applications for such films exist. Gallium arsenide is used in some integrated circuits (ICs) and photovoltaic devices. Amorphous polysilicon is used in photovoltaic devices. Certain carbides and nitrides confer wear-resistance. Polymerization by CVD, perhaps the most versatile of all applications, allows for super-thin coatings which possess some very desirable qualities, such as lubricity, hydrophobicity and weather-resistance to name a few. CVD of metal-organic frameworks, a class of crystalline nanoporous materials, has recently been demonstrated. Applications for these films are anticipated in gas sensing and low-k dielectrics CVD techniques are adventageous for membrane coatings as well, such as those in desalination or water treatment, as these coatings can be sufficiently uniform (conformal) and thin that they do not clog membrane pores.

Commercially Important Materials Prepared by CVD

Polysilicon

Polycrystalline silicon is deposited from trichlorosilane ($SiHCl_3$) or silane (SiH_4), using the following reactions:

$$SiHCl_3 \rightarrow Si + Cl_2 + HCl$$

$$SiH_4 \rightarrow Si + 2\ H_2$$

This reaction is usually performed in LPCVD systems, with either pure silane feedstock, or a solution of silane with 70–80% nitrogen. Temperatures between 600 and 650 °C and pressures between 25 and 150 Pa yield a growth rate between 10 and 20 nm per minute. An alternative process uses a hydrogen-based solution. The hydrogen reduces the growth rate, but the temperature is raised to 850 or even 1050 °C to compensate. Polysilicon may be grown directly with doping, if gases such as phosphine, arsine or diborane are added to the CVD chamber. Diborane increases the growth rate, but arsine and phosphine decrease it.

Silicon Dioxide

Silicon dioxide (usually called simply "oxide" in the semiconductor industry) may be deposited by several different processes. Common source gases include silane and oxygen, dichlorosilane ($SiCl_2H_2$) and nitrous oxide (N_2O), or tetraethylorthosilicate (TEOS; $Si(OC_2H_5)_4$). The reactions are as follows:

$$SiH_4 + O_2 \rightarrow SiO_2 + 2\ H_2$$

$$SiCl_2H_2 + 2\ N_2O \rightarrow SiO_2 + 2\ N_2 + 2\ HCl$$

$$Si(OC_2H_5)_4 \rightarrow SiO_2 + \text{byproducts}$$

The choice of source gas depends on the thermal stability of the substrate; for instance, aluminium is sensitive to high temperature. Silane deposits between 300 and 500 °C, dichlorosilane at around 900 °C, and TEOS between 650 and 750 °C, resulting in a layer of *low- temperature oxide* (LTO). However, silane produces a lower-quality oxide than the other methods (lower dielectric strength, for instance), and it deposits nonconformally. Any of these reactions may be used in LPCVD, but the silane reaction is also done in APCVD. CVD oxide invariably has lower quality than thermal oxide, but thermal oxidation can only be used in the earliest stages of IC manufacturing.

Oxide may also be grown with impurities (alloying or "doping"). This may have two purposes. During further process steps that occur at high temperature, the impurities may diffuse from the oxide into adjacent layers (most notably silicon) and dope them. Oxides containing 5–15% impurities by mass are often used for this purpose. In addition, silicon dioxide alloyed with phosphorus pentoxide ("P-glass") can be used to smooth out uneven surfaces. P-glass softens and reflows at temperatures above 1000 °C. This process requires a phosphorus concentration of at least 6%, but concentrations above 8% can corrode aluminium. Phosphorus is deposited from phosphine gas and oxygen:

$$4\ PH_3 + 5\ O_2 \rightarrow 2\ P_2O_5 + 6\ H_2$$

Glasses containing both boron and phosphorus (borophosphosilicate glass, BPSG) undergo viscous flow at lower temperatures; around 850 °C is achievable with glasses containing around 5 weight % of both constituents, but stability in air can be difficult to achieve. Phosphorus oxide in high concentrations interacts with ambient moisture to produce phosphoric acid. Crystals of BPO_4 can also precipitate from the flowing glass on cooling; these crystals are not readily etched in the standard reactive plasmas used to pattern oxides, and will result in circuit defects in integrated circuit manufacturing.

Besides these intentional impurities, CVD oxide may contain byproducts of the deposition. TEOS produces a relatively pure oxide, whereas silane introduces hydrogen impurities, and dichlorosilane introduces chlorine.

Lower temperature deposition of silicon dioxide and doped glasses from TEOS using ozone rather than oxygen has also been explored (350 to 500 °C). Ozone glasses have excellent conformality but tend to be hygroscopic – that is, they absorb water from the air due to the incorporation of silanol (Si-OH) in the glass. Infrared spectroscopy and mechanical strain as a function of temperature are valuable diagnostic tools for diagnosing such problems.

Silicon Nitride

Silicon nitride is often used as an insulator and chemical barrier in manufacturing ICs. The following two reactions deposit silicon nitride from the gas phase:

$$3\ SiH_4 + 4\ NH_3 \rightarrow Si_3N_4 + 12\ H_2$$

$$3\ SiCl_2H_2 + 4\ NH_3 \rightarrow Si_3N_4 + 6\ HCl + 6\ H_2$$

Silicon nitride deposited by LPCVD contains up to 8% hydrogen. It also experiences strong tensile stress, which may crack films thicker than 200 nm. However, it has higher resistivity and dielectric strength than most insulators commonly available in microfabrication (10^{16} Ω·cm and 10 MV/cm, respectively).

Another two reactions may be used in plasma to deposit SiNH:

$$2\ SiH_4 + N_2 \rightarrow 2\ SiNH + 3\ H_2$$

$$SiH_4 + NH_3 \rightarrow SiNH + 3\ H_2$$

These films have much less tensile stress, but worse electrical properties (resistivity 10^6 to 10^{15} Ω·cm, and dielectric strength 1 to 5 MV/cm).

Metals

CVD for tungsten is achieved from tungsten hexafluoride (WF_6), which may be deposited in two ways:

$$WF_6 \rightarrow W + 3\ F_2$$

$$WF_6 + 3\ H_2 \rightarrow W + 6\ HF$$

Other metals, notably aluminium and copper, can be deposited by CVD. As of 2010, commercially cost-effective CVD for copper did not exist, although volatile sources exist, such as $Cu(hfac)_2$. Copper is typically deposited by electroplating. Aluminum can be deposited from triisobutylaluminium (TIBAL) and related organoaluminium compounds.

CVD for molybdenum, tantalum, titanium, nickel is widely used. These metals can form useful silicides when deposited onto silicon. Mo, Ta and Ti are deposited by LPCVD, from their pentachlorides. Nickel, molybdenum, and tungsten can be deposited at low temperatures from their carbonyl precursors. In general, for an arbitrary metal *M*, the chloride deposition reaction is as follows:

$$2\, MCl_5 + 5\, H_2 \rightarrow 2\, M + 10\, HCl$$

whereas the carbonyl decomposition reaction can happen spontaneously under thermal treatment or acoustic cavitation and is as follows:

$$M(CO)_n \rightarrow M + n\, CO$$

the decomposition of metal carbonyls is often violently precipitated by moisture or air, where oxygen reacts with the metal precursor to form metal or metal oxide along with carbon dioxide.

Niobium(V) oxide layers can be produced by the thermal decomposition of niobium(V) ethoxide with the loss of diethyl ether according to the equation:

$$2\, Nb(OC_2H_5)_5 \rightarrow Nb_2O_5 + 5\, C_2H_5OC_2H_5$$

Graphene

Many variations of CVD can be utilized to synthesize graphene. Although many advancements have been made, the processes listed below are not commercially viable yet.

- Carbon source

The most popular carbon source used to produce graphene is methane gas. Less popular choices include petroleum asphalt, notable for being inexpensive but more difficult to work with.

- Use of catalyst

The use of catalyst is viable in changing the physical process of graphene production. Notable examples include iron nanoparticles, nickel foam, and gallium vapor. These catalysts can either be used in situ during graphene buildup, or situated at some distance away at the deposition area. Some catalysts require another step to remove them from the sample material.

The direct growth of high-quality, large single-crystalline domains of graphene on a dielectric substrate is of vital importance for applications in electronics and optoelectronics. Combining the advantages of both catalytic CVD and the ultra-flat dielectric substrate, gaseous catalyst-assisted CVD paves the way for synthesizing high-quality graphene for device applications while avoiding the transfer process.

- Physical conditions

Physical conditions such as surrounding pressure, temperature, carrier gas, and chamber material play a big role in production of graphene.

Most systems use LPCVD with pressures ranging from 1 to 1500 Pa. However, some still use APCVD. Low pressures are used more commonly as they help prevent unwanted reactions and produce more uniform thickness of deposition on the substrate.

On the other hand, temperatures used range from 800–1050 °C. High temperatures translate to an increase of the rate of reaction. Caution has to be exercised as high temperatures do pose higher danger levels in addition to greater energy costs.

- Carrier gas

Hydrogen gas and inert gases such as argon are flowed into the system. These gases act as a carrier, enhancing surface reaction and improving reaction rate, thereby increasing deposition of graphene onto the substrate.

- Chamber material

Standard quartz tubing and chambers are used in CVD of graphene. Quartz is chosen because it has a very high melting point and is chemically inert. In other words, quartz does not interfere with any physical or chemical reactions regardless of the conditions.

- Methods of analysis of results

Raman spectroscopy, X-ray spectroscopy, transmission electron microscopy (TEM), and scanning electron microscopy (SEM) are used to examine and characterize the graphene samples.

Raman spectroscopy is used to characterize and identify the graphene particles; X-ray spectroscopy is used to characterize chemical states; TEM is used to provide fine details regarding the internal composition of graphene; SEM is used to examine the surface and topography.

Sometimes, atomic force microscopy (AFM) is used to measure local properties such as friction and magnetism.

Cold wall CVD technique can be used to study the underlying surface science involved in graphene nucleation and growth as it allows unprecedented control of process parameters like gas flow rates, temperature and pressure as demonstrated in a recent study. The study was carried out in a home-built vertical cold wall system utilizing resistive heating by passing direct current through the substrate. It provided conclusive insight into a typical surface-mediated nucleation and growth mechanism involved in two-dimensional materials grown using catalytic CVD under conditions sought out in the semiconductor industry.

Graphene Nanoribbon

In spite of graphene's exciting electronic and thermal properties, it is unsuitable as a transistor for future digital devices, due to the absence of a bandgap between the conduction and valence bands. This makes it impossible to switch between on and off states with respect to electron flow. Scaling things down, graphene nanoribbons of less than 10 nm in width do exhibit electronic bandgaps and are therefore potential candidates for digital devices. Precise control over their dimensions, and hence electronic properties, however, represents a challenging goal, and the ribbons typically possess rough edges that are detrimental to their performance.

Diamond

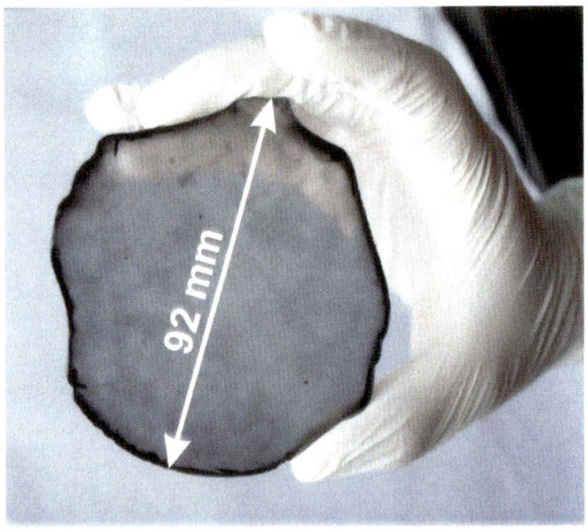

Free-standing single-crystal CVD diamond disc

Colorless gem cut from diamond grown by chemical vapor deposition

CVD can be used to produce a synthetic diamond by creating the circumstances necessary for carbon atoms in a gas to settle on a substrate in crystalline form. CVD of diamonds has received much attention in the materials sciences because it allows many new applications that had previously been considered too expensive. CVD diamond growth typically occurs under low pressure (1–27 kPa; 0.145–3.926 psi; 7.5–203 Torr) and involves feeding varying amounts of gases into a chamber, energizing them and providing conditions for diamond growth on the substrate. The gases always include a carbon source, and typically include hydrogen as well, though the amounts used vary greatly depending on the type of diamond being grown. Energy sources include hot filament, microwave power, and arc discharges, among others. The energy source is intended to generate a plasma in which the gases are broken down and more complex chemistries occur. The actual chemical process for diamond growth is still under study and is complicated by the very wide variety of diamond growth processes used.

Using CVD, films of diamond can be grown over large areas of substrate with control over the properties of the diamond produced. In the past, when high pressure high temperature (HPHT) techniques were used to produce a diamond, the result was typically very small free standing diamonds of varying sizes. With CVD diamond growth areas of greater than fifteen centimeters (six inches) diameter have been achieved and much larger areas are likely to be successfully coated with diamond in the future. Improving this process is key to enabling several important applications.

The growth of diamond directly on a substrate allows the addition of many of diamond's important qualities to other materials. Since diamond has the highest thermal conductivity of any bulk material, layering diamond onto high heat producing electronics (such as optics and transistors) allows the diamond to be used as a heat sink. Diamond films are being grown on valve rings, cutting tools, and other objects that benefit from diamond's hardness and exceedingly low wear rate. In each case the diamond growth must be carefully done to achieve the necessary adhesion onto the substrate. Diamond's very high scratch resistance and thermal conductivity, combined with a lower coefficient of thermal expansion than Pyrex glass, a coefficient of friction close to that of Teflon (polytetrafluoroethylene) and strong lipophilicity would make it a nearly ideal non-stick coating for cookware if large substrate areas could be coated economically.

CVD growth allows one to control the properties of the diamond produced. In the area of diamond growth, the word "diamond" is used as a description of any material primarily made up of sp3-bonded carbon, and there are many different types of diamond included in this. By regulating the processing parameters—especially the gases introduced, but also including the pressure the system is operated under, the temperature of the diamond, and the method of generating plasma—many different materials that can be considered diamond can be made. Single crystal diamond can be made containing various dopants. Polycrystalline diamond consisting of grain sizes from several nanometers to several micrometers can be grown. Some polycrystalline diamond grains are surrounded by thin, non-diamond carbon, while others are not. These different factors affect the diamond's hardness, smoothness, conductivity, optical properties and more.

Chalcogenides

Commercially, mercury cadmium telluride is of continuing interest for detection of infrared radiation. Consisting of an alloy of CdTe and HgTe, this material can be prepared from the dimethyl derivatives of the respective elements.

Molecular Beam Epitaxy (MBE)

The source material and the substrate are kept in anultra high vacuum chamber. The source material is heated and the vapors are taken out of a cavity such that they take the form of a molecular beam. This collimated beam is then aimed at the substrate which is kept at some temperature. In this process, source material is placed in a crucible which is heated to an optimum temperature. Thus evaporating the material and resulting in thermal energy molecular beams. The growth conditions can be controlled more precisely as compared to other techniques as a consequence of control over beam flux. Moreover, due to the presence of ultra high vacuum, various in situ analytical processes can be carried out to check and control of the process. MBE is carried out in conditions far away from thermal equilibrium since it is a vacuum based process. Schematic of MBE setup is shown in figure.

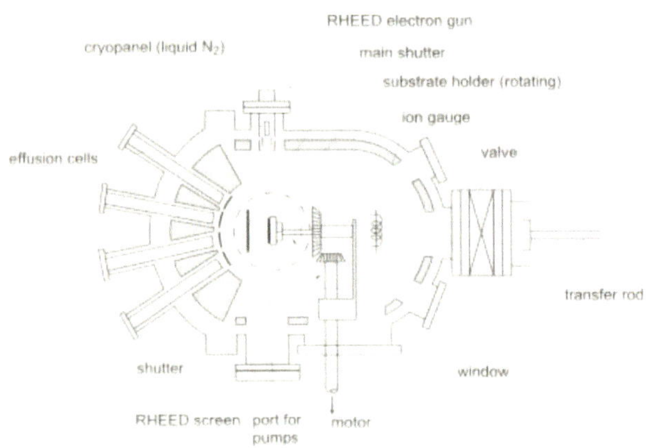

Schematic of MBE setup

Method

Molecular beam epitaxy takes place in high vacuum or ultra-high vacuum (10^{-8}–10^{-12} Torr). The most important aspect of MBE is the deposition rate (typically less than 3,000 nm per hour) that allows the films to grow epitaxially. These deposition rates require proportionally better vacuum to achieve the same impurity levels as other deposition techniques. The absence of carrier gases as well as the ultra high vacuum environment result in the highest achievable purity of the grown films.

In solid source MBE, elements such as gallium and arsenic, in ultra-pure form, are heated in separate quasi-Knudsen effusion cells or electron beam evaporators until they begin to slowly sublime. The gaseous elements then condense on the wafer, where they may react with each other. In the example of gallium and arsenic, single-crystal gallium arsenide is formed. When evaporation sources such as copper or gold are used, the gaseous elements impinging on the surface may be adsorbed (after a time window where the impinging atoms will hop around the surface) or reflected. Atoms on the surface may also desorb. Controlling the temperature of the source will control the rate of material impinging on the substrate surface and the temperature of the substrate will affect the rate of hopping or desorption. The term "beam" means that evaporated atoms do not interact with each other or vacuum chamber gases until they reach the wafer, due to the long mean free paths of the atoms.

During operation, reflection high energy electron diffraction (RHEED) is often used for monitoring the growth of the crystal layers. A computer controls shutters in front of each furnace, allowing precise control of the thickness of each layer, down to a single layer of atoms. Intricate structures of layers of different materials may be fabricated this way. Such control has allowed the development of structures where the electrons can be confined in space, giving quantum wells or even quantum dots. Such layers are now a critical part of many modern semiconductor devices, including semiconductor lasers and light-emitting diodes.

In systems where the substrate needs to be cooled, the ultra-high vacuum environment within the growth chamber is maintained by a system of cryopumps, and cryopanels, chilled using liquid nitrogen or cold nitrogen gas to a temperature close to 77 kelvins (−196 degrees Celsius). Cryogenic temperatures act as a sink for impurities in the vacuum, so vacuum levels need to be several orders of magnitude better to deposit films under these conditions. In other systems, the wafers on which the crystals are grown may be mounted on a rotating platter which can be heated to several hundred degrees Celsius during operation.

Molecular beam epitaxy is also used for the deposition of some types of organic semiconductors. In this case, molecules, rather than atoms, are evaporated and deposited onto the wafer. Other variations include gas-source MBE, which resembles chemical vapor deposition.

MBE systems can also be modified accordingly to the needs. Oxygen sources, for examples, can be incorporated for depositing oxide materials for advanced electronic, magnetic and optical applications, as well as for fundamental research.

Quantum Nanostructures

One of the most accomplished achievements of MBE is the nano-structures that permitted the formation of atomically flat and abrupt hetero-interfaces. Such structures have played an unprecedented role in expanding the knowledge of physics and electronics. Most recently the construction of nanowires and quantum structures built within them that allow for information processing and the possible integration with on-chip applications for quantum communication and computing. These heterostructure nanowire lasers are only possible to build using advance MBE techniques, allowing for monolithically integration on silicon , and picosecond signal processing.

Asaro-Tiller-Grinfeld Instability

The Asaro-Tiller-Grinfeld (ATG) instability, also known as the Grinfeld instability, is an elastic instability often encountered during molecular beam epitaxy. If there is a mismatch between the lattice sizes of the growing film and the supporting crystal, elastic energy will be accumulated in the growing film. At some critical height, the free energy of the film can be lowered if the film breaks into isolated islands, where the tension can be relaxed laterally. The critical height depends on the Young's modulus, mismatch size, and surface tension.

Some applications for this instability have been researched, such as the self-assembly of quantum dots. This community uses the name of Stranski–Krastanov growth for ATG.

Liquid Phase Epitaxy (LPE)

Cooling of a solution containing the material to be deposited leads to the precipitation of the solute on the substrate. The system is cooled below the saturation temperature. It is similar to seeded growth bulk crystal growth techniques but takes place at relatively lower temperatures. However, unlike MBE, it occurs under equilibrium conditions. It is a comparatively safer and cheaper technique.

In addition to this, there are certain physical vapor deposition techniques to grow thin films.

1. Thermal Evaporation:

   The material to be deposited (target) is heated and evaporated by resistive heating method under vacuum resulting in a thin film at the substrate.

2. Electron Beam Evaporation:

   If the heating of the target material is carried out by focused electron beams, the process is known as electron beam evaporation.

3. Sputtering:

   The target material to be deposited is bombarded by high energy ions of inert gas such as Argon which can be generated using RF or DC source. The target atoms are ejected out due to momentum transfer and finally settled at the substrate resulting in the formation of a thin film.

Types

Homoepitaxy is a kind of epitaxy performed with only one material, in which a crystalline film is grown on a substrate or film of the same material. This technology is used to grow a film which is more pure than the substrate and to fabricate layers having different doping levels. In academic literature, homoepitaxy is often abbreviated to "homoepi".

Heteroepitaxy is a kind of epitaxy performed with materials that are different from each other. In heteroepitaxy, a crystalline film grows on a crystalline substrate or film of a different material. This technology is often used to grow crystalline films of materials for which crystals cannot otherwise be obtained and to fabricate integrated crystalline layers of different materials. Examples include silicon on sapphire, gallium nitride (GaN) on sapphire, aluminium gallium indium phosphide (AlGaInP) on gallium arsenide (GaAs) or diamond or iridium, and graphene on hexagonal boron nitride.

Heterotopotaxy is a process similar to heteroepitaxy except that thin-film growth is not limited to two-dimensional growth; the substrate is similar only in structure to the thin-film material.

Pendeo-epitaxy is a process in which the heteroepitaxial film is growing vertically and laterally at the same time. In 2D crystal heterostructure, graphenen nanoribbons embedded in hexagonal boron nitride are an example for pendeo-epitaxy. Epitaxy is used in silicon-based manufacturing processes for bipolar junction transistors (BJTs) and modern complementary metal–oxide–semiconductors (CMOS), but it is particularly important for compound semiconductors such as gallium arsenide. Manufacturing issues include control of the amount and uniformity of the deposition's resistivity and thickness, the cleanliness and purity of the surface and the chamber atmosphere, the prevention of the typically much more highly doped substrate wafer's diffusion of dopant to the new layers, imperfections of the growth process, and protecting the surfaces during the manufacture and handling.

Applications

Epitaxy is used in nanotechnology and in semiconductor fabrication. Indeed, epitaxy is the only affordable method of high quality crystal growth for many semiconductor materials. In surface science, epitaxy is used to create and study monolayer and multilayer films of adsorbed organic molecules on single crystalline surfaces. Adsorbed molecules form ordered structures on atomically flat terraces of single crystalline surfaces and can directly be observed via scanning tunnelling microscopy. In contrast, surface defects and their geometry have significant influence on the adsorption of organic molecules

Methods

Epitaxial silicon is usually grown using vapor-phase epitaxy (VPE), a modification of chemical vapor deposition. Molecular-beam and liquid-phase epitaxy (MBE and LPE) are also used, mainly for compound semiconductors. Solid-phase epitaxy is used primarily for crystal-damage healing.

Vapor-phase

Silicon is most commonly deposited by doping with silicon tetrachloride and hydrogen at approximately 1200 to 1250 °C:

$$SiCl_{4(g)} + 2H_{2(g)} \leftrightarrow Si_{(s)} + 4HCl_{(g)}$$

This reaction is reversible, and the growth rate depends strongly upon the proportion of the two source gases. Growth rates above 2 micrometres per minute produce polycrystalline silicon, and negative growth rates (etching) may occur if too much hydrogen chloride byproduct is present. (In fact, hydrogen chloride may be added intentionally to etch the wafer.) An additional etching reaction competes with the deposition reaction:

$$SiCl_{4(g)} + Si_{(s)} \leftrightarrow 2SiCl_{2(g)}$$

Silicon VPE may also use silane, dichlorosilane, and trichlorosilane source gases. For instance, the silane reaction occurs at 650 °C in this way:

$$SiH_4 \rightarrow Si + 2H_2$$

This reaction does not inadvertently etch the wafer, and takes place at lower temperatures than deposition from silicon tetrachloride. However, it will form a polycrystalline film unless tightly controlled, and it allows oxidizing species that leak into the reactor to contaminate the epitaxial layer with unwanted compounds such as silicon dioxide.

VPE is sometimes classified by the chemistry of the source gases, such as hydride VPE and metalorganic VPE.

Liquid-phase

Liquid-phase epitaxy (LPE) is a method to grow semiconductor crystal layers from the melt on solid substrates. This happens at temperatures well below the melting point of the deposited semiconductor. The semiconductor is dissolved in the melt of another material. At conditions that are close to the equilibrium between dissolution and deposition, the deposition of the semiconductor crystal on the substrate is relatively fast and uniform. The most used substrate is indium phosphide (InP). Other substrates like glass or ceramic can be applied for special applications. To facilitate nucleation, and to avoid tension in the grown layer the thermal expansion coefficient of substrate and grown layer should be similar.

Centrifugal liquid-phase epitaxy is used commercially to make thin layers of silicon, germanium, and gallium arsenide. Centrifugally formed film growth is a process used to form thin layers of materials by using a centrifuge. The process has been used to create silicon for thin-film solar cells and far-infrared photodetectors. Temperature and centrifuge spin rate are used to control layer growth. Centrifugal LPE has the capability to create dopant concentration gradients while the solution is held at constant temperature.

Solid-phase

Solid-phase epitaxy (SPE) is a transition between the amorphous and crystalline phases of a material. It is usually done by first depositing a film of amorphous material on a crystalline substrate. The substrate is then heated to crystallize the film. The single crystal substrate serves as a template for crystal growth. The annealing step used to recrystallize or heal silicon layers amorphized during ion implantation is also considered one type of Solid Phase Epitaxy. The Impurity segregation and redistribution at the growing crystal-amorphous layer interface during this process is used to incorporate low-solubility dopants in metals and Silicon.

Molecular-beam Epitaxy

In molecular beam epitaxy (MBE), a source material is heated to produce an evaporated beam of particles. These particles travel through a very high vacuum (10^{-8} Pa; practically free space) to the substrate, where they condense. MBE has lower throughput than other forms of epitaxy. This technique is widely used for growing periodic groups III, IV, and V semiconductor crystals.

Doping

An epitaxial layer can be doped during deposition by adding impurities to the source gas, such as arsine, phosphine, or diborane. The concentration of impurity in the gas phase determines its concentration in the deposited film. As in chemical vapor deposition (CVD), impurities change the deposition rate. Additionally, the high temperatures at which CVD is performed may allow dopants to diffuse into the growing layer from other layers in the wafer ("out-diffusion"). Also, dopants in the source gas, liberated by evaporation or wet etching of the surface, may diffuse into the epitaxial layer ("autodoping"). The dopant profiles of underlying layers change as well, however not as significantly.

Minerals

Rutile epitaxial on hematite nearly 6 cm long. Bahia, Brazil

In mineralogy, epitaxy is the overgrowth of one mineral on another in an orderly way, such that certain crystal directions of the two minerals are aligned. This occurs when some planes in the lattices of the overgrowth and the substrate have similar spacings between atoms.

If the crystals of both minerals are well formed so that the directions of the crystallographic axes are clear then the epitaxic relationship can be deduced just by a visual inspection.

Sometimes many separate crystals form the overgrowth on a single substrate, and then if there is epitaxy all the overgrowth crystals will have a similar orientation. The reverse, however, is not necessarily true. If the overgrowth crystals have a similar orientation there is probably an epitaxic relationship, but it is not certain.

Some authors consider that overgrowths of a second generation of the same mineral species should also be considered as epitaxy, and this is common terminology for semiconductor scientists who induce epitaxic growth of a film with a different doping level on a semiconductor substrate of the same material. For naturally produced minerals, however, the International Mineralogical Association (IMA) definition requires that the two minerals be of different species.

Another man-made application of epitaxy is the making of artificial snow using silver iodide, which is possible because hexagonal silver iodide and ice have similar cell dimensions.

Isomorphic Minerals

Minerals that have the same structure (isomorphic minerals) may have epitaxic relations. An example is albite $NaAlSi_3O_8$ on microcline $KAlSi_3O_8$. Both these minerals are triclinic, with space group 1, and with similar unit cell parameters, a = 8.16 Å, b = 12.87 Å, c = 7.11 Å, $\alpha = 93.45°$, $\beta = 116.4°$, $\gamma = 90.28°$ for albite and a = 8.5784 Å, b = 12.96 Å, c = 7.2112 Å, $\alpha = 90.3°$, $\beta = 116.05°$, $\gamma = 89°$ for microcline.

Polymorphic Minerals

Minerals that have the same composition but different structures (polymorphic minerals) may also have epitaxic relations. Examples are pyrite and marcasite, both FeS_2, and sphalerite and wurtzite, both ZnS.

Rutile on hematite, from Novo Horizonte, Bahia, Northeast Region, Brazil

Rutile on Hematite

Hematite pseudomorph after magnetite, with terraced epitaxial faces. La Rioja, Argentina

Some pairs of minerals that are not related structurally or compositionally may also exhibit epitaxy. A common example is rutile TiO_2 on hematite Template: Fe. Rutile is tetragonal and hematite is trigonal, but there are directions of similar spacing between the atoms in the plane of rutile (perpendicular to the a axis) and the plane of hematite (perpendicular to the c axis). In epitaxy these directions tend to line up with each other, resulting in the axis of the rutile overgrowth being parallel to the c axis of hematite, and the c axis of rutile being parallel to one of the axes of hematite.

Hematite on Magnetite

Another example is hematite $Fe^{3+}{}_2O_3$ on magnetite $Fe^{2+}\ Fe^{3+}{}_2O_4$. The magnetite structure is based on close-packed oxygen anions stacked in an ABC-ABC sequence. In this packing the close-packed layers are parallel to (a plane that symmetrically "cuts off" a corner of a cube). The hematite structure is based on close-packed oxygen anions stacked in an AB-AB sequence, which results in a crystal with hexagonal symmetry.

If the cations were small enough to fit into a truly close-packed structure of oxygen anions then the spacing between the nearest neighbour oxygen sites would be the same for both species. The radius of the oxygen ion, however, is only 1.36 Å and the Fe cations are big enough to cause some variations. The Fe radii vary from 0.49 Å to 0.92 Å, depending on the charge (2+ or 3+) and the coordination number. Nevertheless, the O spacings are similar for the two minerals hence hematite can readily grow on the faces of magnetite, with hematite parallel to magnetite.

Growth Techniques to Fabricate a pn Junction

There are various techniques to fabricate a pn junction and they are explained in detail as follows-

Grown junction:

It is similar to the Czochralski method where a seed crystal melt is used to produce single crystal.

However, in this method, one kind of dopant (either n type or p type) is introduced into the melt during the crystal growth. After some time, the other dopant is added into the melt with a concentration higher than the concentration of the initial impurity. This abrupt change in the impurity results in the formation of a pn junction. So, a single crystal consists of p and n type semiconductors.

Alloyed junction:

An alloy is created between a doped semiconductor and a metal containing other type of dopant. For e.g. a small piece of metal containing the p-type impurity is placed on a n-type semiconductor and heated at a high temperature in inert atmosphere such that the metal melts and a part of n and p type impurities are alloyed. The molten portion is then cooled and solidified forming a pn junction.

Diffused junction:

The p-type/n-type semiconductor is heated at a high temperature in the presence of gases containing ntype/ p-type impurities. The dopant atoms diffuse into the bulk crystal as a result of the concentration gradient. Some of the host atoms leave their lattice sites at high temperatures and create vacancies which may be occupied by the dopant atoms. SiO2 layer is used as a mask to protect different regions of the semiconductor so that various geometries can be created. Diffusion is carried out under two steps. In the first step, known as the pre deposition step, the impurities are introduced upto a certain depth (around one tenth of a micron). Once these impurities have been introduced, they are embedded deeper into the semiconductor, without more impurities being added. This is done by removing the source of dopants and passing oxygen gas at high temperatures leading to the formation of oxide layer on the surface which seals the impurities. This second step is known as drive in diffusion. This is the most common method used in IC fabrication.

Ion implanted junction:

The dopant ions are accelerated at very high speed and bombarded on the surface of the semiconductor. These ions replace the semiconductor atoms upon collision upto a certain depth where a pn junction is created. Before the bombardment of ions, SiO2 layer is deposited on the top of the semiconductor and certain regions are etched. The ion energies are typically in the range 30-300 keV. When the ion displaces the atom from its lattice site, the material becomes amorphous in that region and can be crystallized by annealing i.e. heating at a certain temperature. The concentration of impurities can be controlled precisely in this case.

When p and n-type semiconductors are brought in contact, holes diffuse from p side to n side due to lesser concentration of holes in the n type semiconductor. Similarly, the electrons from the n type semiconductor diffuse from the n-side to the p-side. This movement of electrons and holes leads to the flow of diffusion current across the junction. These holes and electrons leave behind negatively charged acceptor ions and positively charged donor ions in the p and n regions respectively. As a result, a negative charge region is created on p side and a positive charge region is created on the n side near the junction. The space charge region created on both sides of the junction resist further movement of the charge carriers because of the potential barrier. Electrons crossing from n side to p side are repelled by the negative charge whereas holes crossing from p side to n side are repelled by the positive charge. Also, an electric field is created resulting in a drift current. In equilibrium, the currents balance each other and net current across the junction is zero. Now, when a forward bias is applied to the pn junction, the depletion layer width decreases and current flows across the junction. On application of reverse bias, the depletion layer width increases and only small reverse current flows.

Energy Band Diagram of pn Junction and Formation of Depletion Region

As evident from figure, when a pn junction is formed the condition of alignment of fermi level leads to band bending and the valence and conduction bands in an n-type material are at slightly lower energy levels than the valence and conduction bands in a p-type material. This may also be due to difference in the atomic characteristics of the pentavalent and trivalent impurity atoms. There will be a possibility of the diffusion of the free electrons in the n region occupying the upper part of the conduction band across the junction (they do not require any additional energy) and temporarily become free electrons in the lower part of the p-region conduction band.

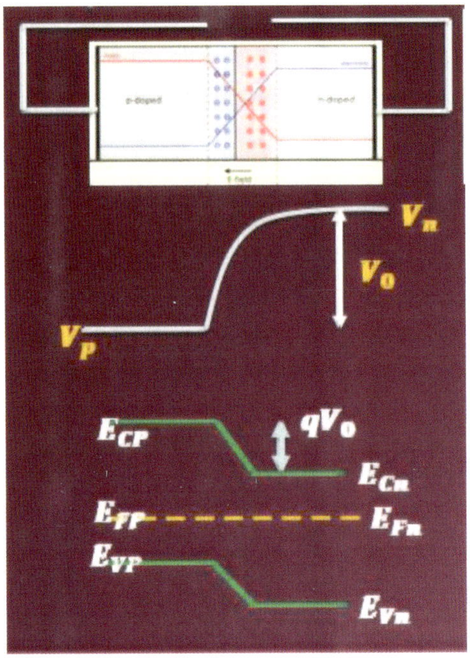

Energy band diagrams illustrating the formation of pn junction and depletion region

As the diffusion of the electrons continues, the depletion region is formed consisting of immobile charge carriers due to the recombination of electrons from n-side and holes from p-side. An energy diagram for a pn junction at the instant of depletion region formation is shown in figure. Soon, there are no electrons left in the n-region conduction band with enough energy to move across the junction to the p-region conduction band, as indicated by the alignment of the top of the n-region conduction band and the bottom of the p-region conduction band as shown in figure. At this point, the junction is at equilibrium; and the formation of depletion region is complete because diffusion has ceased. There is an energy gradient across the depletion region which acts as an "energy hill" denoted by Vo in figure which can electron must have in order to climb to get to the p region.

References

- (Polish), (English), Paweł Tomaszewski, ""Jan Czochralski i jego metoda" (ang.Jan Czochralski and his method), Oficyna Wydawnicza ATUT, Wrocław–Kcynia 2003, ISBN 83-89247-27-5

- Neurotechnology Group, Berlin Institute of Technology, IEEE Xplore Digital Library. "Regression Methods for Virtual Metrology of Layer Thickness in Chemical Vapor Deposition." January 17, 2014. Retrieved November 9, 2015
- Georg Müller (1988) Crystal growth from the melt Springer-Verlag, Science 138 pages ISBN 3-540-18603-4, ISBN 978-3-540-18603-8
- Custer, J. S.; Polman, A.; Van Pinxteren, H. M. (1994). "Erbium in crystal silicon: Segregation and trapping during solid phase epitaxy of amorphous silicon". Journal of Applied Physics. 75 (6): 2809. Bibcode:-1994JAP....75.2809C. doi:10.1063/1.356173
- Aleksic, Jalena; Zielke, Paul; Szymczyk, Janusz A.; et al. (2002). "Temperature and Flow Visualization in a Simulation of the Czochralski Process Using Temperature-Sensitive Liquid Crystals". Ann. N.Y. Acad. Sci. 972: 158. Bibcode:2002NYASA.972..158A. doi:10.1111/j.1749-6632.2002.tb04567.x
- Maruyama, Toshiro (1994). "Electrochromic Properties of Niobium Oxide Thin Films Prepared by Chemical Vapor Deposition". Journal of the Electrochemical Society. 141 (10): 2868. doi:10.1149/1.2059247
- Custer, J.S.; Polman, A.; Pinxteren, H. M. (15 March 1994). "Erbium in crystal silicon: Segregation and trapping during solid phase epitaxy of amorphous silicon". Journal of Applied Physics. 75(6)
- Luque, A.; Sala, G.; Palz, Willeke; Santos, G. dos; Helm, P. (2012). Tenth E.C. Photovoltaic Solar Energy Conference: Proceedings of the International Conference, held at Lisbon, Portugal, 8–12 April 1991. Springer. p. 694.
- James D. Plummer, Michael D. Deal, and Peter B. Griffin, Silicon VLSI Technology, Prentice Hall, 2000, ISBN 0-13-085037-3 pp. 126–27
- Bridgman, Percy W. (1925). "Certain Physical Properties of Single Crystals of Tungsten, Antimony, Bismuth, Tellurium, Cadmium, Zinc, and Tin". Proceedings of the American Academy of Arts and Sciences. 60 (6): 305–383. doi:10.2307/25130058. JSTOR . 25130058
- Das, Shantanu; Drucker, Jeff (28 May 2018). "Pre-coalescence scaling of graphene island sizes". Journal of Applied Physics. 123(20): 205306. Bibcode:2018JAP...123t5306D. doi:10.1063/1.5021341
- Hans J. Scheel; Peter Capper; Peter Rudolph (25 October 2010). Crystal Growth Technology: Semiconductors and Dielectrics. John Wiley & Sons. pp. 177–178. ISBN 978-3-527-32593-1

4

Diode and its Types

A two-terminal electronic component, which has low electrical resistance in one direction of flow and high resistance in the other, is known as a diode. They are of different types, some of which are discussed in this chapter, like Zener diode, tunnel diode, Impatt diode, Light Emitting Diode (LED), and Gunn diode.

The word diode can be explained as 'Di' means two and 'ode' is obtained from electrode. A diode has two terminals or electrodes (one connected to P-type and the other to the N-type). A P-N junction diode is formed either by using any semiconductor such as Si or Ge. The P and N type regions are referred to as anode and cathode respectively. Its circuit symbol is shown in fig. In its symbol, arrowhead indicates the direction of current flow.

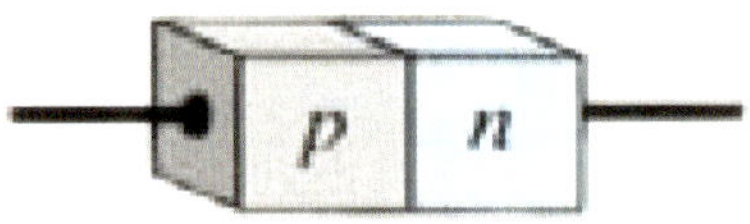

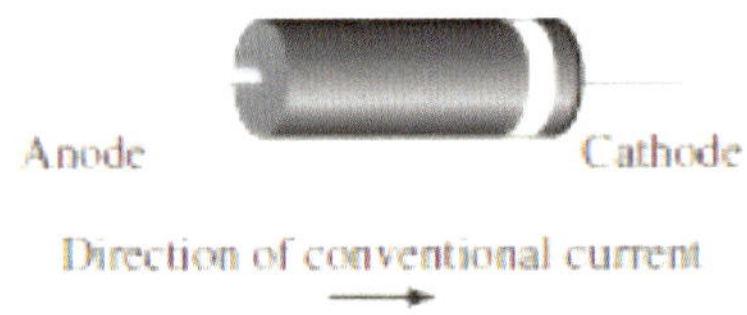

A semiconductor diode has two terminals or electrodes that act like an on-off switch. When the diode is "on", it acts as a short circuit and allows current to flow through it. When it is "off", it behaves like an open circuit and passes no current. The two terminals are different and are marked as plus and minus in Figure 1. If the polarity of the applied voltage matches that of the diode (forward bias), then the diode turns "on".

When the applied voltage polarity is opposite (reverse bias), it turns "off". Of course this is the theoretical behavior of an ideal diode, but it can be seen as a good approximation for a real diode. Diodes are widely used in power electronic circuits for conversion of electric power. The basic application of diode in power electronics is as AC to DC converters commonly known as rectifiers which provide fixed output voltage.

Main Functions

The most common function of a diode is to allow an electric current to pass in one direction (called the diode's *forward* direction), while blocking it in the opposite direction (the *reverse* direction). As such, the diode can be viewed as an electronic version of a check valve. This unidirectional behavior is called rectification, and is used to convert alternating current (ac) to direct current (dc). Forms of rectifiers, diodes can be used for such tasks as extracting modulation from radio signals in radio receivers.

However, diodes can have more complicated behavior than this simple on–off action, because of their nonlinear current-voltage characteristics. Semiconductor diodes begin conducting electricity only if a certain threshold voltage or cut-in voltage is present in the forward direction (a state in which the diode is said to be *forward-biased*). The voltage drop across a forward-biased diode varies only a little with the current, and is a function of temperature; this effect can be used as a temperature sensor or as a voltage reference.

A semiconductor diode's current–voltage characteristic can be tailored by selecting the semiconductor materials and the doping impurities introduced into the materials during manufacture. These techniques are used to create special-purpose diodes that perform many different functions. For example, diodes are used to regulate voltage (Zener diodes), to protect circuits from high voltage surges (avalanche diodes), to electronically tune radio and TV receivers (varactor diodes), to generate radio-frequency oscillations (tunnel diodes, Gunn diodes, IMPATT diodes), and to produce light (light-emitting diodes). Tunnel, Gunn and IMPATT diodes exhibit negative resistance, which is useful in microwave and switching circuits.

Diodes, both vacuum and semiconductor, can be used as shot-noise generators.

Vacuum Tube Diodes

In 1873, Frederick Guthrie observed that a grounded, white hot metal ball brought in close proximity to an electroscope would discharge a positively charged electroscope, but not a negatively charged electroscope.

In 1880, Thomas Edison observed unidirectional current between heated and unheated elements in a bulb, later called Edison effect, and was granted a patent on application of the phenomenon for use in a dc voltmeter.

About 20 years later, John Ambrose Fleming (scientific adviser to the Marconi Company and former Edison employee) realized that the Edison effect could be used as a radio detector. Fleming patented the first true thermionic diode, the Fleming valve, in Britain on November 16, 1904 (followed by U.S. Patent 803,684 in November 1905).

Throughout the vacuum tube era, valve diodes were used in in almost all electronics such as radios, televisions, sound systems and instrumentation. They slowly lost market share beginning in the late 1940s due to selenium rectifier technology and then to semiconductor diodes during the 1960s. Today they are still used in a few high power applications where their ability to withstand transient voltages and their robustness gives them an advantage over semiconductor devices, and in musical instrument and audiophile applications.

Solid-state Diodes

In 1874, German scientist Karl Ferdinand Braun discovered the "unilateral conduction" across a contact between a metal and a mineral. Indian scientist Jagadish Chandra Bose was the first to use a crystal for detecting radio waves in 1894. The crystal detector was developed into a practical device for wireless telegraphy by Greenleaf Whittier Pickard, who invented a silicon crystal detector in 1903 and received a patent for it on November 20, 1906. Other experimenters tried a variety of other minerals as detectors. Semiconductor principles were unknown to the developers of these early rectifiers. During the 1930s understanding of physics advanced and in the mid 1930s researchers at Bell Telephone Laboratories recognized the potential of the crystal detector for application in microwave technology. Researchers at Bell Labs, Western Electric, MIT, Purdue and in the UK intensively developed point-contact diodes (*crystal rectifiers* or *crystal diodes*) during World War II for application in radar. After World War II, AT&T used these in their microwave towers that criss-crossed the United States, and many radar sets use them even in the 21st century. In 1946, Sylvania began offering the 1N34 crystal diode. During the early 1950s, junction diodes were developed.

Rectifiers

Although all diodes *rectify*, the term *rectifier* is usually applied to diodes intended for power supply application in order to differentiate them from diodes intended for small signal circuits.

Vacuum Tube Diodes

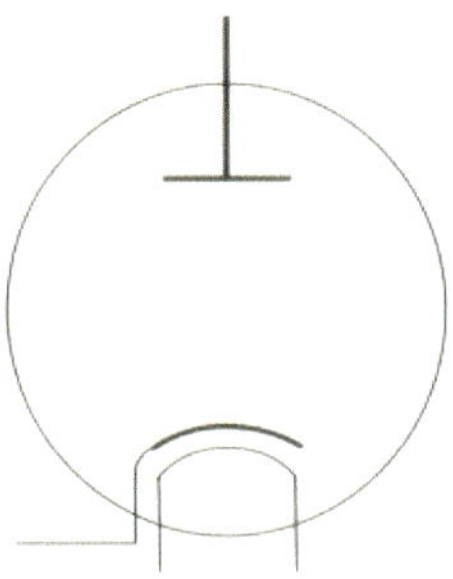

The symbol for an indirectly heated vacuum tube diode.
From top to bottom, the element names are: *plate*, *cathode*, and *heater*.

A thermionic diode is a thermionic-valve device consisting of a sealed, evacuated glass or metal envelope containing two electrodes: a cathode and a plate. The cathode is either *indirectly heated* or *directly heated*. If indirect heating is employed, a heater is included in the envelope.

In operation, the cathode is heated to red heat (800–1000 °C). A directly heated cathode is made of tungsten wire and is heated by current passed through it from an external voltage source. An indirectly heated cathode is heated by infrared radiation from a nearby heater that is formed of Nichrome wire and supplied with current provided by an external voltage source.

The operating temperature of the cathode causes it to release electrons into the vacuum, a process called thermionic emission. The cathode is coated with oxides of alkaline earth metals, such as barium and strontium oxides. These have a low work function, meaning that they more readily emit electrons than would the uncoated cathode.

The plate, not being heated, does not emit electrons; but is able to absorb them.

The alternating voltage to be rectified is applied between the cathode and the plate. When the plate voltage is positive with respect to the cathode, the plate electrostatically attracts the electrons from the cathode, so a current of electrons flows through the tube from cathode to plate. When the plate voltage is negative with respect to the cathode, no electrons are emitted by the plate, so no current can pass from the plate to the cathode.

Semiconductor Diodes

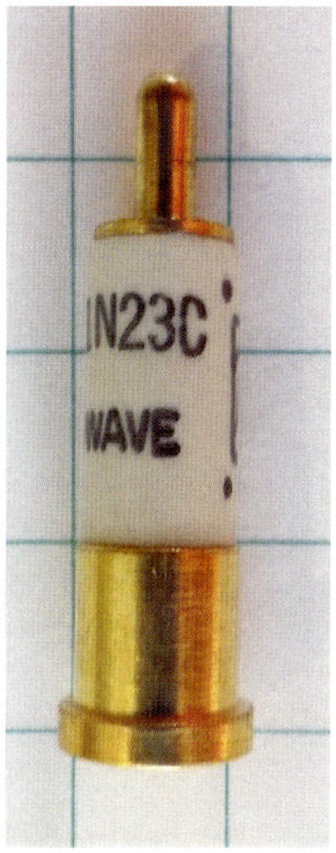

Point contact diode (*crystal rectifier* or *crystal diode*), type 1N23C. Grid one quarter inch.

Point-contact Diodes

Point-contact diodes were developed starting in the 1930s, out of the early crystal detector technology, and are generally used in the 3 to 30 gigahertz range. Point-contact diodes use a small diameter metal wire in contact with a semiconductor crystal, and are of either *non-welded* contact type or *welded contact* type. Non-welded contact construction utilizes the Schottky barrier principle. The metal side is the pointed end of a small diameter wire that is in contact with the semiconductor crystal. In the welded contact type, a small P region is formed in the otherwise N type crystal around the metal point during manufacture by momentarily passing a relatively large current through the device. Point contact diodes generally exhibit lower capacitance, higher forward resistance and greater reverse leakage than junction diodes.

Junction Diodes

p–n Junction Diode

A p–n junction diode is made of a crystal of semiconductor, usually silicon, but germanium and gallium arsenide are also used. Impurities are added to it to create a region on one side that contains negative charge carriers (electrons), called an n-type semiconductor, and a region on the other side that contains positive charge carriers (holes), called a p-type semiconductor. When the n-type and p-type materials are attached together, a momentary flow of electrons occur from the n to the p side resulting in a third region between the two where no charge carriers are present. This region is called the depletion region because there are no charge carriers (neither electrons nor holes) in it. The diode's terminals are attached to the n-type and p-type regions. The boundary between these two regions, called a p–n junction, is where the action of the diode takes place. When a sufficiently higher electrical potential is applied to the P side (the anode) than to the N side (the cathode), it allows electrons to flow through the depletion region from the N-type side to the P-type side. The junction does not allow the flow of electrons in the opposite direction when the potential is applied in reverse, creating, in a sense, an electrical check valve.

Schottky Diode

Another type of junction diode, the Schottky diode, is formed from a metal–semiconductor junction rather than a p–n junction, which reduces capacitance and increases switching speed.

Current–voltage Characteristic

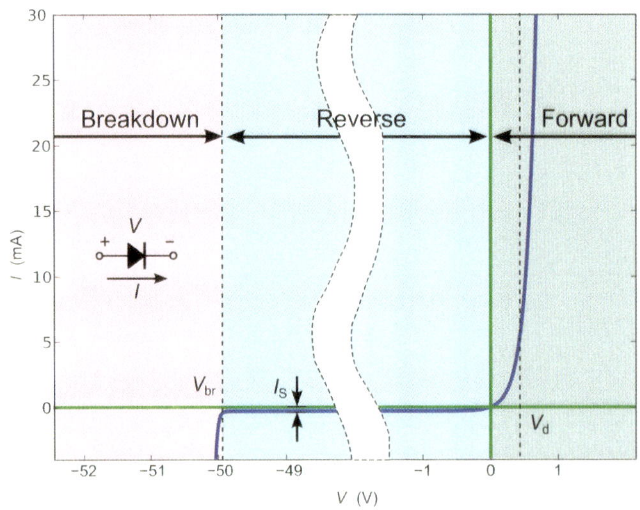

I–V (current vs. voltage) characteristics of a p–n junction diode

A semiconductor diode's behavior in a circuit is given by its current–voltage characteristic, or I–V graph. The shape of the curve is determined by the transport of charge carriers through the so-called *depletion layer* or *depletion region* that exists at the p–n junction between differing semiconductors. When a p–n junction is first created, conduction-band (mobile) electrons from the N-doped region diffuse into the P-doped region where there is a large population of holes (vacant places for electrons) with which the electrons "recombine". When a mobile electron recombines with a hole, both hole and electron vanish, leaving behind an immobile positively charged donor (dopant) on the N side and negatively charged acceptor (dopant) on the P side. The region around the p–n junction becomes depleted of charge carriers and thus behaves as an insulator.

However, the width of the depletion region (called the depletion width) cannot grow without limit. For each electron–hole pair recombination made, a positively charged dopant ion is left behind in the N-doped region, and a negatively charged dopant ion is created in the P-doped region. As recombination proceeds and more ions are created, an increasing electric field develops through the depletion zone that acts to slow and then finally stop recombination. At this point, there is a "built-in" potential across the depletion zone.

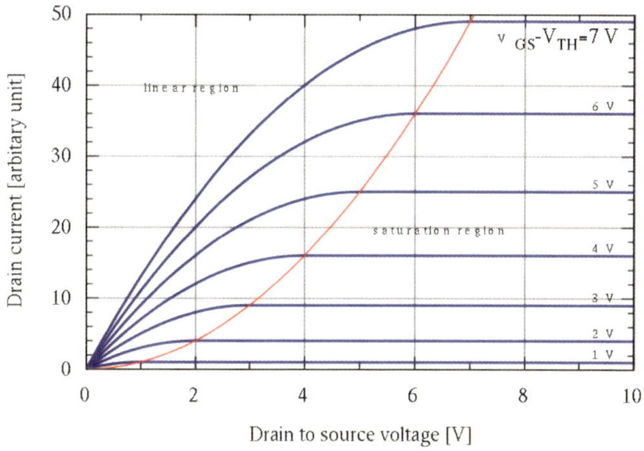

Reverse Bias

If an external voltage is placed across the diode with the same polarity as the built-in potential, the depletion zone continues to act as an insulator, preventing any significant electric current flow (unless electron–hole pairs are actively being created in the junction by, for instance, light). This is called the *reverse bias* phenomenon.

Forward Bias

However, if the polarity of the external voltage opposes the built-in potential, recombination can once again proceed, resulting in a substantial electric current through the p–n junction (i.e. substantial numbers of electrons and holes recombine at the junction). For silicon diodes, the built-in potential is approximately 0.7 V (0.3 V for germanium and 0.2 V for Schottky). Thus, if an external voltage greater than and opposite to the built-in voltage is applied, a current will flow and the diode is said to be "turned on" as it has been given an external *forward bias*. The diode is commonly said to have a forward "threshold" voltage, above which it conducts and below which conduction stops. However, this is only an approximation as the forward characteristic is smooth.

A diode's I–V characteristic can be approximated by four regions of operation:

- At very large reverse bias, beyond the peak inverse voltage or PIV, a process called reverse breakdown occurs that causes a large increase in current (i.e., a large number of electrons and holes are created at, and move away from the p–n junction) that usually damages the device permanently. The avalanche diode is deliberately designed for use in that manner. In the Zener diode, the concept of PIV is not applicable. A Zener diode contains a heavily doped p–n junction allowing electrons to tunnel from the valence band of the p-type material to the conduction band of the n-type material, such that the reverse voltage is "clamped" to a known value (called the *Zener voltage*), and avalanche does not occur. Both devices, however, do have a limit to the maximum current and power they can withstand in the clamped reverse-voltage region. Also, following the end of forward conduction in any diode, there is reverse current for a short time. The device does not attain its full blocking capability until the reverse current ceases.

- For a bias less than the PIV, the reverse current is very small. For a normal P–N rectifier diode, the reverse current through the device in the micro-ampere (μA) range is very low. However, this is temperature dependent, and at sufficiently high temperatures, a substantial amount of reverse current can be observed (mA or more).
- With a small forward bias, where only a small forward current is conducted, the current–voltage curve is exponential in accordance with the ideal diode equation. There is a definite forward voltage at which the diode starts to conduct significantly. This is called the *knee voltage* or *cut-in voltage* and is equal to the barrier potential of the p-n junction. This is a feature of the exponential curve, and appears sharper on a current scale more compressed than in the diagram shown here.
- At larger forward currents the current-voltage curve starts to be dominated by the ohmic resistance of the bulk semiconductor. The curve is no longer exponential, it is asymptotic to a straight line whose slope is the bulk resistance. This region is particularly important for power diodes. The diode can be modeled as an ideal diode in series with a fixed resistor.

In a small silicon diode operating at its rated currents, the voltage drop is about 0.6 to 0.7 volts. The value is different for other diode types—Schottky diodes can be rated as low as 0.2 V, germanium diodes 0.25 to 0.3 V, and red or blue light-emitting diodes (LEDs) can have values of 1.4 V and 4.0 V respectively.

At higher currents the forward voltage drop of the diode increases. A drop of 1 V to 1.5 V is typical at full rated current for power diodes.

Shockley Diode Equation

The *Shockley ideal diode equation* or the *diode law* (named after the bipolar junction transistor co-inventor William Bradford Shockley) gives the I–V characteristic of an ideal diode in either forward or reverse bias (or no bias). The following equation is called the *Shockley ideal diode equation* when n, the ideality factor, is set equal to 1 :

$$I = I_S \left(e^{\frac{V_D}{nV_T}} - 1 \right)$$

where

I is the diode current,

I_S is the reverse bias saturation current (or scale current),

V_D is the voltage across the diode,

V_T is the thermal voltage, and

n is the *ideality factor*, also known as the *quality factor* or sometimes *emission coefficient*. The ideality factor n typically varies from 1 to 2 (though can in some cases be higher), depending on the fabrication process and semiconductor material and is set equal to 1 for the case of an "ideal" diode (thus the n is sometimes omitted). The ideality factor was added to account for imperfect junctions as observed in real transistors. The factor mainly accounts for carrier recombination as the charge carriers cross the depletion region.

The thermal voltage V_T is approximately 25.85 mV at 300 K, a temperature close to "room temperature" commonly used in device simulation software. At any temperature it is a known constant defined by:

$$V_T = \frac{kT}{q},$$

where *k* is the Boltzmann constant, *T* is the absolute temperature of the p–n junction, and *q* is the magnitude of charge of an electron (the elementary charge).

The reverse saturation current, I_S, is not constant for a given device, but varies with temperature; usually more significantly than V_T, so that V_D typically decreases as *T* increases.

The *Shockley ideal diode equation* or the *diode law* is derived with the assumption that the only processes giving rise to the current in the diode are drift (due to electrical field), diffusion, and thermal recombination–generation (R–G) (this equation is derived by setting n = 1 above). It also assumes that the R–G current in the depletion region is insignificant. This means that the *Shockley ideal diode equation* doesn't account for the processes involved in reverse breakdown and photon-assisted R–G. Additionally, it doesn't describe the "leveling off" of the I–V curve at high forward bias due to internal resistance. Introducing the ideality factor, n, accounts for recombination and generation of carriers.

Under *reverse bias* voltages the exponential in the diode equation is negligible, and the current is a constant (negative) reverse current value of $-I_S$. The reverse *breakdown region* is not modeled by the Shockley diode equation.

For even rather small *forward bias* voltages the exponential is very large, since the thermal voltage is very small in comparison. The subtracted '1' in the diode equation is then negligible and the forward diode current can be approximated by

$$I = I_S e^{\frac{V_D}{nV_T}}$$

The use of the diode equation in circuit problems is illustrated in the article on diode modeling.

Small-signal Behavior

At forward voltages less than the saturation voltage, the voltage versus current characteristic curve of most diodes is not a straight line. The current can be approximated by $I = I_S e^{\frac{V_D}{nV_T}}$ as mentioned in the previous section.

In detector and mixer applications, the current can be estimated by a Taylor's series. The odd terms can be omitted because they produce frequency components that are outside the pass band of the mixer or detector. Even terms beyond the second derivative usually need not be included because they are small compared to the second order term. The desired current component is approximately proportional to the square of the input voltage, so the response is called *square law* in this region.

Reverse-recovery Effect

Following the end of forward conduction in a p–n type diode, a reverse current can flow for a short time. The device does not attain its blocking capability until the mobile charge in the junction is depleted.

The effect can be significant when switching large currents very quickly. A certain amount of "reverse recovery time" t_r (on the order of tens of nanoseconds to a few microseconds) may be required to remove the reverse recovery charge Q_r from the diode. During this recovery time, the diode can actually conduct in the reverse direction. This might give rise to a large constant current in the reverse direction for a short time while the diode is reverse biased. The magnitude of such a reverse current is determined by the operating circuit (i.e., the series resistance) and the diode is said to be in the storage-phase. In certain real-world cases it is important to consider the losses that are incurred by this non-ideal diode effect. However, when the slew rate of the current is not so severe (e.g. Line frequency) the effect can be safely ignored. For most applications, the effect is also negligible for Schottky diodes.

The reverse current ceases abruptly when the stored charge is depleted; this abrupt stop is exploited in step recovery diodes for generation of extremely short pulses.

Types of Semiconductor Diode

Normal (p–n) diodes, which operate as described above, are usually made of doped silicon or germanium. Before the development of silicon power rectifier diodes, cuprous oxide and later selenium was used. Their low efficiency required a much higher forward voltage to be applied (typically 1.4 to 1.7 V per "cell", with multiple cells stacked so as to increase the peak inverse voltage rating for application in high voltage rectifiers), and required a large heat sink (often an extension of the diode's metal substrate), much larger than the later silicon diode of the same current ratings would require. The vast majority of all diodes are the p–n diodes found in CMOS integrated circuits, which include two diodes per pin and many other internal diodes.

Avalanche diodes

> These are diodes that conduct in the reverse direction when the reverse bias voltage exceeds the breakdown voltage. These are electrically very similar to Zener diodes (and are often mistakenly called Zener diodes), but break down by a different mechanism: the *avalanche effect*. This occurs when the reverse electric field applied across the p–n junction causes a wave of ionization, reminiscent of an avalanche, leading to a large current. Avalanche diodes are designed to break down at a well-defined reverse voltage without being destroyed. The difference between the avalanche diode (which has a reverse breakdown above about 6.2 V) and the Zener is that the channel length of the former exceeds the mean free path of the electrons, resulting in many collisions between them on the way through the channel. The only practical difference between the two types is they have temperature coefficients of opposite polarities.

Constant current diodes

These are actually JFETs with the gate shorted to the source, and function like a two-terminal current-limiting analog to the voltage-limiting Zener diode. They allow a current through them to rise to a certain value, and then level off at a specific value. Also called *CLDs, constant-current diodes, diode-connected transistors*, or *current-regulating diodes*.

Crystal rectifiers or crystal diodes

These are point-contact diodes. The 1N21 series and others are used in mixer and detector applications in radar and microwave receivers. The 1N34A is another example of a crystal diode.

Esaki or tunnel diodes

These have a region of operation showing negative resistance caused by quantum tunneling, allowing amplification of signals and very simple bistable circuits. Because of the high carrier concentration, tunnel diodes are very fast, may be used at low (mK) temperatures, high magnetic fields, and in high radiation environments. Because of these properties, they are often used in spacecraft.

Gunn diodes

These are similar to tunnel diodes in that they are made of materials such as GaAs or InP that exhibit a region of negative differential resistance. With appropriate biasing, dipole domains form and travel across the diode, allowing high frequency microwave oscillators to be built.

Light-emitting diodes (LEDs)

In a diode formed from a direct band-gap semiconductor, such as gallium arsenide, charge carriers that cross the junction emit photons when they recombine with the majority carrier on the other side. Depending on the material, wavelengths (or colors) from the infrared to the near ultraviolet may be produced. The forward potential of these diodes depends on the wavelength of the emitted photons: 2.1 V corresponds to red, 4.0 V to violet. The first LEDs were red and yellow, and higher-frequency diodes have been developed over time. All LEDs produce incoherent, narrow-spectrum light; "white" LEDs are actually combinations of three LEDs of a different color, or a blue LED with a yellow scintillator coating. LEDs can also be used as low-efficiency photodiodes in signal applications. An LED may be paired with a photodiode or phototransistor in the same package, to form an opto-isolator.

Laser diodes

When an LED-like structure is contained in a resonant cavity formed by polishing the parallel end faces, a laser can be formed. Laser diodes are commonly used in optical storage devices and for high speed optical communication.

Thermal diodes

This term is used both for conventional p–n diodes used to monitor temperature because of their varying forward voltage with temperature, and for Peltier heat pumps for thermoelectric heating and cooling. Peltier heat pumps may be made from semiconductor, though they do not have any rectifying junctions, they use the differing behaviour of charge carriers in N and P type semiconductor to move heat.

Perun's diodes

This is a special type of voltage-surge protection diode. It is characterized by the symmetrical voltage-current characteristic, similar to DIAC. It has much faster response time however, that's why it is used in demanding applications.

Photodiodes

All semiconductors are subject to optical charge carrier generation. This is typically an undesired effect, so most semiconductors are packaged in light blocking material. Photodiodes are intended to sense light(photodetector), so they are packaged in materials that allow light to pass, and are usually PIN (the kind of diode most sensitive to light). A photodiode can be used in solar cells, in photometry, or in optical communications. Multiple photodiodes may be packaged in a single device, either as a linear array or as a two-dimensional array. These arrays should not be confused with charge-coupled devices.

PIN diodes

A PIN diode has a central un-doped, or *intrinsic*, layer, forming a p-type/intrinsic/n-type structure. They are used as radio frequency switches and attenuators. They are also used as large-volume, ionizing-radiation detectors and as photodetectors. PIN diodes are also used in power electronics, as their central layer can withstand high voltages. Furthermore, the PIN structure can be found in many power semiconductor devices, such as IGBTs, power MOSFETs, and thyristors.

Schottky diodes

Schottky diodes are constructed from a metal to semiconductor contact. They have a lower forward voltage drop than p–n junction diodes. Their forward voltage drop at forward currents of about 1 mA is in the range 0.15 V to 0.45 V, which makes them useful in voltage clamping applications and prevention of transistor saturation. They can also be used as low loss rectifiers, although their reverse leakage current is in general higher than that of other diodes. Schottky diodes are majority carrier devices and so do not suffer from minority carrier storage problems that slow down many other diodes—so they have a faster reverse recovery than p–n junction diodes. They also tend to have much lower junction capacitance than p–n diodes, which provides for high switching speeds and their use in high-speed circuitry and RF devices such as switched-mode power supply, mixers, and detectors.

Super barrier diodes

Super barrier diodes are rectifier diodes that incorporate the low forward voltage drop of the Schottky diode with the surge-handling capability and low reverse leakage current of a normal p–n junction diode.

Gold-doped diodes

As a dopant, gold (or platinum) acts as recombination centers, which helps a fast recombination of minority carriers. This allows the diode to operate at signal frequencies, at the expense of a higher forward voltage drop. Gold-doped diodes are faster than other p–n diodes (but not as fast as Schottky diodes). They also have less reverse-current leakage than Schottky diodes (but not as good as other p–n diodes). A typical example is the 1N914.

Snap-off or Step recovery diodes

The term *step recovery* relates to the form of the reverse recovery characteristic of these devices. After a forward current has been passing in an SRD and the current is interrupted or reversed, the reverse conduction will cease very abruptly (as in a step waveform). SRDs can, therefore, provide very fast voltage transitions by the very sudden disappearance of the charge carriers.

Stabistors or *Forward Reference Diodes*

The term *stabistor* refers to a special type of diodes featuring extremely stable forward voltage characteristics. These devices are specially designed for low-voltage stabilization applications requiring a guaranteed voltage over a wide current range and highly stable over temperature.

Transient voltage suppression diode (TVS)

These are avalanche diodes designed specifically to protect other semiconductor devices from high-voltage transients. Their p–n junctions have a much larger cross-sectional area than those of a normal diode, allowing them to conduct large currents to ground without sustaining damage.

Varicap or varactor diodes

These are used as voltage-controlled capacitors. These are important in PLL (phase-locked loop) and FLL (frequency-locked loop) circuits, allowing tuning circuits, such as those in television receivers, to lock quickly on to the frequency. They also enabled tunable oscillators in early discrete tuning of radios, where a cheap and stable, but fixed-frequency, crystal oscillator provided the reference frequency for a voltage-controlled oscillator.

Zener diodes

These can be made to conduct in reverse bias (backward), and are correctly termed reverse breakdown diodes. This effect, called Zener breakdown, occurs at a precisely defined voltage, allowing the diode to be used as a precision voltage reference. The term Zener diode is colloquially applied to several types of breakdown diodes, but strictly speaking Zener diodes have a breakdown voltage of below 5 volts, whilst avalanche diodes are used for breakdown voltages above that value. In practical voltage reference circuits, Zener and switching diodes are connected in series and opposite directions to balance the temperature coefficient response of the diodes to near-zero. Some devices labeled as high-voltage

Zener diodes are actually avalanche diodes. Two (equivalent) Zeners in series and in reverse order, in the same package, constitute a transient absorber (or Transorb, a registered trademark).

Other uses for semiconductor diodes include the sensing of temperature, and computing analog logarithms.

Graphic Symbols

The symbol used to represent a particular type of diode in a circuit diagram conveys the general electrical function to the reader. There are alternative symbols for some types of diodes, though the differences are minor. The triangle in the symbols points to the forward direction, i.e. in the direction of conventional current flow.

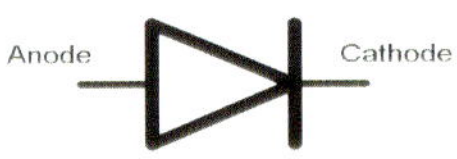

Diode

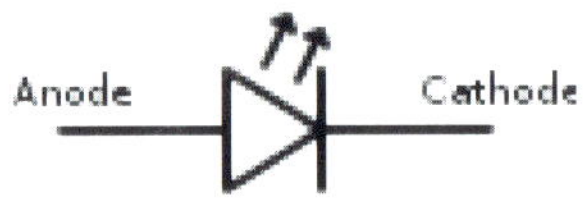

Light-emitting diode (LED)

Photodiode

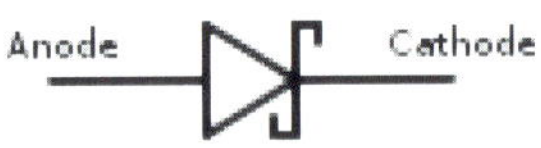

Schottky diode

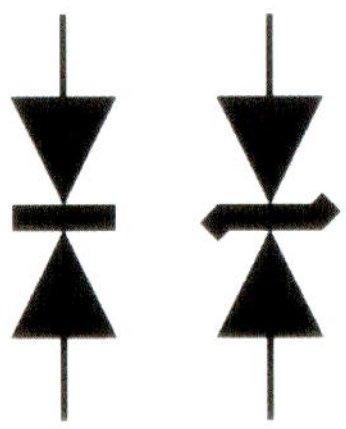
Transient-voltage-suppression diode (TVS)

Tunnel diode

Varicap

Zener diode

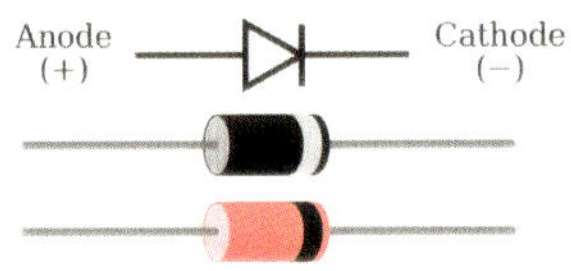

Typical diode packages in same alignment as diode symbol. Thin bar depicts the cathode.

Numbering and Coding Schemes

There are a number of common, standard and manufacturer-driven numbering and coding schemes for diodes; the two most common being the EIA/JEDEC standard and the European Pro Electron standard:

EIA/JEDEC

The standardized 1N-series numbering *EIA370* system was introduced in the US by EIA/JEDEC (Joint Electron Device Engineering Council) about 1960. Most diodes have a 1-prefix designation (e.g., 1N4003). Among the most popular in this series were: 1N34A/1N270 (germanium signal), 1N914/1N4148 (silicon signal), 1N400x (silicon 1A power rectifier), and 1N580x (silicon 3A power rectifier).

JIS

The JIS semiconductor designation system has all semiconductor diode designations starting with "1S".

Pro Electron

The European Pro Electron coding system for active components was introduced in 1966 and comprises two letters followed by the part code. The first letter represents the semiconductor material used for the component (A = germanium and B = silicon) and the second letter represents the general function of the part (for diodes, A = low-power/signal, B = variable capacitance, X = multiplier, Y = rectifier and Z = voltage reference); for example:

- AA-series germanium low-power/signal diodes (e.g., AA119)
- BA-series silicon low-power/signal diodes (e.g., BAT18 silicon RF switching diode)
- BY-series silicon rectifier diodes (e.g., BY127 1250V, 1A rectifier diode)
- BZ-series silicon Zener diodes (e.g., BZY88C4V7 4.7V Zener diode)

Other common numbering / coding systems (generally manufacturer-driven) include:

- GD-series germanium diodes (e.g., GD9) – this is a very old coding system
- OA-series germanium diodes (e.g., OA47) – a coding sequence developed by Mullard, a UK company

As well as these common codes, many manufacturers or organisations have their own systems too – for example:

- HP diode 1901-0044 = JEDEC 1N4148
- UK military diode CV448 = Mullard type OA81 = GEC type GEX23

Related Devices

- Rectifier
- Transistor
- Thyristor or silicon controlled rectifier (SCR)
- TRIAC
- DIAC
- Varistor

In optics, an equivalent device for the diode but with laser light would be the Optical isolator, also known as an Optical Diode, that allows light to only pass in one direction. It uses a Faraday rotator as the main component.

Applications

Radio Demodulation

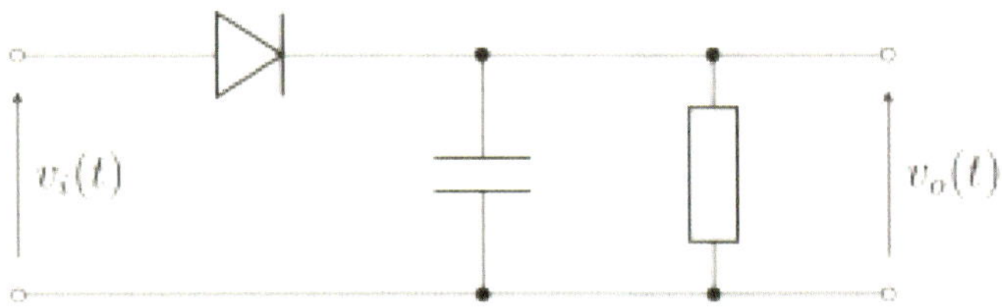

A simple envelope demodulator circuit.

The first use for the diode was the demodulation of amplitude modulated (AM) radio broadcasts. The history of this discovery is treated in depth in the radio article. In summary, an AM signal consists of alternating positive and negative peaks of a radio carrier wave, whose amplitude or envelope is proportional to the original audio signal. The diode rectifies the AM radio frequency signal, leaving only the positive peaks of the carrier wave. The audio is then extracted from the rectified carrier wave using a simple filter and fed into an audio amplifier or transducer, which generates sound waves.

In microwave and millimeter wave technology, beginning in the 1930s, researchers improved and miniaturized the crystal detector. Point contact diodes (*crystal diodes*) and Schottky diodes are used in radar, microwave and millimeter wave detectors.

Power Conversion

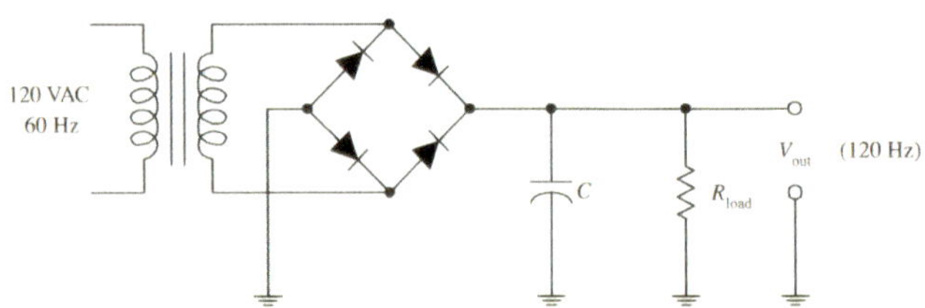

Schematic of basic ac-to-dc power supply

Rectifiers are constructed from diodes, where they are used to convert alternating current (ac) electricity into direct current (dc). Automotive alternators are a common example, where the diode, which rectifies the AC into dc, provides better performance than the commutator or earlier, dynamo. Similarly, diodes are also used in *Cockcroft–Walton voltage multipliers* to convert ac into higher ac voltages.

Over-voltage Protection

Diodes are frequently used to conduct damaging high voltages away from sensitive electronic devices. They are usually reverse-biased (non-conducting) under normal circumstances. When the voltage rises above the normal range, the diodes become forward-biased (conducting). For example, diodes are used in (stepper motor and H-bridge) motor controller and relay circuits to de-energize coils rapidly without the damaging voltage spikes that would otherwise occur. (A diode used in such an application is called a flyback diode). Many integrated circuits also incorporate diodes on the connection pins to prevent external voltages from damaging their sensitive transistors. Specialized diodes are used to protect from over-voltages at higher power.

Logic Gates

Diodes can be combined with other components to construct AND and OR logic gates. This is referred to as diode logic.

Ionizing Radiation Detectors

In addition to light, semiconductor diodes are sensitive to more energetic radiation. In electronics, cosmic rays and other sources of ionizing radiation cause noise pulses and single and multiple bit errors. This effect is sometimes exploited by particle detectors to detect radiation. A single particle of radiation, with thousands or millions of electron volts of energy, generates many charge carrier pairs, as its energy is deposited in the semiconductor material. If the depletion layer is large enough to catch the whole shower or to stop a heavy particle, a fairly accurate measurement of the particle's energy can be made, simply by measuring the charge conducted and without the complexity of a magnetic spectrometer, etc. These semiconductor radiation detectors need efficient and uniform charge collection and low leakage current. They are often cooled by liquid nitrogen. For longer-range (about a centimetre) particles, they need a very large depletion depth and large area. For short-range particles, they need any contact or un-depleted semiconductor on at least one surface to be very thin. The back-bias voltages are near breakdown (around a thousand volts per centimetre). Germanium and silicon are common materials. Some of these detectors sense position as well as energy. They have a finite life, especially when detecting heavy particles, because of radiation damage. Silicon and germanium are quite different in their ability to convert gamma rays to electron showers.

Semiconductor detectors for high-energy particles are used in large numbers. Because of energy loss fluctuations, accurate measurement of the energy deposited is of less use.

Temperature Measurements

A diode can be used as a temperature measuring device, since the forward voltage drop across the diode depends on temperature, as in a silicon bandgap temperature sensor. From the Shockley ideal diode equation given above, it might *appear* that the voltage has a *positive* temperature coefficient (at a constant current), but usually the variation of the reverse saturation current term is more significant than the variation in the thermal voltage term. Most diodes therefore have a *negative* temperature coefficient, typically −2 mV/°C for silicon diodes. The temperature coefficient is approximately constant for temperatures above about 20 kelvins. Some graphs are given for 1N400x series, and CY7 cryogenic temperature sensor.

Current Steering

Diodes will prevent currents in unintended directions. To supply power to an electrical circuit during a power failure, the circuit can draw current from a battery. An uninterruptible power supply may use diodes in this way to ensure that current is only drawn from the battery when necessary. Likewise, small boats typically have two circuits each with their own battery/batteries: one used for engine starting; one used for domestics. Normally, both are charged from a single alternator, and a heavy-duty split-charge diode is used to prevent the higher-charge battery (typically the engine battery) from discharging through the lower-charge battery when the alternator is not running.

Diodes are also used in electronic musical keyboards. To reduce the amount of wiring needed in electronic musical keyboards, these instruments often use keyboard matrix circuits. The keyboard controller scans the rows and columns to determine which note the player has pressed. The problem with matrix circuits is that, when several notes are pressed at once, the current can flow backwards through the circuit and trigger "phantom keys" that cause "ghost" notes to play. To avoid triggering unwanted notes, most keyboard matrix circuits have diodes soldered with the switch under each key of the musical keyboard. The same principle is also used for the switch matrix in solid-state pinball machines.

Waveform Clipper

Diodes can be used to limit the positive or negative excursion of a signal to a prescribed voltage.

Clamper

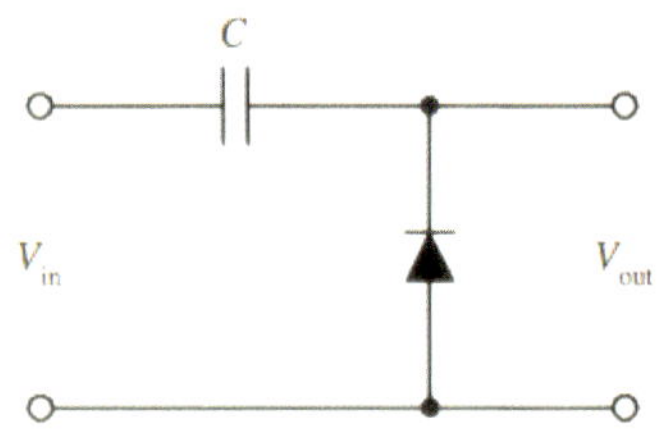

This simple diode clamp will clamp the negative peaks of the incoming waveform to the common rail voltage

A diode clamp circuit can take a periodic alternating current signal that oscillates between positive and negative values, and vertically displace it such that either the positive, or the negative peaks occur at a prescribed level. The clamper does not restrict the peak-to-peak excursion of the signal, it moves the whole signal up or down so as to place the peaks at the reference level.

Zener Diode

A Zener diode is a special type of device designed to operate in the zener breakdown region. Zener diodes acts like normal p-n junction diodes under forward biased condition. When forward biased voltage is applied to the zener diode it allows large amount of electric current and blocks only a small amount of electric current.

Zener diode is heavily doped than the normal p-n junction diode. Hence, it has very thin depletion region. A zener diodes allow more electric current than the normal p-n junction diodes.

Zener diode symbol

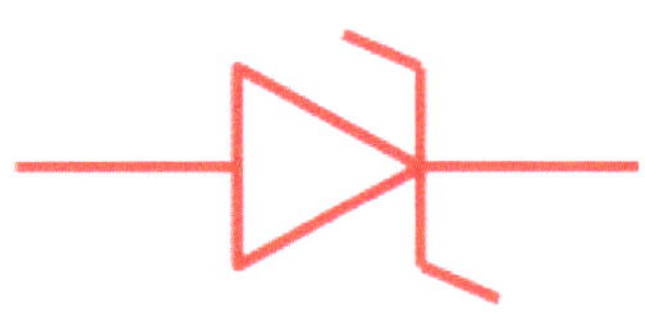

Zener diode allows electric current in forward direction like a normal diode but also allows electric current in the reverse direction if the applied reverse voltage is greater than the zener voltage. Zener diode is always connected in reverse direction because it is specifically designed to work in reverse direction.

Operation

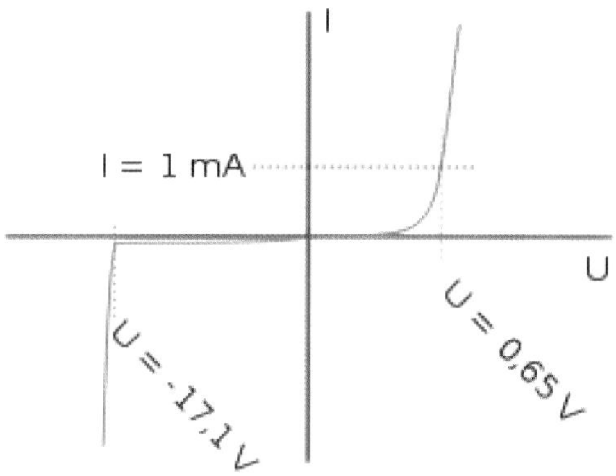

Current-voltage characteristic of a Zener diode with a breakdown voltage of 17 volts. Notice the change of voltage scale between the forward biased (positive) direction and the reverse biased (negative) direction.

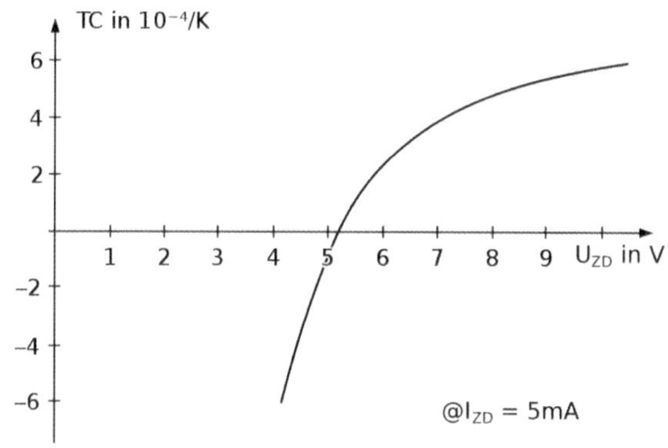

Temperature coefficient of Zener voltage against nominal Zener voltage.

A conventional solid-state diode allows significant current if it is reverse-biased above its reverse breakdown voltage. When the reverse bias breakdown voltage is exceeded, a conventional diode is subject to high current due to avalanche breakdown. Unless this current is limited by circuitry, the diode may be permanently damaged due to overheating. A Zener diode exhibits almost the same properties, except the device is specially designed so as to have a reduced breakdown voltage, the so-called Zener voltage. By contrast with the conventional device, a reverse-biased Zener diode exhibits a controlled breakdown and allows the current to keep the voltage across the Zener diode close to the Zener breakdown voltage. For example, a diode with a Zener breakdown voltage of 3.2 V exhibits a voltage drop of very nearly 3.2 V across a wide range of reverse currents. The Zener diode is therefore ideal for applications such as the generation of a reference voltage (e.g. for an amplifier stage), or as a voltage stabilizer for low-current applications.

Another mechanism that produces a similar effect is the avalanche effect as in the avalanche diode. The two types of diode are in fact constructed the same way and both effects are present in diodes of this type. In silicon diodes up to about 5.6 volts, the Zener effect is the predominant effect and shows a marked negative temperature coefficient. Above 5.6 volts, the avalanche effect becomes predominant and exhibits a positive temperature coefficient.

In a 5.6 V diode, the two effects occur together, and their temperature coefficients nearly cancel each other out, thus the 5.6 V diode is useful in temperature-critical applications. An alternative, which is used for voltage references that need to be highly stable over long periods of time, is to use a Zener diode with a temperature coefficient (TC) of +2 mV/°C (breakdown voltage 6.2–6.3 V) connected in series with a forward-biased silicon diode (or a transistor B-E junction) manufactured on the same chip. The forward-biased diode has a temperature coefficient of −2 mV/°C, causing the TCs to cancel out.

Modern manufacturing techniques have produced devices with voltages lower than 5.6 V with negligible temperature coefficients, but as higher-voltage devices are encountered, the temperature coefficient rises dramatically. A 75 V diode has 10 times the coefficient of a 12 V diode.

Zener and avalanche diodes, regardless of breakdown voltage, are usually marketed under the umbrella term of "Zener diode".

Under 5.6 V, where the Zener effect dominates, the IV curve near breakdown is much more rounded, which calls for more care in targeting its biasing conditions. The IV curve for Zeners above 5.6 V (being dominated by Avalanche), is much sharper at breakdown.

Construction

The Zener diode's operation depends on the heavy doping of its p-n junction. The depletion region formed in the diode is very thin (<1 μm) and the electric field is consequently very high (about 500 kV/m) even for a small reverse bias voltage of about 5 V, allowing electrons to tunnel from the valence band of the p-type material to the conduction band of the n-type material.

At the atomic scale, this tunnelling corresponds to the transport of valence band electrons into the empty conduction band states; as a result of the reduced barrier between these bands and high electric fields that are induced due to the relatively high levels of doping on both sides. The breakdown voltage can be controlled quite accurately in the doping process. While tolerances within 0.07% are available, the most widely used tolerances are 5% and 10%. Breakdown voltage for commonly available Zener diodes can vary widely from 1.2 volts to 200 volts.

For diodes that are lightly doped the breakdown is dominated by the avalanche effect rather than the Zener effect. Consequently, the breakdown voltage is higher (over 5.6 V) for these devices.

Surface Zeners

The emitter-base junction of a bipolar NPN transistor behaves as a Zener diode, with breakdown voltage at about 6.8 V for common bipolar processes and about 10 V for lightly doped base regions in BiCMOS processes. Older processes with poor control of doping characteristics had the variation of Zener voltage up to ±1 V, newer processes using ion implantation can achieve no more than ±0.25 V. The NPN transistor structure can be employed as a *surface Zener diode*, with collector and emitter connected together as its cathode and base region as anode. In this approach the base doping profile usually narrows towards the surface, creating a region with intensified electric field where the avalanche breakdown occurs. The hot carriers produced by acceleration in the intense field sometime shoot into the oxide layer above the junction and become trapped there. The accumulation of trapped charges can then cause 'Zener walkout', a corresponding change of the Zener voltage of the junction. The same effect can be achieved by radiation damage.

The emitter-base Zener diodes can handle only smaller currents as the energy is dissipated in the base depletion region which is very small. Higher amount of dissipated energy (higher current for longer time, or a short very high current spike) causes thermal damage to the junction and/or its contacts. Partial damage of the junction can shift its Zener voltage. Total destruction of the Zener junction by overheating it and causing migration of metallization across the junction ("spiking") can be used intentionally as a 'Zener zap' antifuse.

Subsurface Zeners

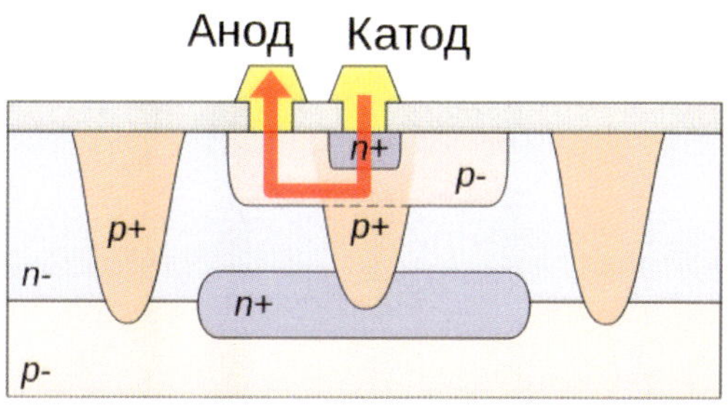

Buried zener structure (labels at top: "Anode", "Cathode")

A subsurface Zener diode, also called 'buried Zener', is a device similar to the Surface Zener, but with the avalanche region located deeper in the structure, typically several micrometers below the oxide. The hot carriers then lose energy by collisions with the semiconductor lattice before reaching the oxide layer and cannot be trapped there. The Zener walkout phenomenon therefore does not occur here, and the buried Zeners have voltage constant over their entire lifetime. Most buried Zeners have breakdown voltage of 5–7 volts. Several different junction structures are used.

Uses

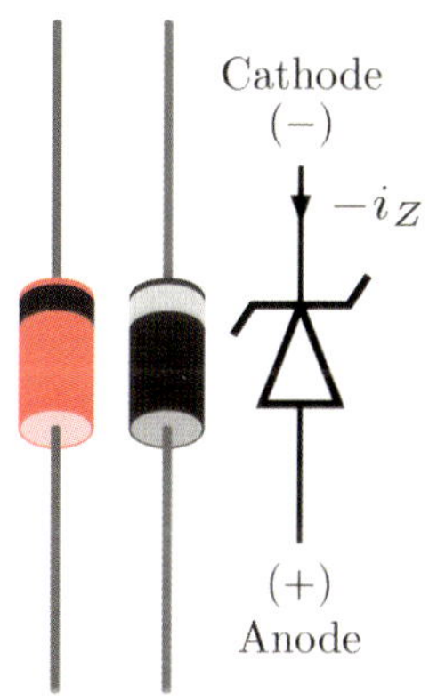

Zener diode shown with typical packages. *Reverse* current $-i_Z$ is shown.

Zener diodes are widely used as voltage references and as shunt regulators to regulate the voltage across small circuits. When connected in parallel with a variable voltage source so that it is reverse biased, a Zener diode conducts when the voltage reaches the diode's reverse breakdown voltage. From that point on, the relatively low impedance of the diode keeps the voltage across the diode at that value.

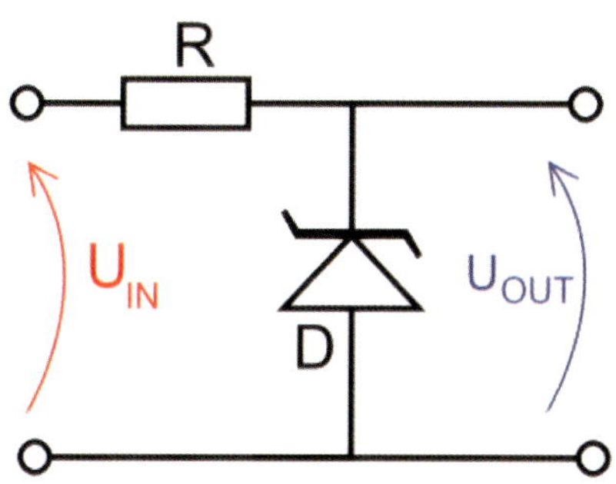

In this circuit, a typical voltage reference or regulator, an input voltage, U_{in}, is regulated down to a stable output voltage U_{out}. The breakdown voltage of diode D is stable over a wide current range and holds U_{out} relatively constant even though the input voltage may fluctuate over a fairly wide range. Because of the low impedance of the diode when operated like this, resistor R is used to limit current through the circuit.

In the case of this simple reference, the current flowing in the diode is determined using Ohm's law and the known voltage drop across the resistor R;

$$I_{\text{diode}} = \frac{U_{\text{in}} - U_{\text{out}}}{R}$$

The value of R must satisfy two conditions:

- R must be small enough that the current through D keeps D in reverse breakdown. For example, the common BZX79C5V6 device, a 5.6 V 0.5 W Zener diode, has a recommended reverse current of 5 mA. If insufficient current exists through D, then U_{out} is unregulated and less than the nominal breakdown voltage (this differs to voltage-regulator tubes where the output voltage will be higher than nominal and could rise as high as U_{in}). When calculating R, allowance must be made for any current through the external load, not shown in this diagram, connected across U_{out}.
- R must be large enough that the current through D does not destroy the device. If the current through D is I_D, its breakdown voltage V_B and its maximum power dissipation P_{max} correlate as such: $I_D V_B < P_{\max}$.

A load may be placed across the diode in this reference circuit, and as long as the Zener stays in reverse breakdown, the diode provides a stable voltage source to the load. Zener diodes in this configuration are often used as stable references for more advanced voltage regulator circuits.

Shunt regulators are simple, but the requirements that the ballast resistor be small enough to avoid excessive voltage drop during worst-case operation (low input voltage concurrent with high load current) tends to leave a lot of current flowing in the diode much of the time, making for a fairly wasteful regulator with high quiescent power dissipation, only suitable for smaller loads.

These devices are also encountered, typically in series with a base-emitter junction, in transistor stages where selective choice of a device centered around the avalanche or Zener point can be used to introduce compensating temperature co-efficient balancing of the transistor p–n junction. An example of this kind of use would be a DC error amplifier used in a regulated power supply circuit feedback loop system.

Zener diodes are also used in surge protectors to limit transient voltage spikes.

Another application of the Zener diode is the use of noise caused by its avalanche breakdown in a random number generator.

Waveform Clipper

Two Zener diodes facing each other in series will act to clip both halves of an input signal. Waveform clippers can be used not only to reshape a signal, but also to prevent voltage spikes from affecting circuits that are connected to the power supply.

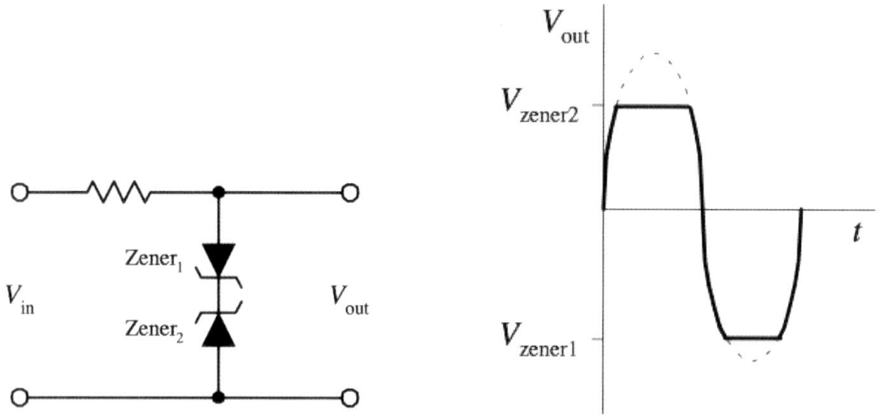

Examples of a Waveform Clipper

Voltage Shifter

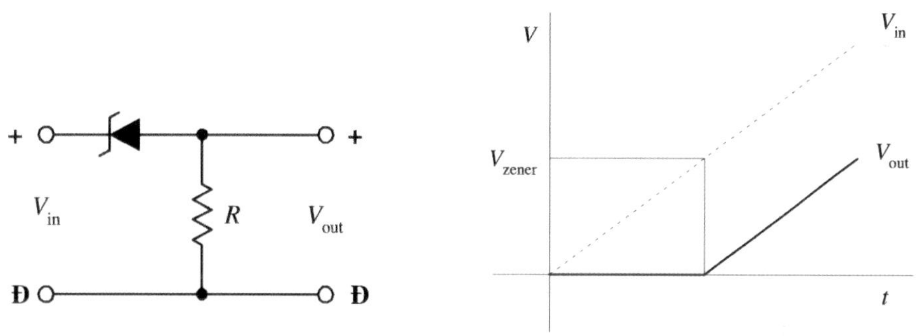

Examples of a Voltage Shifter

A Zener diode can be applied to a circuit with a resistor to act as a voltage shifter. This circuit lowers the output voltage by a quantity that is equal to the Zener diode's breakdown voltage.

Voltage Regulator

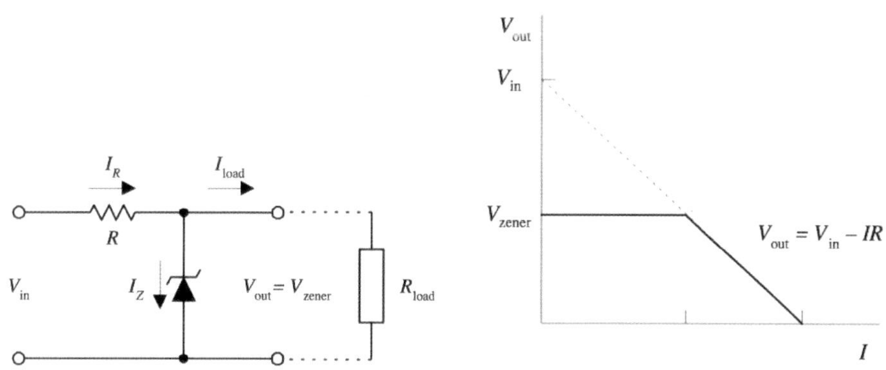

Examples of a Voltage Regulator

A Zener diode can be applied in a voltage regulator circuit to regulate the voltage applied to a load, such as in a linear regulator.

Tunnel Diode

In the tunnel diode, the semiconductor materials used in forming a junction are doped to the extent of one-thousand impurity atoms for ten-million semiconductor atoms. This heavy doping produces an extremely narrow depletion. Also because of the heavy doping, a tunnel diode exhibits an unusual current-voltage characteristic curve as compared with that of an ordinary junction diode as shown in Figure.

At zero bias, there are no states available for tunneling, while with increasing positive bias, more and more such states exist. We thus expect the current to rise with voltage. The current will have a maximum when the band edge on one side coincides with the Fermi level on the other, since for voltages beyond this point, the electrons will have fewer states to tunnel to. Ideally, the tunneling current should therefore go to zero, while in practise tunneling is still possible through a small number of states in the forbidden gap. At the normal cut-in voltage of the p-n-junction diode, thermionic excitation "over the top" of the potential barrier will occur, and the current will again rise, as in the normal 1-V-characteristic of a diode. The end result will be a characteristic as shown in Figure, with a region where the differential conductance is negative. The "valley" current, Iv, is determined by the excess current due to the gap states. It is in general advantageous to achieve a large ratio Ip / Iv.

Operation of tunnel diode depends on the tunneling effect. In an open circuit i.e. when no bias is applied, Fermi level EF in the p-side is at the same energy as the Fermi level EF in the n-side as shown in Figure (a). It can be seen from energy band diagram of tunnel diode in zero bias mode, that there are no filled states on one side of the junction which are at the same energy as empty allowed states on the other side.

Thus there is no flow of charge in either direction across the junction and current is zero.

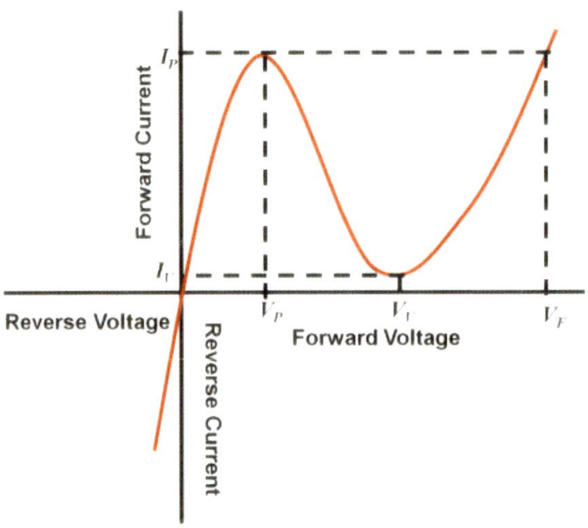

I-V characteristics of a tunnel diode

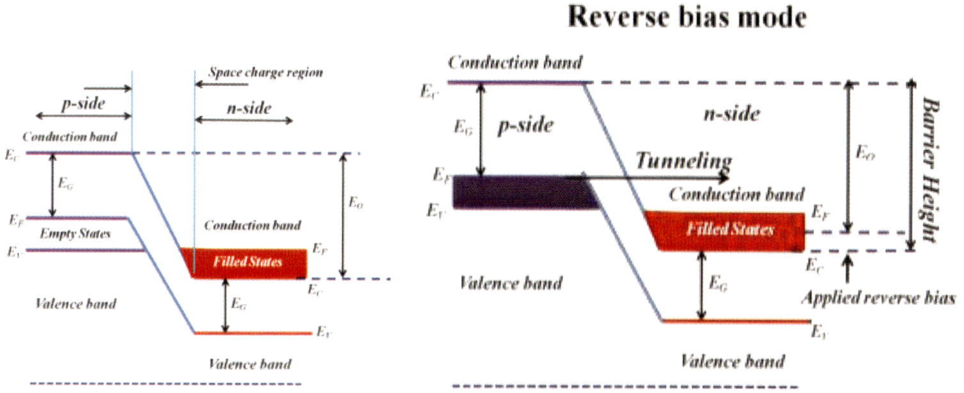

Energy band diagram of a heavily doped p-n diode
(a) under open circuit condition and (b) with an applied reverse bias

Reverse Bias Mode: When reverse bias is applied, the height of the barrier is increased above the open circuit value EO. Hence, n-side levels must shift downward with respect to the p-side level as shown in Figure (b). Now some energy states in the valence band of the p-side lies at the same level as allowed empty states in the conduction band of the n-side, thus the electrons will tunnel from p to n side leading to reverse diode current. If reverse bias is increased further then tunneling and reverse current increases as shown in Figure. When Tunnel diodes are used in the reverse direction they are called back diodes and can act as fast rectifiers with zero offset voltage and extreme linearity for power signals.

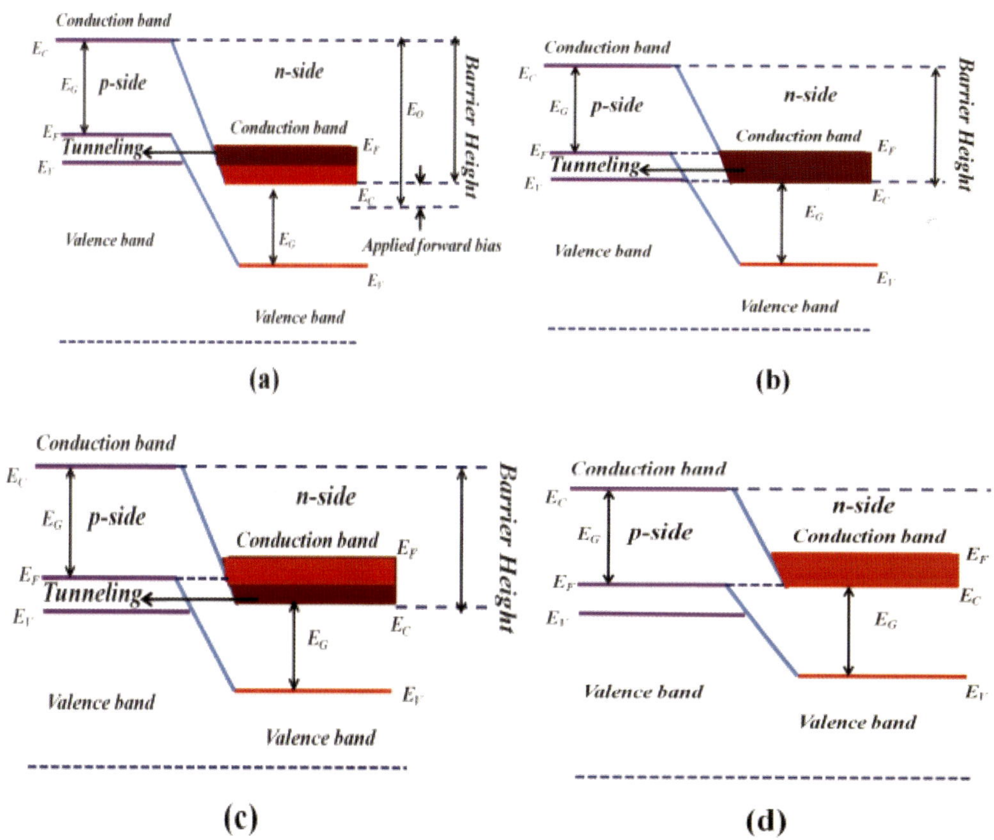

Change in energy band structure of heavily doped p-n diode with increasing forward bias ((a) to (d)).

Forward Bias Mode: When forward bias is applied, the height of the barrier is decreased below the open circuit value EO. Hence, n-side levels must shift upward with respect to the p-side level as shown in Figure (a). It can be seen that there are occupied states in the conduction band of the n-side which are at the same energy level as allowed empty states in the valence band of the p-side. Thus electrons tunnel from the n-side to p-side leading to forward current. As the forward bias is further increased to VP (shown in Figure), the maximum number of electrons tunnel from occupied states of n-side to empty states of the p-side, leading to maximum peak current value IP (Figure (b)). But if the forward bias still increases (> VP), the tunneling current decreases (shown in Figure). In this condition the occupied states of n-sides are shifting towards upwards with respect to empty states of p-sides which reduce the tunneling of electron resulting in negative resistance condition (Figure (c)). When operated in the negative region used as oscillator. When forward bias further increases to VV (shown in Figure), there are no empty allowed states are present on one side of the junction at the same energy as occupied states on the other side, thus tunneling current drops to zero (Figure (d)). On increasing the forward bias voltage (> VV) the tunnel diode starts conducting as normal p-n diode giving rise to forward current again (shown in Figure).

Tunnel diodes of the type described above can be fabricated from any of a number of semiconductor materials. They continued to be used for many years, but at the present time have been superceded by other devices in almost all applications. One primary use was as low-noise amplifiers, where the simplicity and low noise temperature of the device made it attractive. A disadvantage was that dopants in some diodes would slowly diffuse through the device, activated by the high electric field this made the reliability less than desired. The mechanical structure of the device also made it quite delicate. Tunnel diodes thus might have been relegated to the category of obsolete devices, were it not for new advances in fabrication techniques, which have recently made a new version of the tunnel diode feasible with considerably improved characteristics: the "resonant tunneling device", or RTD.

Technical Comparisons

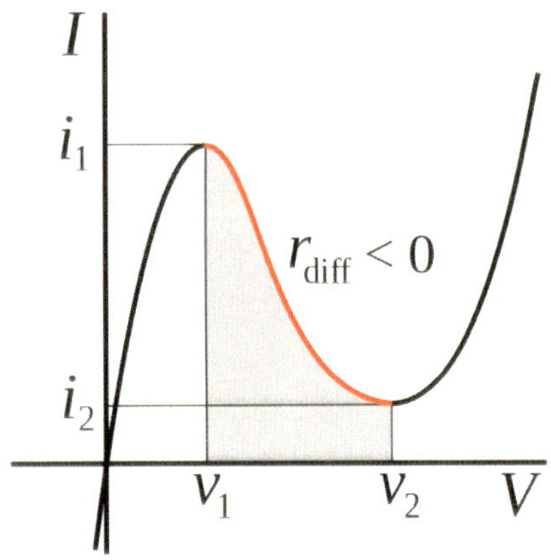

IV curve similar to a tunnel diode characteristic curve.
It has negative differential resistance in the shaded voltage region, between v_1 and v_2.

In a conventional semiconductor diode, conduction takes place while the p–n junction is forward biased and blocks current flow when the junction is reverse biased. This occurs up to a point known as the "reverse breakdown voltage" at which point conduction begins (often accompanied by destruction of the device). In the tunnel diode, the dopant concentrations in the p and n layers are increased to a level such that the reverse breakdown voltage becomes zero and the diode conducts in the reverse direction. However, when forward-biased, an effect occurs called quantum mechanical tunneling which gives rise to a region in its voltage-current behavior where an *increase* in forward voltage is accompanied by a *decrease* in forward current. This negative resistance region can be exploited in a solid state version of the dynatron oscillator which normally uses a tetrode thermionic valve (vacuum tube).

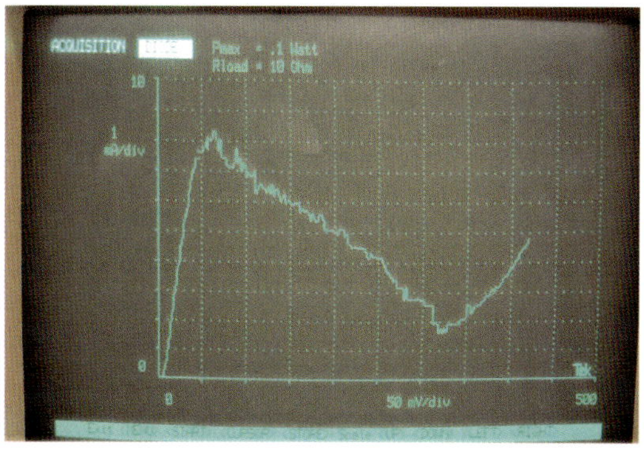

I-V curve of 10mA germanium tunnel diode, taken on a Tektronix model 571 curve tracer.

The tunnel diode showed great promise as an oscillator and high-frequency threshold (trigger) device since it operated at frequencies far greater than the tetrode could: well into the microwave bands. Applications for tunnel diodes included local oscillators for UHF television tuners, trigger circuits in oscilloscopes, high-speed counter circuits, and very fast-rise time pulse generator circuits. The tunnel diode can also be used as a low-noise microwave amplifier. Since its discovery, more conventional semiconductor devices have surpassed its performance using conventional oscillator techniques. For many purposes, a three-terminal device, such as a field-effect transistor, is more flexible than a device with only two terminals. Practical tunnel diodes operate at a few milliamperes and a few tenths of a volt, making them low-power devices. The Gunn diode has similar high frequency capability and can handle more power.

Tunnel diodes are also more resistant to ionizing radiation than other diodes. This makes them well suited to higher radiation environments such as those found in space.

Longevity

Tunnel diodes are notable for their longevity, with devices made in the 1960s still functioning. Writing in *Nature*, Esaki and coauthors state that semiconductor devices in general are extremely stable, and suggest that their shelf life should be "infinite" if kept at room temperature. They go on to report that a small-scale test of 50-year-old devices revealed a "gratifying confirmation of the diode's longevity". As noticed on some samples of Esaki diodes, the gold plated iron pins can in fact corrode and short out to the case. This can usually be diagnosed and treated with simple peroxide/vinegar technique normally used for repairing phone PCBs and the diode inside normally still works.

These components are susceptible to damage by overheating, and thus special care is needed when soldering them. Surplus Russian units are also reliable and often can be purchased for a few pence despite original cost being in the £30-50 range. The units typically sold are GaAs based and have a Ipk/Iv ratio of 5:1 at around 1-20mA Ipk so should be protected against overcurrent.

Resonant Tunneling Devices

The RTD was proposed by Tsu and Esaki (1973) but its realization as a microwave device had to await the advanced state of MBE and OMCVD material growth technology of the 1980's. Chang, Esaki and Tsu measured a small negative differential conductance (Chang et al., 1974). A few years later, Sollner et al. (1983) demonstrated for the first time that this device can have a negative differential conductance which is large enough for practical device applications. The early work of Sollner's group also made it clear, as was expected, that the intrinsic speed of the tunneling process was extremely fast - of the order of 0.1 picoseconds. Resonant tunnel diodes have been investigated intensely since 1983 as a result of the potential for development of very high frequency devices based on the RTD structure. Such devices are at the present time superior to the conventional tunnel diode.

Fabrication of RTDs

The barriers through which electrons tunnel in the RTD are formed by growing alternating, very thin, layers of semiconductors with different bandgaps, rather than relying on thin ~n junction depletion layers, as in the original tunnel diode. A typical structure is shown in Figure.

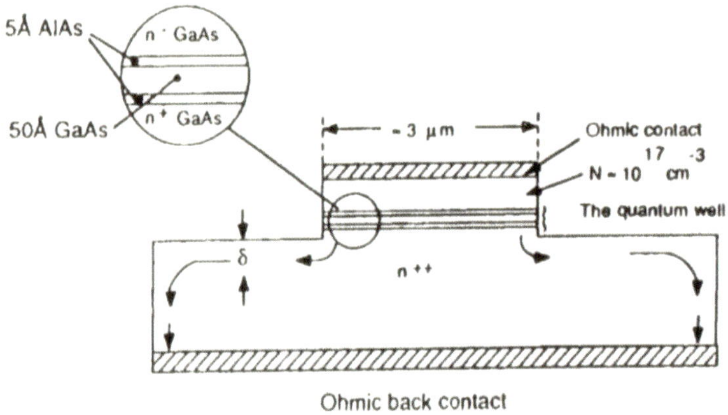

Cross-section of a typical RTD device.

It consists of two layers of AlAs, (bandgap 2.16 eV) of thickness 1.5 nm, on either side of a 5.0 nm thick layer of GaAs, (bandgap 1.43 eV). Thicker layers of n-type GaAs, on either side of the above layers are intended to provide a low resistance path from the ohmic contact (at the top) and the highly doped substrate to the tunneling structure. Electrons in the GaAs layer will find themselves surrounded by two fairly high potential barriers (Figure (a)), and will form quasi-stationary states, according to elementary quantum mechanics. The number of states, and their location, will depend on the thickness and height of the barriers, as well as the width of the "well". Tunneling through the barriers will be enhanced when electrons outside the barriers have an energy equal to one of the energy levels in the well – theoretically a transmission probability of 1 is predicted (Figure (b)).

If the voltage is increased further, the energy of the electrons at the cathode no longer matches that of those inside the well, and the tunneling stops (Figure (c)) and is shown in the I-V-characteristic of Figure (d) results. There is a useful analogy to the transmission of electron waves through the RTD structure with the transmission of electromagnetic waves through a Fabry-Perot resonator (used in optics or quasi-optics), or a microwave resonator such as a waveguide transmission cavity with coupling holes at either end, as indicated in the insert to Figure. In the microwave case, 100% transmission also occurs at resonance if the cavity is lossless. It is assumed in the RTD case (as for the electromagnetic case) that the electron wave-function is coherent through the tunneling process, i.e. the scattering probability inside the well must not be too high for the device to work.

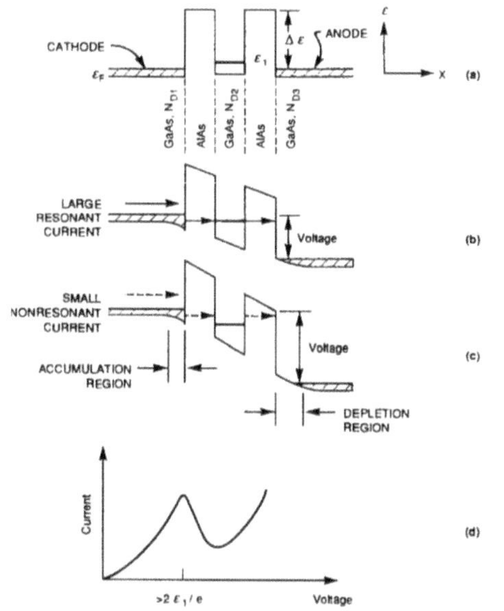

Description of the tunneling process in an RTD device

Equivalent Circuit Model

The first equivalent circuit model used for the RTD was the one which has been applied to the pn junction tunnel diode as shown in Figure.

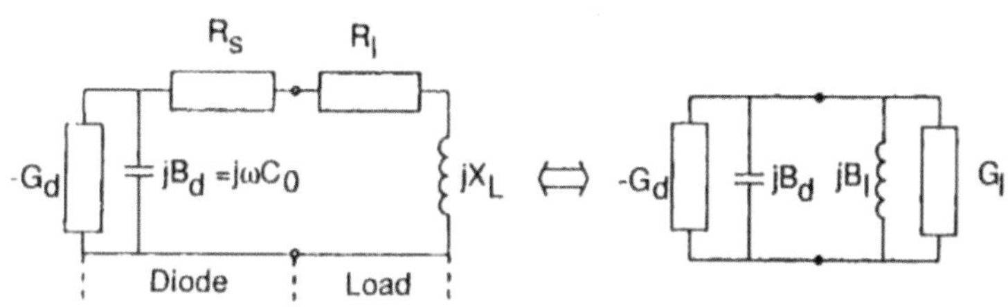

Two versions of the equivalent circuit of an RTD device.

A load with an inductive component has been added, and oscillations will occur at a frequency for which the total admittance of the circuit is zero. The oscillation frequency can easily be derived to be:

$$f_{ase} = \frac{1}{(2\pi C_0)} \times \sqrt{\frac{G}{(R_L + R_s}} - G^2$$

The highest frequency (and of course zero power out) is obtained for the extreme case of RL = 0 and G = 1/2Rs. This frequency is:

$$f_{max} = \frac{1}{2\pi} \times \frac{1}{(2R_s C_O)}$$

The above expression for f_{max} is the one usually given in the tunnel diode. The recent work on RTDs has demonstrated that the RTD can not be described by the simple equivalent circuit in Figure, however. Most importantly, a depletion region is formed "down-stream" from the double-barrier, and a substantial part of the terminal voltage may be dropped across this region, as shown in Figure.

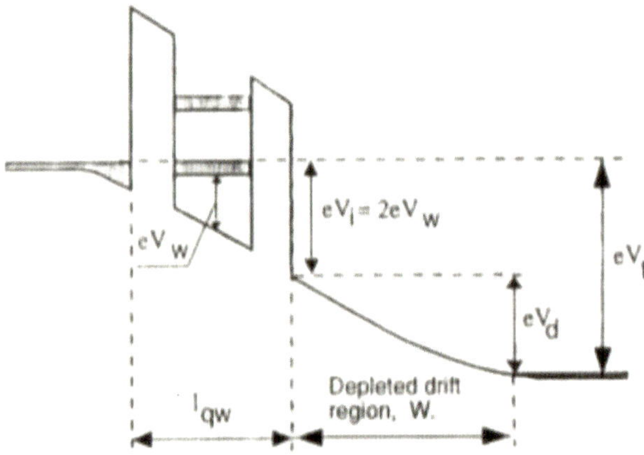

Energy band diagram of an RTD device with a bias voltage applied.

A thin accumulation layer is also located on the opposite side of the double-barrier, the "cathode". The width of the depletion layer can be varied by changing the doping and width of the n-type layer on the anode side.

IMPATT Diode

IMPATT (Impact Avalanche Transit Time) devices are solid state oscillators or amplifiers for microwave and millimeter wave frequencies up to above 200 GHz, generally with high power output (for a solid state device). Efficiencies as high as 25% have recently been obtained with GaAII devices, but the traditional IMPATT is a silicon device with up to 15% efficiency. The basic principle involves a 1800 phase-delay of the current with respect to the voltage, which clearly will make the device exhibit a negative resistance. Shockley (1954) had introduced this general idea, not yet involving impact avalanching to produce the phase-shift. A number of different p-n-junction type structures are used today. During operation, the diode is undergoing (controlled) reverse breakdown. The history of the development of the IMPATT device has some similarities to that of the Gunn-device. Both were being developed in parallel in the early 1960's, for example one Bell Laboratories group worked simultaneously on both devices (DeLoach, 1976). In the case of the IMPATT, there was also a theoretical prediction of one version of the IMPATT operation, published in 1958 by W.T. Read (Read, 1958). The event described above, resulted in the first publication regarding an IMPATT type oscillator (Johnston, DeLoach and Cohen, 1965). Only a couple of months later, Read-type IMPATT operation was also observed (Lee et al., 1965).

Device Structure

The IMPATT diode family includes many different junctions and metal semiconductor devices. The first IMPATT oscillation was obtained from a simple silicon p-n junction diode biased into a reverse avalanche break down and mounted in a microwave cavity. Because of the strong dependence of the ionization coefficient on the electric field, most of the electron–hole pairs are generated in the high field region. The generated electron immediately moves into the N region, while the generated holes drift across the P region. The time required for the hole to reach the contact constitutes the transit time delay.

The original proposal for a microwave device of the IMPATT type was made by Read. The Read diode consists of two regions (i) The Avalanche region (a region with relatively high doping and high field) in which avalanche multiplication occurs and (ii) the drift region (a region with essentially intrinsic doping and constant field) in which the generated holes drift towards the contact. A similar device can be built with the configuration in which electrons generated from the avalanche multiplication drift through the intrinsic region.

An IMPATT diode generally is mounted in a microwave package. The diode is mounted with its low–field region close to a silicon heatsink so that the heat generated at the diode junction can be readily dissipated. Similar microwave packages are used to house other microwave devices.

The IMPATT diode operates over a narrow frequency band, and diode internal dimensions must correlate with the desired operating frequency. An IMPATT oscillator can be tuned by adjusting the resonant frequency of the coupled circuit, and also by varying the current in the diode; this can be used for frequency modulation.

Principle of Operation

If a free electron with a sufficient energy strikes a silicon atom, it can break the covalent bond of silicon and liberate an electron from the covalent bond. If the electron liberated gains energy by being in an electric field and liberates other electrons from other covalent bonds then this process can cascade very quickly into a chain reaction producing a large number of electrons and a large current flow. This phenomenon is called impact avalanche.

At breakdown, the n – region is punched through and forms the avalanche region of the diode. The high resistivity region is the drift zone through which the avalanche generated electrons move toward the anode.

Consider a dc bias V_B, just short of that required to cause breakdown, applied to the diode. Let an AC voltage of sufficiently large magnitude be superimposed on the dc bias, such that during the positive cycle of the AC voltage, the diode is driven deep into the avalanche breakdown. At t=0, the AC voltage is zero, and only a small pre-breakdown current flows through the diode. As t increases, the voltage goes above the breakdown voltage and secondary electron-hole pairs are produced by impact ionization. As long as the field in the avalanche region is maintained above the breakdown field, the electron-hole concentration grows exponentially with t. Similarly this concentration decays exponentially with time when the field is reduced below breakdown voltage during the negative swing of the AC voltage. The holes generated in the avalanche region disappear in the p+ region and are collected by the cathode. The electrons are injected into the i – zone where they drift toward the n+ region. Then, the field in the avalanche region reaches its maximum value and the population of the electron-hole pairs starts building up. At this time, the ionization coefficients have their maximum values. The generated electron concentration does not follow the electric field instantaneously because it also depends on the number of electron-hole pairs already present in the avalanche region. Hence, the electron concentration at this point will have a small value. Even after the field has passed its maximum value, the electron-hole concentration continues to grow because the secondary carrier generation rate still remains above its average value. For this reason, the electron concentration in the avalanche region attains its maximum value at, when the field has dropped to its average value. Thus, it is clear that the avalanche region introduces a 90° phase shift between the AC signal and the electron concentration in this region.

With a further increase in t, the AC voltage becomes negative, and the field in the avalanche region drops below its critical value. The electrons in the avalanche region are then injected into the drift zone which induces a current in the external circuit which has a phase opposite to that of the AC voltage. The AC field, therefore, absorbs energy from the drifting electrons as they are decelerated by the decreasing field. It is clear that an ideal phase shift between the diode current and the AC signal is achieved if the thickness of the drift zone is such that the bunch of electron is collected at the n^+ - anode at the moment the AC voltage goes to zero. This condition is achieved by making the length of the drift region equal to the wavelength of the signal. This situation produces an additional phase shift of 90° between the AC voltage and the diode current.

Operation of IMPATT Devices-Physical Discussion

The Impact Ionization Process

Impact ionization is the process in a material by which one energetic charge carrier can lose energy by the creation of other charge carriers. For example, in semiconductors, an electron (or hole) with enough kinetic energy can knock a bound electron out of its bound state (in the valence band) and promote it to a state in the conduction band, creating an electron-hole pair. For carriers to have sufficient kinetic energy a sufficiently large electric field must be applied, in essence requiring a sufficiently large voltage but not necessarily a large current.

If this occurs in a region of high electrical field then it can result in avalanche breakdown. This process is exploited in avalanche diodes, by which a small optical signal is amplified before entering an external electronic circuit. In an avalanche photodiode the original charge carrier is created by the absorption of a photon.

In some sense, impact ionization is the reverse process to Auger recombination.

Avalanche photodiodes (APD) are used in optical receivers before the signal is given to the receiver circuitry the photon is multiplied with the photocurrent and this increases the sensitivity of the receiver since photocurrent is multiplied before encountering of the thermal noise associated with the receiver circuit.

Reverse break-down is most conveniently first discussed for a p^+-n diode, usually implemented with a p^+-n-n+ structure.

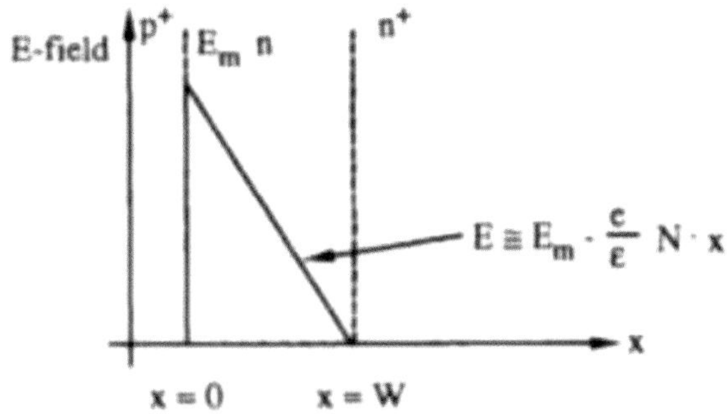

Electric field versus position in a p⁺ -n-diode.

The corresponding energy band diagram for large reverse bias is given in Figure. The reverse current consists primarily of holes which diffuse up to the potential barrier from the n-side, and then quickly traverse the junction, accelerated by the very strong electric field there. A smaller number of electrons cross the junction in the opposite direction, smaller because the p-side is more highly doped, and becauseit is the minority carriers which constitute the reverse direction current. The holes will become energetic enough to be able to ionize an electron pair, somewhere in the depletion layer. This process requires about 1.5 times the bandgap energy, i.e. about 1.6 eV for silicon.

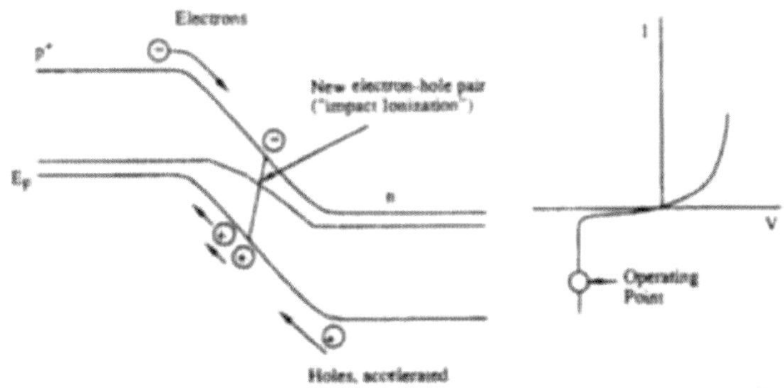

Energy bands of a p⁺ -n-junction in reverse bias

The impact ionization will be able to sustain an avalanche if the probability that a carrier will cause an ionization event on its transit of the depletion region is equal to 1.0 or higher, and under these circumstances, the process will start to grow rapidly (in time) in the manner of an avalanche. The avalanche condition can be expressed:

$$\int_{o}^{W} \alpha(E)dx = 1$$

Here, a is the probability of impact ionization occurring in unit length of travel, or the "ionization rate". The units are usually cm -1. To be more realistic, we must recognize that the ionization rates for electrons and holes are not same. Also, the ionization rates depend very strongly on the electric field. A typical dependence is:

$$\alpha(E) = \alpha_0.E^6$$

A generalized avalanche condition can be derived for the case of unequal rates for electrons and holes

$$\int_0^W \beta \exp\left[-\int_0^m (\beta-\alpha)dx'\right]dx = 1$$

Define the average ionization rate for electrons and holes:

$$<\alpha> \cong \beta xp\left[-\int_o^m (\beta-\alpha)dx'\right]$$

Which leads to: $\int_0^w <\alpha> dx = 1$

By using (S.2b) and (S.2c), the effective a can be shown to be:

Here $$\alpha_{eft} = (k-1)\frac{\alpha}{\ell nk}$$

$$K(E) = \frac{\beta(E)}{\alpha(E)}.$$

This value for < a > can be used with very accurate results for GaAs.

Due to the strong dependence of a on electric field, we find that the avalanching occurs almost exclusively very close to the peak of the electric field (in Figure). This region of the diode is termed the "avalanche region". In a device with field distribution as in Figure, the holes generated will drift to the left into the contact and will be out of the picture as far as device operation is concerned, whereas the electrons will drift to the right through the so called "drift region" of the device, see Figure.

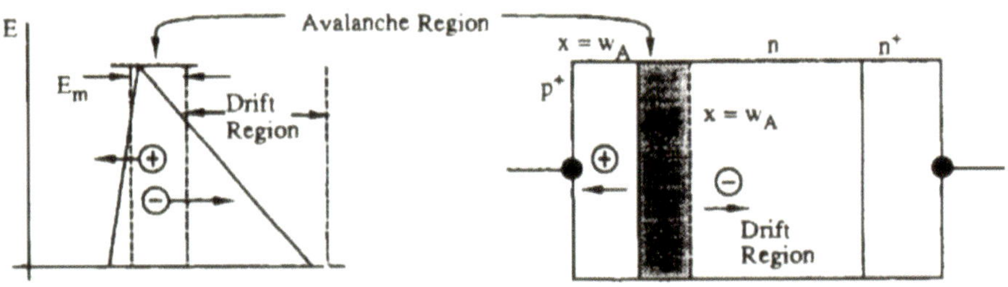

Definition of avalanche and drift regions in a p^+-n IMPATT diode.

Break- Down Voltage

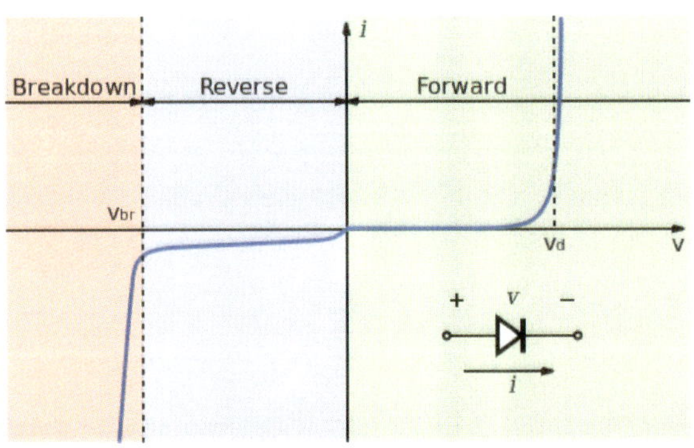

Diode I-V diagram

Breakdown voltage is a parameter of a diode that defines the largest reverse voltage that can be applied without causing an exponential increase in the leakage current in the diode. As long as the current is limited, exceeding the breakdown voltage of a diode does no harm to the diode. In fact, Zener diodes are essentially just heavily doped normal diodes that exploit the breakdown voltage of a diode to provide regulation of voltage levels.

Rectifier diodes (semiconductor or tube/valve) may have several voltage ratings, such as the Peak Inverse Voltage (PIV) across the diode, and the maximum RMS input voltage to the rectifier circuit (which will be much less).

Many small-signal transistors need to have any breakdown currents limited to much lower values to avoid excessive heating. To avoid damage to the device, and to limit the effects excessive leakage current may have on the surrounding circuit, the following bipolar transistor Maximum Ratings are often specified:

- V_{CEO} (sometimes written BV_{CEO} or $V_{(BR)CEO}$ – the maximum voltage between collector and emitter that can be safely applied (and with no more than some specified leakage current, often) when no *circuit at the base* of the transistor is there to remove collector-base leakage. Typical values: 20 Volts to as high as 700 Volts; very early Germanium point-contact transistors such as the OC10 had values around 5 Volts or less.
- V_{CBO} – the maximum collector-to-base voltage, *with emitter open-circuit*. Typical values 25 to 1200 Volts.
- V_{CER} – the maximum voltage rating between collector and emitter with some specified resistance (or less) between base and emitter. A more realistic rating for real-world circuits than the open-base or open-emitter scenarios above.
- V_{EBO} – the maximum reverse voltage on the base with respect to the emitter.
- V_{CES} – collector to emitter rating when base is shorted to emitter; equivalent to V_{CER} when $R = 0$;

Field Effect transistors have similar maximum ratings, the most important one for Junction FETs is the gate-drain rating.

Some devices may also have a *maximum rate of change* of voltage specified.

Reverse Break-Down Voltage

It is convenient to observe that the maximum field of the triangular field distribution in a p-n-junction diode at breakdown is fairly independent of the doping and the type of junction. The explanation for this fact is again the very high E-field dependence of the ionization coefficient, Q. Thus, when the field has reached a certain magnitude, then it becomes very easy to satisfy the avalanche condition when the integral is taken over typical junction widths, and the field for which this happens is basically a universal constant for the material. The reverse break-down voltage can now be derived in terms of the maximum E-field at break-down, Em. The results are for the standard cases of a one-sided (p+ - n) diode, and twosided diode, respectively,

$$V_B = \frac{1}{2} E_M W = \frac{\varepsilon_s E_m^2}{2eN_B}$$

$$V_B = \frac{1}{2} E_M W = \frac{\varepsilon_s E_m^2}{eN_B}$$

Note that the break-down voltage is inversely proportional to the doping in both cases. Typical breakdown voltages in the 10 to 100 volt range are obtained if the doping is of the order of 10^{16} to 10^{17}cm-3 (on the low side in one-sided junctions). These widths are 10-20% of the width of the entire depletion region, justifying the model, which assumes that the drift region occupies most of the depletion region.

Saturation Velocity

Saturation velocity is the maximum velocity a charge carrier in a semiconductor, generally an electron, attains in the presence of very high electric fields.When this happens, the semiconductor is said to be in a state of velocity saturation. Charge carriers normally move at an average drift speed proportional to the electric field strength they experience temporally. The proportionality constant is known as mobility of the carrier, which is a material property. A good conductor would have a high mobility value for its charge carrier, which means higher velocity, and consequently higher current values for a given electric field strength. There is a limit though to this process and at some high field value, a charge carrier can not move any faster, having reached its saturation velocity, due to mechanisms that eventually limit the movement of the carriers in the material.

As the applied electric field increases from that point, the carrier velocity no longer increases because the carriers lose energy through increased levels of interaction with the lattice, by emitting phonons and even photons as soon as the carrier energy is large enough to do so.

Field Effect Transistors

Saturation velocity is a very important parameter in the design of semiconductor devices, especially field effect transistors, which are basic building blocks of almost all modern integrated circuits. Typical values of saturation velocity may vary greatly for different materials, for example for Si it is in the order of 1×10^7 cm/s, for GaAs 1.2×10^7 cm/s, while for 6H-SiC, it is near 2×10^7 cm/s. Typical electric field strengths at which carrier velocity saturates is usually on the order of 10-100 kV/cm. Both saturation field and the saturation velocity of a semiconductor material are typically strong function of impurities, crystal defects and temperature.

Small Scale Devices

For extremely small scale devices, where the high-field regions may be comparable or smaller than the average mean free path of the charge carrier, one can observe velocity overshoot, or hot electron effects which has become more important as the transistor geometries continually decrease to enable design of faster, larger and more dense integrated circuits. The regime where the two terminals between which the electron moves is much smaller than the mean free path, is sometimes referred as ballistic transport. There have been numerous attempts in the past to build transistors based on this principle without much success. Nevertheless, developing field of nanotechnology, and new materials such as Carbon nanotubes and graphene, offers new hope.

Negative Differential Resistivity

Though in a semiconductor such as Si saturation velocity of a carrier is same as the peak velocity of the carrier, for some other materials with more complex energy band structures, this is not true. In GaAs or InP for example the carrier drift velocity reaches to a maximum as a function of field and then it begins to actually decrease as the electric field applied is increased further. Carriers which have gained enough energy are kicked up to a different conduction band which presents a lower drift velocity and eventually a lower saturation velocity in these materials. This results in an overall decrease of current for higher voltage until all electrons are in the "slow" band and this is the principle behind operation of a Gunn diode, which can display negative differential resistivity. Due to the transfer of electrons to a different conduction band involved, such devices, usually single terminal, are referred to as Transferred electron devices, or TEDs.

Design Considerations

When designing semiconductor devices, especially on a sub-micrometre scale as used in modern microprocessors, velocity saturation is an important design characteristic. Velocity saturation greatly affects the voltage transfer characteristics of a field-effect transistor, which is the basic device used in most integrated circuits. If a semiconductor device enters velocity saturation, an increase in voltage applied to the device will not cause a linear increase in current as would be expected by Ohm's law. Instead, the current may only increase by a small amount, or not at all. It is possible to take advantage of this result when trying to design a device that will pass a constant current regardless of the voltage applied, a current limiter in effect.

Saturation velocity is a very important parameter in the design of semiconductor devices, especially field effect transistors, which are basic building blocks of almost all modern integrated circuits. Typical values of saturation velocity may vary greatly for different materials, for example for Si it is in the order of 1×10^7 cm/s, for GaAs 1.2×10^7 cm/s, while for 6H-SiC, it is near 2×10^7 cm/s. Typical electric field strengths at which carrier velocity saturates is usually on the order of 10-100 kV/cm. Both saturation field and the saturation velocity of a semiconductor material are typically strong function of impurities, crystal defects and temperature.

Saturated Drift Velocity

The saturation velocity decreases as the temperature increases. It is important that the velocity is almost independent of the electric field for the high fields which exist in IMPATT devices - this means that we can regard the carrier velocity as essentially constant and equal to the saturation velocity anywhere in the drift-region. The saturation velocity of electrons and holes in GaAs is somewhat lower. The saturation velocity is essentially the same for electrons and holes. Its temperature-dependence is given by:

$$vs(T) = 6.12\times10^{6}\left[0.268+\frac{262}{273+T}-\frac{1.21\times10^{4}}{(273+T)^{2}}\right]$$

where is the cm/sec and T is in °C.

Displacement Current

Due to the large amounts of rapidly moving electric charge in an IMPATT device, it is necessary to take the displacement current into account. The avalanche gives rise to the "injected" current which is localised to the avalanche region. Due to the nature of the avalanche process, the injected current is concentrated toward the end of the first half period of the AC-voltage across the device. The time dependence of the injected charge is therefore similar to a short pulse. As this pulse of charge drifts through the drift region, we can model it with a constant charge which moves through a capacitor. In general, the total current is continuous, and is represented by a displacement current in any plane of the capacitor which does not contain the charge, as shown in figure . If the charge is moving with constant velocity, then the total current is also constant in time until the charge has traversed the capacitor. We can calculate the average current when the charge, q, moves across a capacitor of width, W, in a time at, as

$$I = \frac{q}{\Delta t} = \frac{q.v}{w}$$

If the velocity of the charge is equal to the saturated velocity, vs then the current becomes

$$I = \frac{q.v_s}{w}$$

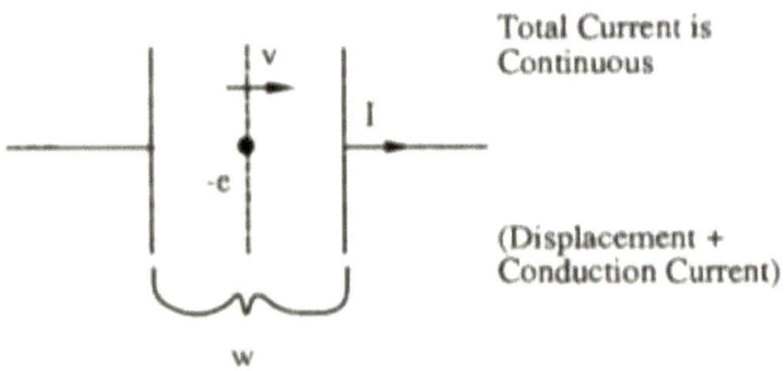

An electron moving between two metal plates

Diagram of p$^+$- n Diode IMPATT Device Operation

We are now ready to put the above facts together in a diagram (Figure). We assume that the E-field consists of a DC component, adjusted so that reverse break-down just occurs, and an AC component, which is added uniformly across the device (compare the Gunn-device case, where the electric field could not be uniform).

The structure of the device is shown in Figure. The E-field and voltage are positive when they accelerate carriers in the positive z-direction, and the current is positive in the same direction as the electric field.

Referring to Figure, the AC electric field is shown in the lowest set of curves, as a function of time. In the top of the diagram, we see in the first "frame" the DC electric field distribution. As the AC field component grows positive, the avalanche is initiated. The growth of the charge produced by the avalanche is pictured in the second set of frames from the top. In the second frame (t = T/4), the AC field is at its maximum, but the maximum avalanche charge injection does not occur until the third frame (t = T/2), just before the AC voltage goes negative and drives the total voltage below the breakdown voltage. In other words, the avalanche continues to grow as long as the electric field is larger than Em. As the AC voltage becomes more negative, the avalanche injected charge drifts through the drift region, and the device current, shown in the next set of frames, is constant with time until the charge reaches the right contact. If we compare the two lowest sets of frames, we notice that the current has an approximately square wave-form, being off for the first half period, and constant and positive during the second half period, which represents the negative resistance which we sought.

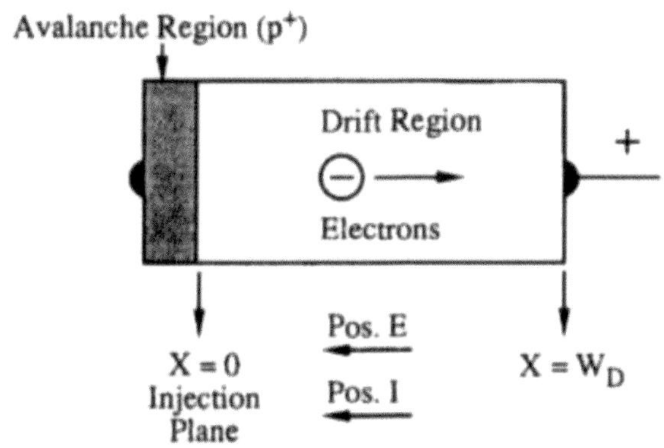

Definition of positive directions for electron velocity, electric field, and current in an IMPATT device.

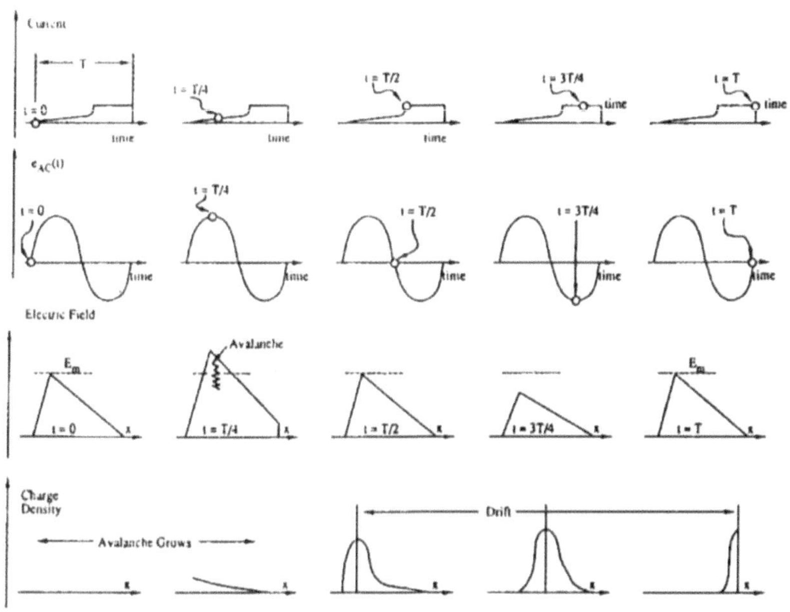

Operation of an IMPATT device, illu6trating the electric field, charge density, current and A C voltage for the device

The negative resistance arises as a result of three effects:

1) The delay of the maximum of the avalanche until it just turns off, i.e. by T/2 with respect to t =0.

2) The transit time delay of T /2 for the drifting charges in the drift region.

3) The effect of the displacement current, which produces a square current pulse from a short pulse of injected charge. This factor effectively speeds up the onset of the current in the contact by T /2.

Taking into account all of the three above effects, we find a total delay of T/2 of the current with respect to the voltage. Specifically, we see that the correct value of the transit time delay in the drift-region should be about T/2. We can also calculate the estimated frequency of operation as follows (note that there is a factor of two ratio of the transit times, in periods, for the IMPATT device compared with a transit-time mode Gunn device.):

$$\tau_t = T/2 \equiv \frac{1}{2f} = \frac{W_D}{v_s} \Rightarrow f = \frac{v_s}{2W_D}$$

As an example, a 10 GHz IMPATT device using silicon should have a drift region length of about 5 micron. We also find it useful to introduce the concept oftransit angle, defined as:

$$\theta_D = w \times \frac{W_D}{v_s} \equiv w \times r_t$$

Doping Profiles for IMPATT Diodes

A great variety of doping profiles have been used for IMPATT diodes and a few are illustrated in figures below. The recent trend has been toward structures with very narrow (higher efficiency) avalanche regions, such as the profile originally proposed by Read (1958). The "lo-hi-Io" structure has a uniform field in the avalanche region, while the narrow clump of (donor) charge allows the field to decrease to values well below Em immediately as the carriers enter the drift region, preventing any avalanche there. Another popular structure is the double-drift one, which allows both electrons and holes to drift through separate drift regions.

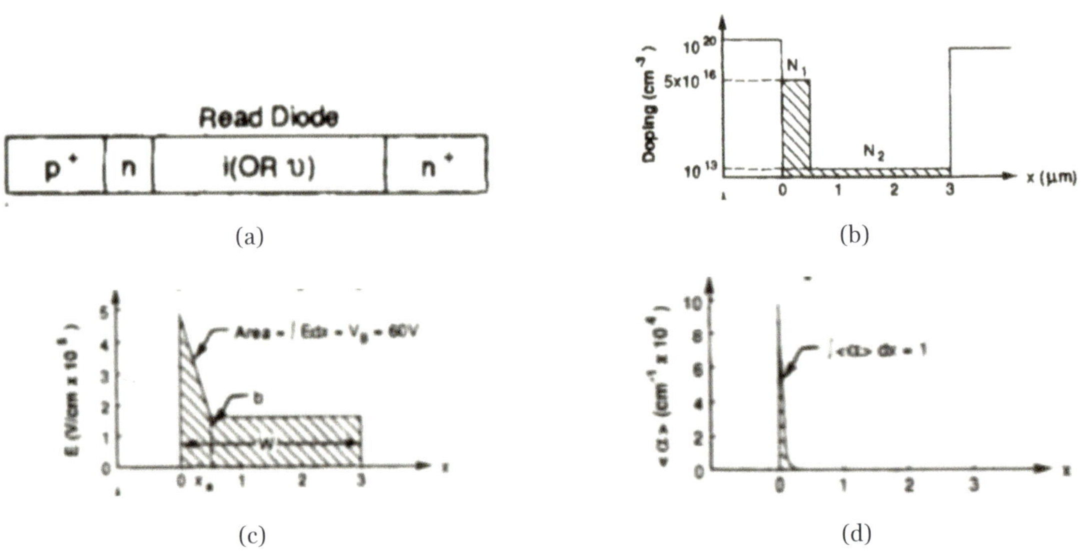

(a) Schematic of IMPATT diode, (b) Distribution of doping, (c) electric field and (d) ionization probability in a Read type IMPATT diode.

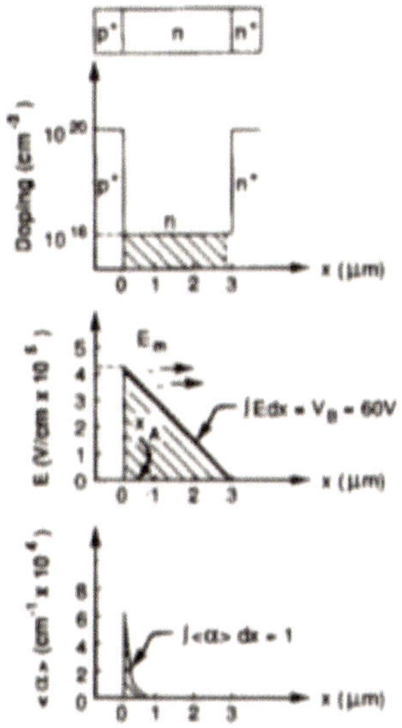

Distribution of doping, electric field and ionization probability in a p^+-n (single drift) IMPATT diode.

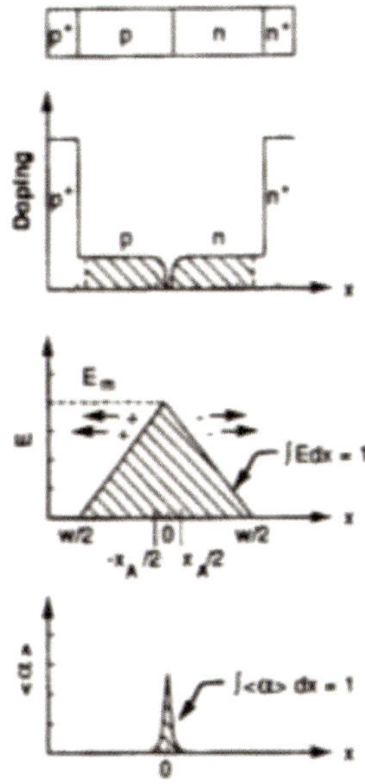

Distribution of doping, electric field and ionization probability in a p-n^+ (single drift) IMPATT diode.

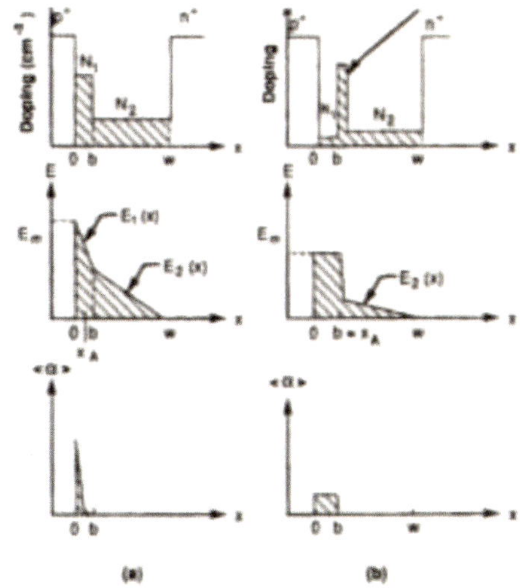

Distribution of doping, electric field and ionization probability in (a) an IMPATT diode device with Hi-Lo structure, (b) and IMPAAT device with Lo-Hi-Lo structure

Light Emitting Diode

We all are familiar with the LEDs as it is very widely used indicator or as an optical source. But now a days it's application domain is changing from indicators to lighting applications. It has become possible because of recent developments in fabrication technology of LEDs. Reliable, cheaper power LEDs are available now a days.

These are nothing but p-n junction diodes which emit light when forward biased. The circuit symbol and actual LED device is shown below. The outward arrows in the circuit symbol symbolize the radiated light.

Observe the actual device carefully. It has two terminals having different leg length. The longer one is 'Anode' terminal and shorter one is 'Cathode' terminal.

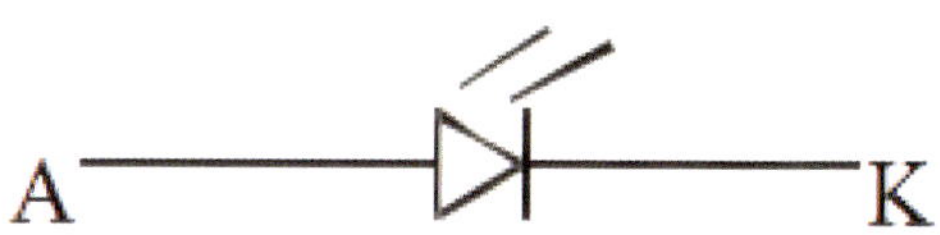

LEDs differ from normal p-n junction diode in working principle, fabrication material, specifications and hence application.

Appearing as practical electronic components in 1962, the earliest LEDs emitted low-intensity infrared light. Infrared LEDs are still frequently used as transmitting elements in remote-control circuits, such as those in remote controls for a wide variety of consumer electronics. The first visible-light LEDs were of low intensity and limited to red. Modern LEDs are available across the visible, ultraviolet, and infrared wavelengths, with very high brightness.

Working of LEDs

Basic difference between rectifier diode and LED is that p and n type semiconductors of LED are heavily doped (more than 1026/m3). So other diodes give off heat when forward biased, while LEDs emit light when forward biased. The process of giving light energy by applying an electrical voltage is called *electro-luminescence*.

When LED is forward biased electron are injected into the n-region and holes into the p-region. Electrons cross the pn junction from the n-type material and recombine with holes in the p-type material. The free electrons are in the conduction band and holes in the valence band. The electron – hole recombination takes place and during this process the energy is released in the form of radiation. Thus, in LED, emitted light comes from the hole-electron recombination.

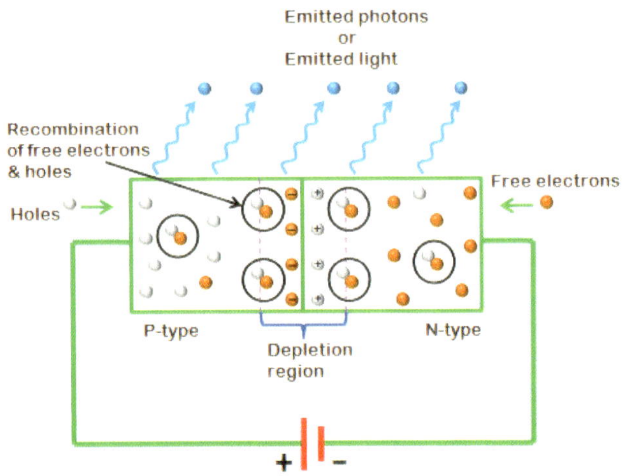

Light Emitting Diode (LED)

The electro-luminescence phenomenon will be more cleared using Energy-band diagram. Free electrons present in the conduction band are at higher energy level than holes in the valence band. Radiative and nonradiative recombinations take place in LED. In all semiconductors these two types of radiations are occurring when forward biased. Only difference is that in some radiative phenomenon is predominant and in some cases nonradiative. Accordingly energy is released either in form of light or heat. In Si or Ge semiconductors greater percentage of energy is in the form of heat and percentage of light energy is less.

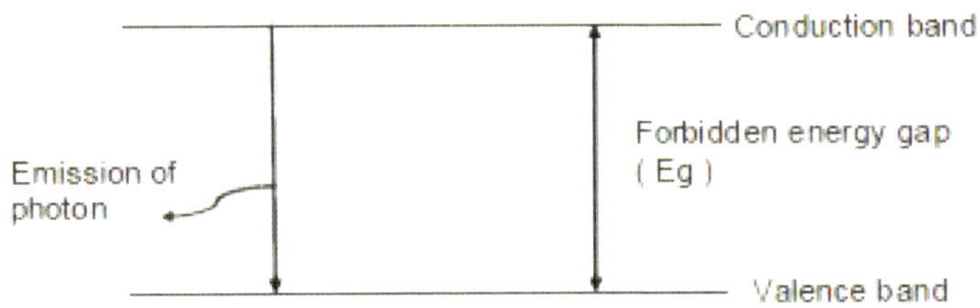

Since semiconductors used for LED fabrication are mostly compounds of Gallium and Indium, behavior of LED is different from normal diode when forward biased. During recombination the light of wavelength λ is emitted where λ is related to the bandgap energy, Eg of semiconductor by formula:

$$Eg = h\upsilon = hC/\lambda$$

where h = plank's constant = $4.14X10^{-15}$ eVs

C = velocity of light = $3\ X10^{8}$ m/s

Thus it is clear from the equation that wavelength of light emitted is dependent on the bandgap energy. If material is changed then Eg changes and hence the color of emitted light.

Under reverse bias no carrier injection takes place and consequently no photon (light) is emitted.

Characteristics of LEDs

Broadly LED characteristics can be categorized as Electrical and Optical.

Electrical Characteristics

Electrical characteristic of LED is important. The forward characteristic of LED is same as that of rectifying diode except for change in forward voltage drop. Typically forward voltage drop across an LED is in between 1.2 V to 3.2 V, while forward voltage drop across diode is 0.2 V (for Ge diode) or 0.6V (for Si diode).

In figure the forward characteristics of several LED materials is shown. One should note that although color of light is determined by the material used, the amount of light produced by an LED is dependent on forward current flowing through the circuit. The current can be controlled by its driving circuit and hence the light intensity.

It is seen from the characteristic that knee voltage for the red LED is the lowest and it increases for shorter wavelength materials.

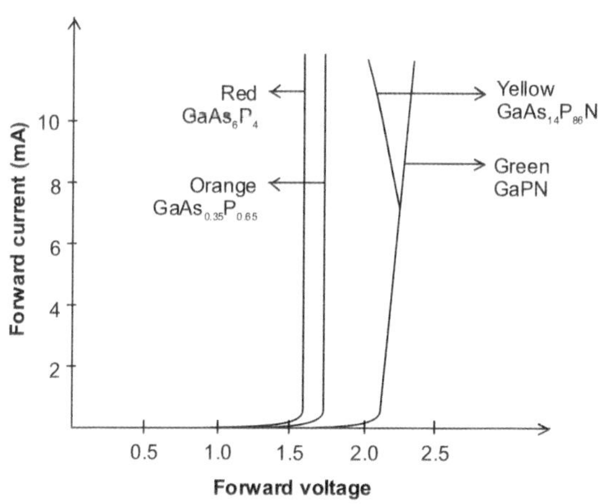

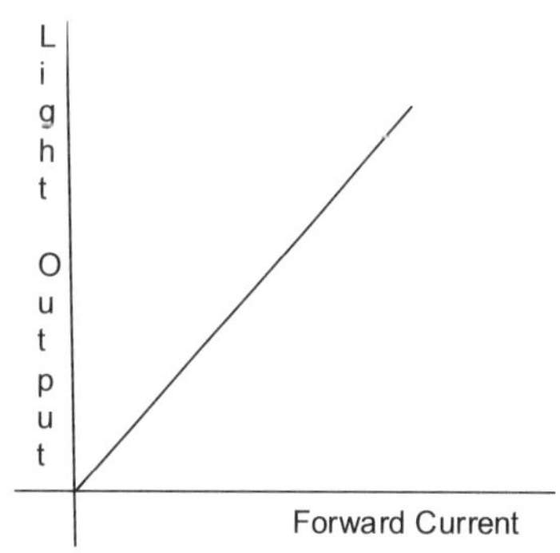

Optical Characteristics

Since now a days LEDs are used for illumination purposes the optical characteristics of LED are also becoming important. There are two important optical characteristics viz. spectral and spatial need to be studied.

- Spectral characteristic: It represents the light output versus wavelength. Following curves show spectral response of colored LEDs in visible range. LEDs have single peak occurring at corresponding color wavelength.

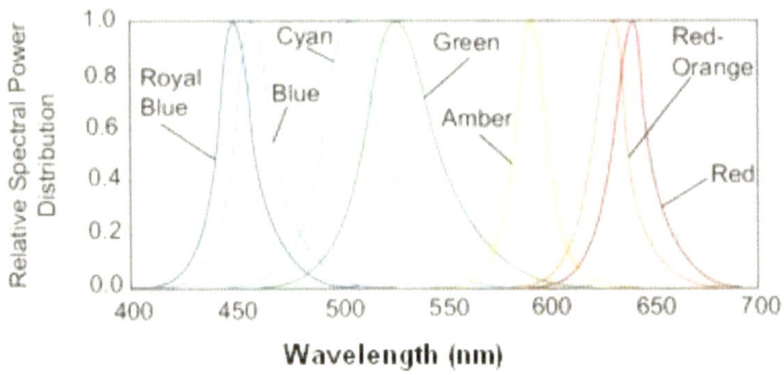

- Spatial characteristics: It gives the radiation pattern of LEDs. It shows the spread of light output with respect to the space. In short it gives directionality of emitted light. The following graph shows typical radiation pattern of LED. The pattern depends upon the lens structure of the LED. The narrower the radiation pattern, the more the light is concentrated in a particular direction.

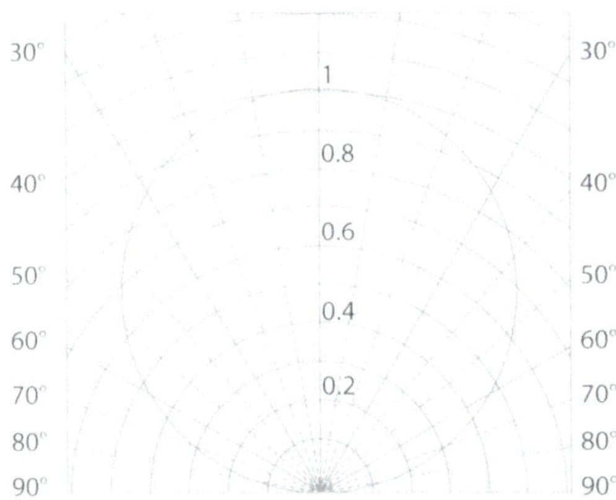

Fabrication of LED

LEDs are specialised form of p-n junction diodes that have been designed to optimise their optical output. As a result the LED structure and fabrication techniques need to ensure that electroluminescence is optimised.

LEDs are basically consists of three main components :

1. p-n junction

2. Electrical contacts

3. Transparent plastic encapsulation

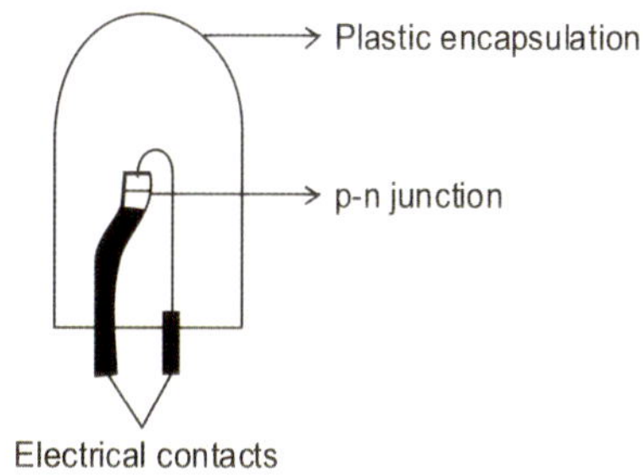

The major part of structure is p-n junction. Active films of the LED structure are normally grown epitaxially - often by liquid phase or vapour phase epitaxy. Common substrates used are compounds of Gallium, Arsenic and Phosphorus. The PN junction can be created by either impurity diffusion, ion implantation, or it can be incorporated during the epitaxial growth phase.

Once a shallow p-n junction is formed, electrical contacts are taken out from both the regions.

The junction is encapsulated in transparent plastic medium. The plastic is molded into an approximate hemishperical shape. Other geometries are also possible which improves the light emission direction and conversion efficiency from electrical to optical.

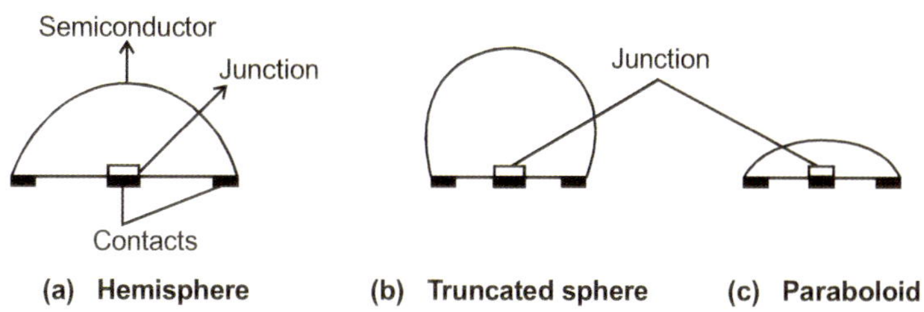

Cross-section of LED

Two basic configurations for the LED structure are :

1] *Surface emitting LED structure:* LED structure emits light perpendicular to the plane of the PN junction.

2] *Edge emitting LED structure:* LED structure emits light in a plane parallel to the junction of the PN junction. In this configuration the light can be confined to a narrow angle.

Materials used for LEDs

Light emitting diodes are available in a wide range of colors with the most common being red, amber, yellow and green and are thus widely used as indicators.

The actual color of a light emitted by LED is determined by the actual semiconductor compound used for fabrication. The different materials emit different wavelength of the light and hence different color. Following table shows the used semiconductor material to produce particular color.

Semicounductor Material	Wavelength of emitted Light (nm)	Color
GaAs	850-940	Infra-Red
$GaAs_{0.6}P_{0.4}$	630-660	Red
$GaAs_{0.25}P_{0.75}$	605-620	Amber
GaAsP.N	585-595	Yellow
AlGaP	550-570	Green
SiC	430-505	Blue
GaInN	450	White

From the above table, it is clear that the main P-type dopant used in the manufacture of Light Emitting Diodes is Gallium (Ga, atomic number 31) and that the main N-type dopant used is Arsenic (As, atomic number 33). Other colors are produced by adding nitrogen or phosphorus impurities.

Thus one should remember that the color of the light emitted by an LED is NOT determined by the coloring of the LED's plastic body although these are slightly colored to both enhance the light output and to indicate its color when not illuminated.

Most light emitting diodes produce just a single output of coloured light however, multi-coloured LEDs are now available that can produce a range of different colours from within a single device. Most of these are actually two or three LEDs fabricated within a single package.

Up to 1970s only colored LEDs were available which were mainly used as indicators. The use of LEDs for signaling applications started about 35 years ago. These are low power LEDs radiating low light intensity. Development of white power LEDs started when high quality GaN crystal was deposited on Sapphire substrate in 1986 and high brightness blue LEDs were developed in 1993 and have been continuously improved. The first white colored OLED was launched around 1990 and made the LEDs more and more interesting to use as light bulbs. White LEDs were developed using blue LEDs combined with yellow phosphors and commercialized in 1996. Nowadays LEDs, colored or white, surpass the incandescent and halogen bulbs and are widely accepted as light bulbs.

Still lot of research is going on to produce high intensity LEDs.

Technology

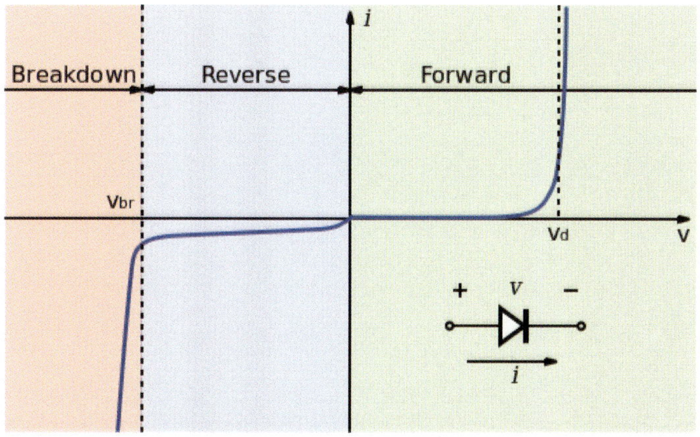

I-V diagram for a diode. An LED begins to emit light when more than 2 or 3 volts is applied. The reverse bias region uses a different vertical scale from the forward bias region to show that the leakage current is nearly constant with voltage until breakdown occurs. In forward bias, the current is small but increases exponentially with voltage.

Physics

The LED consists of a chip of semiconducting material doped with impurities to create a *p-n junction*. As in other diodes, current flows easily from the p-side, or anode, to the n-side, or cathode, but not in the reverse direction. Charge-carriers—electrons and holes—flow into the junction from electrodes with different voltages. When an electron meets a hole, it falls into a lower energy level and releases energy in the form of a photon.

The wavelength of the light emitted, and thus its colour, depends on the band gap energy of the materials forming the *p-n junction*. In silicon or germanium diodes, the electrons and holes usually recombine by a *non-radiative transition*, which produces no optical emission, because these are indirect band gap materials. The materials used for the LED have a direct band gap with energies corresponding to near-infrared, visible, or near-ultraviolet light.

LED development began with infrared and red devices made with gallium arsenide. Advances in materials science have enabled making devices with ever-shorter wavelengths, emitting light in a variety of colours.

LEDs are usually built on an n-type substrate, with an electrode attached to the p-type layer deposited on its surface. P-type substrates, while less common, occur as well. Many commercial LEDs, especially GaN/InGaN, also use sapphire substrate.

Refractive Index

Bare uncoated semiconductors such as silicon exhibit a very high refractive index relative to open air, which prevents passage of photons arriving at sharp angles relative to the air-contacting surface of the semiconductor due to total internal reflection. This property affects both the light-emission efficiency of LEDs as well as the light-absorption efficiency of photovoltaic cells. The refractive index of silicon is set at 3.96 (at 590 nm), while air's refractive index is set at 1.0002926.

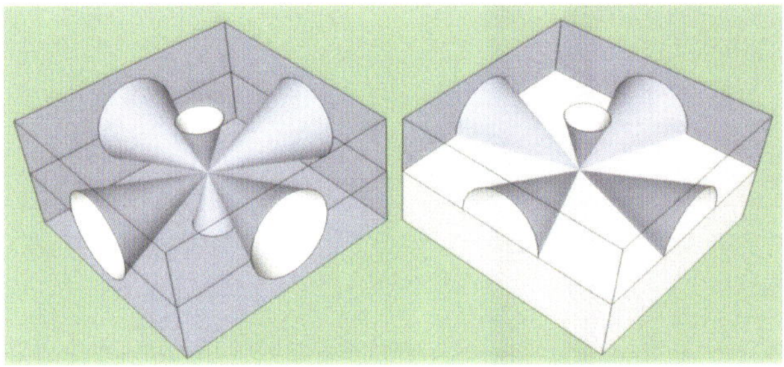

Idealized example of light emission cones in a simple square semiconductor, for a single point-source emission zone.

Most materials used for LED production have very high refractive indices

This means that much of the light is reflected back into the material at the material/air surface interface. Thus, light extraction in LEDs is an important aspect of LED production, subject to much research and development. The light emission cones of a real LED wafer are far more complex than a single point-source light emission. The light emission zone is typically a two-dimensional plane between the wafers.

In general, a flat-surface uncoated LED semiconductor chip emits light only perpendicular to the semiconductor's surface, and a few degrees to the side, in a cone shape referred to as the *light cone*, *cone of light*, or the *escape cone*. The maximum angle of incidence is referred to as the critical angle. When the critical angle is exceeded, photons no longer escape the semiconductor but are, instead, reflected internally inside the semiconductor crystal as if it were a mirror.

Internal reflections can escape through other crystalline faces if the incidence angle is low enough and the crystal is sufficiently transparent to not re-absorb the photon emission. But for a simple square LED with 90-degree angled surfaces on all sides, the faces all act as equal angle mirrors. In this case, most of the light can not escape and is lost as waste heat in the crystal.

A convoluted chip surface with angled facets similar to a jewel or fresnel lens can increase light output by distributing light perpendicular to the chip surface and far to the sides of the photon emission point.

The ideal shape of a semiconductor with maximum light output would be a microsphere with the photon emission occurring at the exact center, with electrodes penetrating to the center to contact at the emission point. All light rays emanating from the center would be perpendicular to the entire surface of the sphere, resulting in no internal reflections. A hemispherical semiconductor would also work, with the flat back-surface serving as a mirror to back-scattered photons.

Transition Coatings

After the doping of the wafer, it is usually cut apart into individual dies. Each die is commonly called a chip.

Many LED semiconductor chips are encapsulated or potted in clear or coloured molded plastic shells. The plastic shell has three purposes:

- Mounting the semiconductor chip in devices is easier to accomplish.
- The tiny fragile electrical wiring is physically supported and protected from damage.
- The plastic acts as a refractive intermediary between the relatively high-index semiconductor and low-index open air.

The third feature helps to boost the light emission from the semiconductor by acting as a diffusing lens, emitting light at a much higher angle of incidence from the light cone than the bare chip would alone.

Efficiency and Operational Parameters

Typical indicator LEDs are designed to operate with no more than 30–60 milliwatts (mW) of electrical power. Around 1999, Philips Lumileds introduced power LEDs capable of continuous use at one watt. These LEDs used much larger semiconductor die sizes to handle the large power inputs. Also, the semiconductor dies were mounted onto metal slugs to allow for greater heat dissipation from the LED die.

One of the key advantages of LED-based lighting sources is high luminous efficacy. White LEDs quickly matched and overtook the efficacy of standard incandescent lighting systems. In 2002, Lumileds made five-watt LEDs available with luminous efficacy of 18–22 lumens per watt (lm/W). For comparison, a conventional incandescent light bulb of 60–100 watts emits around 15 lm/W, and standard fluorescent lights emit up to 100 lm/W.

As of 2012, Philips had achieved the following efficacies for each color. The efficiency values show the physics – light power out per electrical power in. The lumen-per-watt efficacy value includes characteristics of the human eye and is derived using the luminosity function.

	Color	Wavelength range (nm)	Typical efficiency coefficient	Typical efficacy (lm/W)
	Red	$620 < \lambda < 645$	0.39	72
	Red-orange	$610 < \lambda < 620$	0.29	98
	Green	$520 < \lambda < 550$	0.15	93
	Cyan	$490 < \lambda < 520$	0.26	75
	Blue	$460 < \lambda < 490$	0.35	37

In September 2003, a new type of blue LED was demonstrated by Cree. This produced a commercially packaged white light giving 65 lm/W at 20 mA, becoming the brightest white LED commercially available at the time, and more than four times as efficient as standard incandescents. In 2006, they demonstrated a prototype with a record white LED luminous efficacy of 131 lm/W at 20 mA. Nichia Corporation has developed a white LED with luminous efficacy of 150 lm/W at a forward current of 20 mA. Cree's XLamp XM-L LEDs, commercially available in 2011, produce 100 lm/W at their full power of 10 W, and up to 160 lm/W at around 2 W input power. In 2012, Cree announced a white LED giving 254 lm/W, and 303 lm/W in March 2014. Practical general lighting needs high-power LEDs, of one watt or more. Typical operating currents for such devices begin at 350 mA.

These efficiencies are for the light-emitting diode only, held at low temperature in a lab. Since LEDs installed in real fixtures operate at higher temperature and with driver losses, real-world efficiencies are much lower. United States Department of Energy (DOE) testing of commercial LED lamps designed to replace incandescent lamps or CFLs showed that average efficacy was still about 46 lm/W in 2009 (tested performance ranged from 17 lm/W to 79 lm/W).

Efficiency Droop

Efficiency droop is the decrease in luminous efficiency of LEDs as the electric current increases above tens of milliamperes.

This effect was initially thought related to elevated temperatures. Scientists proved the opposite is true: though the life of an LED is shortened, the efficiency droop is less severe at elevated temperatures. The mechanism causing efficiency droop was identified in 2007 as Auger recombination, which was taken with mixed reaction. In 2013, a study confirmed Auger recombination as the cause of efficiency droop.

In addition to being less efficient, operating LEDs at higher electric currents creates higher heat levels, which can compromise LED lifetime. Because of this increased heat at higher currents, high-brightness LEDs have an industry standard of operating at only 350 mA, which is a compromise between light output, efficiency, and longevity.

Possible Solutions

Instead of increasing current levels, luminance is usually increased by combining multiple LEDs in one bulb. Solving the problem of efficiency droop would mean that household LED light bulbs would need fewer LEDs, which would significantly reduce costs.

Researchers at the U.S. Naval Research Laboratory have found a way to lessen the efficiency droop. They found that the droop arises from non-radiative Auger recombination of the injected carriers. They created quantum wells with a soft confinement potential to lessen the non-radiative Auger processes.

Researchers at Taiwan National Central University and Epistar Corp are developing a way to lessen the efficiency droop by using ceramic aluminium nitride (AlN) substrates, which are more thermally conductive than the commercially used sapphire. The higher thermal conductivity reduces self-heating effects.

Lifetime and Failure

Solid-state devices such as LEDs are subject to very limited wear and tear if operated at low currents and at low temperatures. Typical lifetimes quoted are 25,000 to 100,000 hours, but heat and current settings can extend or shorten this time significantly. One way to cool down the LED and use the energy efficiently, is to blink the LED in order to gain biomass for example in microalgae cultivation. It is important to note that these projections are based on a standard test that may not accelerate all the potential mechanisms that can induce failures in LEDs.

The most common symptom of LED (and diode laser) failure is the gradual lowering of light output and loss of efficiency. Sudden failures, although rare, can also occur. Early red LEDs were notable for their short service life. With the development of high-power LEDs, the devices are subjected to higher junction temperatures and higher current densities than traditional devices. This causes stress on the material and may cause early light-output degradation. To quantify useful lifetime in a standardized manner, some suggest using L70 or L50, which are runtimes (typically in thousands of hours) at which a given LED reaches 70% and 50% of initial light output, respectively.

Whereas in most previous sources of light (incandescent lamps, discharge lamps, and those that burn combustible fuel, e.g. candles and oil lamps) the light results from heat, LEDs only operate if they are kept cool enough. The manufacturer commonly specifies a maximum junction temperature of 125 or 150 °C, and lower temperatures are advisable in the interests of long life. At these temperatures, relatively little heat is lost by radiation, which means that the light beam generated by an LED is cool.

The waste heat in a high-power LED (which as of 2015 can be less than half the power that it consumes) is conveyed by conduction through the substrate and package of the LED to a heat sink, which gives up the heat to the ambient air by convection. Careful thermal design is, therefore, essential, taking into account the thermal resistances of the LED's package, the heat sink and the interface between the two. Medium-power LEDs are often designed to solder directly to a printed circuit board that contains a thermally conductive metal layer. High-power LEDs are packaged in large-area ceramic packages that attach to a metal heat sink—the interface being a material with high thermal conductivity (thermal grease, phase-change material, thermally conductive pad, or thermal adhesive).

If an LED-based lamp is installed in an unventilated luminaire, or a luminaire is located in an environment that does not have free air circulation, the LED is likely to overheat, resulting in reduced life or early catastrophic failure. Thermal design is often based on an ambient temperature of 25 °C (77 °F). LEDs used in outdoor applications, such as traffic signals or in-pavement signal lights, and in climates where the temperature within the light fixture becomes very high, could experience reduced output or even failure.

Since LED efficacy is higher at low temperatures, LED technology is well suited for supermarket freezer lighting. Because LEDs produce less waste heat than incandescent lamps, freezer tube lighting use can save on refrigeration costs as well. However, they may be more susceptible to frost and snow buildup than incandescent lamps, so some LED lighting systems have been designed with an added heating circuit. Additionally, research has developed heat sink technologies that transfer heat produced within the junction to appropriate areas of the light fixture.

Specifications of LEDs

While selecting an LED for particular application it is necessary to comprehend the different LED specifications or LED parameters. Some of the major LED specifications are outlined in the paragraphs below.

LED specifications can be categorized as Electrical and Optical.

Electrical Specifications

Since basically LED is a diode, forward voltage and forward current are two important parameters. Low power LEDs are capable of driving forward current upto 100mA while power LEDs can have forward current as high as 1A. Typical forward voltage drop varies between 1.2V to 3V. It is necessary to ensure that the maximum current rating is not exceeded. This could give rise to excessive heat dissipation within the LED chip itself which could result in reduced light output and reduced operating lifetime. The other ratings include:

- Reverse voltage: LEDs are not tolerant to large reverse voltages. They should never be run above their stated maximum reverse voltage, which is normally quite small. If they are then permanent destruction of the device will almost certainly result.

  If there is any chance of a reverse voltage appearing across the LED, then it is always best to build in protection into the circuitry to prevent this. Normally simple diode circuits can be introduced and these will adequately protect any LED.

- Operating and storage junction temperature
- Thermal resistance (junction to ambient)
- Operating life: The light intensity of a LED does diminish gradually with time. This means that a LED has an operational life. This LED specification is of particular importance when a LED or LEDs are to be used for lighting applications. It is not normally as crucial when the LED is used as an indicator - here a catastrophic failure is of greater importance.
- Turn on and turn off time.

For practical relevant data sheets must be referred.

Optical Specifications

In this one has to consider Light color, maximum intensity and radiation pattern of LED.

- Light output: The LED light intensity specification is not of crucial importance for most indicator applications, but with LEDs being used for lighting, this parameter is needed to be able to specify exactly what is needed in many situations. The light output from an LED is quantified as luminous intensity value and is expressed in millicandella, mcd. The luminous intensity value for an LED must be quoted for a given current. Many LEDs operate at currents of around 20mA, but the light output of an LED increases with increase in forward current.

  It is sometimes specified as radiation intensity at particular current in watt/sterdian.

- Operating Wavelength of LED : It signifies color of emitted light. In datasheet directly numerical value of wavelength is given for color LED. In case of white LED, graphical spectral response of device is provided.
- Radiation Pattern: It describes the relative intensity strength in any direction from the light source. Practical light sources have various forms of radiation patterns based on

different emission mechanism, materials used for fabrication of sources and manufacturing techniques. Datasheets of LED manufacturers provide radiation patterns in terms of relative intensity versus angle. The patterns may be symmetrical or nonsymmetrical. In case of symmetric radiation patterns, manufacturers mention view angles.

Types

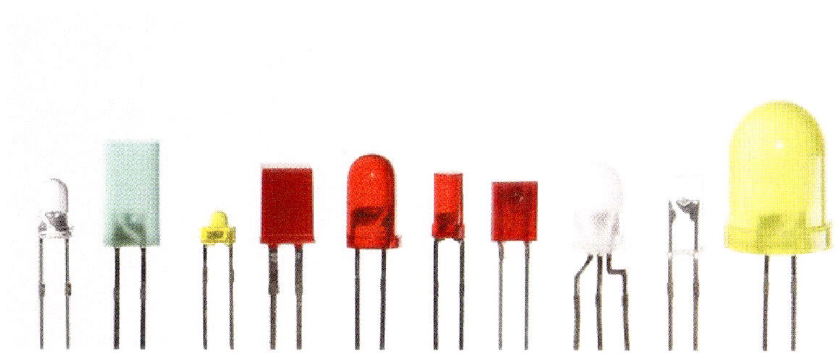

LEDs are produced in a variety of shapes and sizes. The color of the plastic lens is often the same as the actual color of light emitted, but not always. For instance, purple plastic is often used for infrared LEDs, and most blue devices have colorless housings.

The main types of LEDs are miniature, high-power devices, and custom designs such as alphanumeric or multicolor.

Miniature

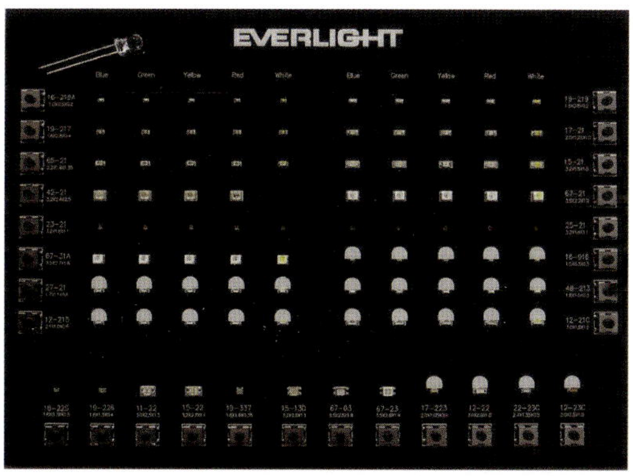

Photo of miniature surface mount LEDs in most common sizes. They can be much smaller than a traditional 5 mm lamp type LED, shown on the upper left corner.

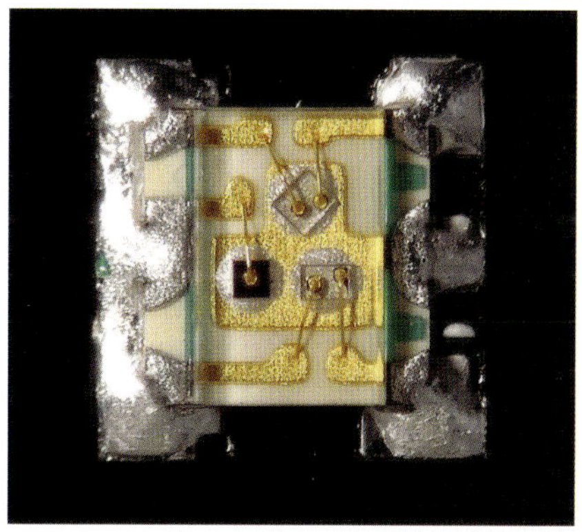

Very small (1.6x1.6x0.35 mm) red, green, and blue surface mount miniature LED package with gold wire bonding details.

These are mostly single-die LEDs used as indicators, and they come in various sizes from 2 mm to 8 mm, through-hole and surface mount packages. They usually do not use a separate heat sink. Typical current ratings range from around 1 mA to above 20 mA. The small size sets a natural upper boundary on power consumption due to heat caused by the high current density and need for a heat sink. Often daisy chained as used in LED tapes.

Common package shapes include round, with a domed or flat top, rectangular with a flat top (as used in bar-graph displays), and triangular or square with a flat top. The encapsulation may also be clear or tinted to improve contrast and viewing angle.

Researchers at the University of Washington have invented the thinnest LED. It is made of two-dimensional (2-D) flexible materials. It is three atoms thick, which is 10 to 20 times thinner than three-dimensional (3-D) LEDs and is also 10,000 times smaller than the thickness of a human hair. These 2-D LEDs are going to make it possible to create smaller, more energy-efficient lighting, optical communication and nano lasers.

There are three main categories of miniature single die LEDs:

Low-current

Typically rated for 2 mA at around 2 V (approximately 4 mW consumption)

Standard

20 mA LEDs (ranging from approximately 40 mW to 90 mW) at around:

1. 1.9 to 2.1 V for red, orange, yellow, and traditional green
2. 3.0 to 3.4 V for pure green and blue
3. 2.9 to 4.2 V for violet, pink, purple and white

Ultra-high-output

20 mA at approximately 2 or 4–5 V, designed for viewing in direct sunlight

5 V and 12 V LEDs are ordinary miniature LEDs that incorporate a suitable series resistor for direct connection to a 5 V or 12 V supply.

High-power

High-power light-emitting diodes attached to an LED star base (Luxeon, Lumileds)

High-power LEDs (HP-LEDs) or high-output LEDs (HO-LEDs) can be driven at currents from hundreds of mA to more than an ampere, compared with the tens of mA for other LEDs. Some can emit over a thousand lumens. LED power densities up to 300 W/cm^2 have been achieved. Since overheating is destructive, the HP-LEDs must be mounted on a heat sink to allow for heat dissipation. If the heat from an HP-LED is not removed, the device fails in seconds. One HP-LED can often replace an incandescent bulb in a flashlight, or be set in an array to form a powerful LED lamp.

Some well-known HP-LEDs in this category are the Nichia 19 series, Lumileds Rebel Led, Osram Opto Semiconductors Golden Dragon, and Cree X-lamp. As of September 2009, some HP-LEDs manufactured by Cree now exceed 105 lm/W.

Examples for Haitz's law—which predicts an exponential rise in light output and efficacy of LEDs over time—are the CREE XP-G series LED, which achieved 105 lm/W in 2009 and the Nichia 19 series with a typical efficacy of 140 lm/W, released in 2010.

AC-driven

LEDs developed by Seoul Semiconductor can operate on AC power without a DC converter. For each half-cycle, part of the LED emits light and part is dark, and this is reversed during the next half-cycle. The efficacy of this type of HP-LED is typically 40 lm/W. A large number of LED elements in series may be able to operate directly from line voltage. In 2009, Seoul Semiconductor released a high DC voltage LED, named as 'Acrich MJT', capable of being driven from AC power with a simple controlling circuit. The low-power dissipation of these LEDs affords them more flexibility than the original AC LED design.

Application-specific Variations

Flashing

Flashing LEDs are used as attention seeking indicators without requiring external electronics. Flashing LEDs resemble standard LEDs but they contain an integrated multivibrator circuit that causes the LED to flash with a typical period of one second. In diffused lens LEDs, this circuit is visible as a small black dot. Most flashing LEDs emit light of one color, but more sophisticated devices can flash between multiple colors and even fade through a color sequence using RGB color mixing.

Bi-color

Bi-color LEDs contain two different LED emitters in one case. There are two types of these. One type consists of two dies connected to the same two leads antiparallel to each other. Current flow in one direction emits one color, and current in the opposite direction emits the other color. The other type consists of two dies with separate leads for both dies and another lead for common anode or cathode so that they can be controlled independently. The most common bi-color combination is red/traditional green, however, other available combinations include amber/traditional green, red/pure green, red/blue, and blue/pure green.

Tri-color

Tri-color LEDs contain three different LED emitters in one case. Each emitter is connected to a separate lead so they can be controlled independently. A four-lead arrangement is typical with one common lead (anode or cathode) and an additional lead for each color.

RGB

RGB LEDs are tri-color LEDs with red, green, and blue emitters, in general using a four-wire connection with one common lead (anode or cathode). These LEDs can have either common positive leads in the case of a common anode LED, or common negative leads in the case of a common cathode LED. Others, however, have only two leads (positive and negative) and have a built-in electronic control unit.

Decorative-multicolor

Decorative-multicolor LEDs incorporate several emitters of different colors supplied by only two lead-out wires. Colors are switched internally by varying the supply voltage.

Alphanumeric

Alphanumeric LEDs are available in seven-segment, starburst, and dot-matrix format. Seven-segment displays handle all numbers and a limited set of letters. Starburst displays can display all letters. Dot-matrix displays typically use 5x7 pixels per character. Seven-segment LED displays were in widespread use in the 1970s and 1980s, but rising use of liquid crystal displays, with their lower power needs and greater display flexibility, has reduced the popularity of numeric and alphanumeric LED displays.

Digital-RGB

Digital RGB Addressable LEDs are RGB LEDs that contain their own "smart" control electronics. In addition to power and ground, these provide connections for data-in, data-out, and sometimes a clock or strobe signal. These are connected in a daisy chain, with the data in of the first LED sourced by a microprocessor, which can control the brightness and color of each LED independently of the others. They are used where a combination of maximum control and minimum visible electronics are needed such as strings for Christmas and LED matrices. Some even have refresh rates in the kHz range, allowing for basic video applications. These devices are also known by their part number (WS2812 being the most common) or a brand name such as NeoPixel

Filament

An LED filament consists of multiple LED chips connected in series on a common longitudinal substrate that forms a thin rod reminiscent of a traditional incandescent filament. These are being used as a low-cost decorative alternative for traditional light bulbs that are being phased out in many countries. The filaments require a rather high voltage to light to nominal brightness, allowing them to work efficiently and simply with mains voltages. Often a simple rectifier and capacitive current limiting are employed to create a low-cost replacement for a traditional light bulb without the complexity of the low voltage, high current converter that single die LEDs need. Usually, they are packaged in a sealed enclosure with a shape similar to lamps they were designed to replace (e.g. a bulb) and filled with inert nitrogen or carbon dioxide gas to remove heat efficiently.

Considerations for use

Power Sources

The current–voltage characteristic of an LED is similar to other diodes, in that the current is dependent exponentially on the voltage. This means that a small change in voltage can cause a large change in current. If the applied voltage exceeds the LED's forward voltage drop by a small amount, the current rating may be exceeded by a large amount, potentially damaging or destroying the LED. The typical solution is to use constant-current power supplies to keep the current below the LED's maximum current rating. Since most common power sources (batteries, mains) are constant-voltage sources, most LED fixtures must include a power converter, at least a current-limiting resistor. However, the high resistance of three-volt coin cells combined with the high differential resistance of nitride-based LEDs makes it possible to power such an LED from such a coin cell without an external resistor.

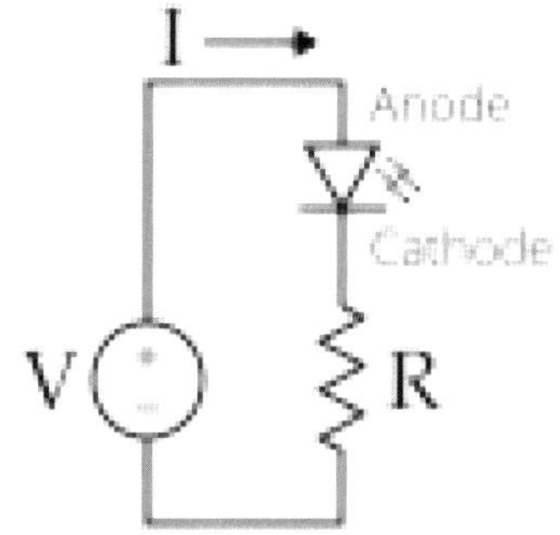

Simple LED circuit with resistor for current limiting

Electrical Polarity

As with all diodes, current flows easily from p-type to n-type material. However, no current flows and no light is emitted if a small voltage is applied in the reverse direction. If the reverse voltage grows large enough to exceed the breakdown voltage, a large current flows and the LED may be damaged. If the reverse current is sufficiently limited to avoid damage, the reverse-conducting LED is a useful noise diode.

Safety and Health

The vast majority of devices containing LEDs are "safe under all conditions of normal use", and so are classified as "Class 1 LED product"/"LED Klasse 1". At present, only a few LEDs—extremely bright LEDs that also have a tightly focused viewing angle of 8° or less—could, in theory, cause temporary blindness, and so are classified as "Class 2". The opinion of the French Agency for Food, Environmental and Occupational Health & Safety (ANSES) of 2010, on the health issues concerning LEDs, suggested banning public use of lamps in the moderate Risk Group 2, especially those with a high blue component, in places frequented by children.

In general, laser safety regulations—and the "Class 1", "Class 2", etc. system—also apply to LEDs.

While LEDs have the advantage over fluorescent lamps that they do not contain mercury, they may contain other hazardous metals such as lead and arsenic. Regarding the toxicity of LEDs when treated as waste, a study published in 2011 stated: "According to federal standards, LEDs are not hazardous except for low-intensity red LEDs, which leached Pb [lead] at levels exceeding regulatory limits (186 mg/L; regulatory limit: 5). However, according to California regulations, excessive levels of copper (up to 3892 mg/kg; limit: 2500), lead (up to 8103 mg/kg; limit: 1000), nickel (up to 4797 mg/kg; limit: 2000), or silver (up to 721 mg/kg; limit: 500) render all except low-intensity yellow LEDs hazardous."

In 2016 a statement of the American Medical Association (AMA) concerning the possible influence of blueish street lighting on the sleep-wake cycle of city-dwellers led to some controversy. So far high-pressure sodium lamps (HPS) with an orange light spectrum were the most efficient light sources commonly used in street-lighting. Now many modern street lamps are equipped with Indium gallium nitride LEDs (InGaN). These are even more efficient and mostly emit blue-rich light with a higher *correlated color temperature (CCT)*. Since light with a high CCT resembles daylight it is thought that this might have an effect on the normal circadian physiology by suppressing melatonin production in the human body. There have been no relevant studies as yet and critics claim exposure levels are not high enough to have a noticeable effect.

Advantages

- Efficiency: LEDs emit more lumens per watt than incandescent light bulbs. The efficiency of LED lighting fixtures is not affected by shape and size, unlike fluorescent light bulbs or tubes.
- Color: LEDs can emit light of an intended color without using any color filters as traditional lighting methods need. This is more efficient and can lower initial costs.
- Size: LEDs can be very small (smaller than 2 mm^2) and are easily attached to printed circuit boards.
- Warmup time: LEDs light up very quickly. A typical red indicator LED achieves full brightness in under a microsecond. LEDs used in communications devices can have even faster response times.
- Cycling: LEDs are ideal for uses subject to frequent on-off cycling, unlike incandescent and fluorescent lamps that fail faster when cycled often, or high-intensity discharge lamps (HID lamps) that require a long time before restarting.
- Dimming: LEDs can very easily be dimmed either by pulse-width modulation or lowering the forward current. This pulse-width modulation is why LED lights, particularly headlights on cars, when viewed on camera or by some people, seem to flash or flicker. This is a type of stroboscopic effect.
- Cool light: In contrast to most light sources, LEDs radiate very little heat in the form of IR that can cause damage to sensitive objects or fabrics. Wasted energy is dispersed as heat through the base of the LED.
- Slow failure: LEDs mostly fail by dimming over time, rather than the abrupt failure of incandescent bulbs.
- Lifetime: LEDs can have a relatively long useful life. One report estimates 35,000 to 50,000 hours of useful life, though time to complete failure may be longer. Fluorescent tubes typically are rated at about 10,000 to 15,000 hours, depending partly on the conditions of use, and incandescent light bulbs at 1,000 to 2,000 hours. Several DOE demonstrations have shown that reduced maintenance costs from this extended lifetime, rather than energy savings, is the primary factor in determining the payback period for an LED product.
- Shock resistance: LEDs, being solid-state components, are difficult to damage with external shock, unlike fluorescent and incandescent bulbs, which are fragile.

- Focus: The solid package of the LED can be designed to focus its light. Incandescent and fluorescent sources often require an external reflector to collect light and direct it in a usable manner. For larger LED packages total internal reflection (TIR) lenses are often used to the same effect. However, when large quantities of light are needed many light sources are usually deployed, which are difficult to focus or collimate towards the same target.

Disadvantages

- Initial price: LEDs are currently slightly more expensive (price per lumen) on an initial capital cost basis, than other lighting technologies. As of March 2014, at least one manufacturer claims to have reached $1 per kilolumen. The additional expense partially stems from the relatively low lumen output and the drive circuitry and power supplies needed.

- Temperature dependence: LED performance largely depends on the ambient temperature of the operating environment – or thermal management properties. Overdriving an LED in high ambient temperatures may result in overheating the LED package, eventually leading to device failure. An adequate heat sink is needed to maintain long life. This is especially important in automotive, medical, and military uses where devices must operate over a wide range of temperatures, which require low failure rates. Toshiba has produced LEDs with an operating temperature range of −40 to 100 °C, which suits the LEDs for both indoor and outdoor use in applications such as lamps, ceiling lighting, street lights, and floodlights.

- Voltage sensitivity: LEDs must be supplied with a voltage above their threshold voltage and a current below their rating. Current and lifetime change greatly with a small change in applied voltage. They thus require a current-regulated supply (usually just a series resistor for indicator LEDs).

- Color rendition: Most cool-white LEDs have spectra that differ significantly from a black body radiator like the sun or an incandescent light. The spike at 460 nm and dip at 500 nm can make the color of objects appear differently under cool-white LED illumination than sunlight or incandescent sources, due to metamerism, red surfaces being rendered particularly poorly by typical phosphor-based cool-white LEDs.

- Area light source: Single LEDs do not approximate a point source of light giving a spherical light distribution, but rather a lambertian distribution. So LEDs are difficult to apply to uses needing a spherical light field; however, different fields of light can be manipulated by the application of different optics or "lenses". LEDs cannot provide divergence below a few degrees. In contrast, lasers can emit beams with divergences of 0.2 degrees or less.

- Electrical polarity: Unlike incandescent light bulbs, which illuminate regardless of the electrical polarity, LEDs only light with correct electrical polarity. To automatically match source polarity to LED devices, rectifiers can be used.

- Blue hazard: There is a concern that blue LEDs and cool-white LEDs are now capable of exceeding safe limits of the so-called blue-light hazard as defined in eye safety specifications such as "ANSI/IESNA RP-27.1–05: Recommended Practice for Photobiological Safety for Lamp and Lamp Systems". However, lighting physicist Sebastien Point and colleagues showed that there is no evidence of a risk in normal use at domestic illuminance for the general population, and that caution is only needed for particular occupationnal situations or for specific populations.

- Light pollution: Because white LEDs, especially those with high color temperature, emit much more short wavelength light than conventional outdoor light sources such as high-pressure sodium vapor lamps, the increased blue and green sensitivity of scotopic vision means that white LEDs used in outdoor lighting cause substantially more sky glow. The American Medical Association warned on the use of high blue content white LEDs in street lighting, due to their higher impact on human health and environment, compared to low blue content light sources (e.g. High-Pressure Sodium, PC amber LEDs, and low CCT LEDs).
- Efficiency droop: The efficiency of LEDs decreases as the electric current increases. Heating also increases with higher currents, which compromises LED lifetime. These effects put practical limits on the current through an LED in high power applications.
- Impact on insects: LEDs are much more attractive to insects than sodium-vapor lights, so much so that there has been speculative concern about the possibility of disruption to food webs.
- Use in winter conditions: Since they do not give off much heat in comparison to incandescent lights, LED lights used for traffic control can have snow obscuring them, leading to accidents.

Applications

LED uses fall into four major categories:

1. Visual signals where light goes more or less directly from the source to the human eye, to convey a message or meaning.
2. Illumination where light is reflected from objects to give visual response of these objects
3. Measuring and interacting with processes involving no human vision.
4. Narrow band light sensors where LEDs operate in a reverse-bias mode and respond to incident light, instead of emitting light.

Indicators and Signs

The low energy consumption, low maintenance and small size of LEDs has led to uses as status indicators and displays on a variety of equipment and installations. Large-area LED displays are used as stadium displays, dynamic decorative displays, and dynamic message signs on freeways. Thin, lightweight message displays are used at airports and railway stations, and as destination displays for trains, buses, trams, and ferries.

One-color light is well suited for traffic lights and signals, exit signs, emergency vehicle lighting, ships' navigation lights or lanterns (chromacity and luminance standards being set under the Convention on the International Regulations for Preventing Collisions at Sea 1972, Annex I and the CIE) and LED-based Christmas lights. In cold climates, LED traffic lights may remain snow-covered. Red or yellow LEDs are used in indicator and alphanumeric displays in environments where night vision must be retained: aircraft cockpits, submarine and ship bridges, astronomy observatories, and in the field, e.g. night time animal watching and military field use.

Red and green LED traffic signals

Automotive applications for LEDs continue to grow.

Because of their long life, fast switching times, and visibility in broad daylight due to their high output and focus, LEDs have been used in brake lights for cars' high-mounted brake lights, trucks, and buses, and in turn signals for some time. However, many vehicles now use LEDs for their rear light clusters. The use in brakes improves safety, due to a great reduction in the time needed to light fully, or faster rise time, up to 0.5 second faster than an incandescent bulb. This gives drivers behind more time to react. In a dual intensity circuit (rear markers and brakes) if the LEDs are not pulsed at a fast enough frequency, they can create a phantom array, where ghost images of the LED appear if the eyes quickly scan across the array. White LED headlamps are beginning to appear. Using LEDs has styling advantages because LEDs can form much thinner lights than incandescent lamps with parabolic reflectors.

Due to the relative cheapness of low output LEDs, they are also used in many temporary uses such as glowsticks, throwies, and the photonic textile Lumalive. Artists have also used LEDs for LED art.

Weather and all-hazards radio receivers with Specific Area Message Encoding (SAME) have three LEDs: red for warnings, orange for watches, and yellow for advisories and statements whenever issued.

Lighting

With the development of high-efficiency and high-power LEDs, it has become possible to use LEDs in lighting and illumination. To encourage the shift to LED lamps and other high-efficiency lighting, the US Department of Energy has created the L Prize competition. The Philips Lighting North America LED bulb won the first competition on August 3, 2011, after successfully completing 18 months of intensive field, lab, and product testing.

LEDs are used as street lights and in other architectural lighting. The mechanical robustness and long lifetime are used in automotive lighting on cars, motorcycles, and bicycle lights. LED light emission may be efficiently controlled by using nonimaging optics principles.

LED street lights are employed on poles and in parking garages. In 2007, the Italian village of Torraca was the first place to convert its entire illumination system to LEDs.

LEDs are used in aviation lighting. Airbus has used LED lighting in its Airbus A320 Enhanced since 2007, and Boeing uses LED lighting in the 787 and 737 Sky Interior. LEDs are also being used now in airport and heliport lighting. LED airport fixtures currently include medium-intensity runway lights, runway centerline lights, taxiway centerline and edge lights, guidance signs, and obstruction lighting.

LEDs are also used as a light source for DLP projectors, and to backlight LCD televisions (referred to as LED TVs) and laptop displays. RGB LEDs raise the color gamut by as much as 45%. Screens for TV and computer displays can be made thinner using LEDs for backlighting.

The lack of IR or heat radiation makes LEDs ideal for stage lights using banks of RGB LEDs that can easily change color and decrease heating from traditional stage lighting, as well as medical lighting where IR-radiation can be harmful. In energy conservation, the lower heat output of LEDs also means air conditioning (cooling) systems have less heat in need of disposal.

LEDs are small, durable and need little power, so they are used in handheld devices such as flashlights. LED strobe lights or camera flashes operate at a safe, low voltage, instead of the 250+ volts commonly found in xenon flashlamp-based lighting. This is especially useful in cameras on mobile phones, where space is at a premium and bulky voltage-raising circuitry is undesirable.

LEDs are used for infrared illumination in night vision uses including security cameras. A ring of LEDs around a video camera, aimed forward into a retroreflective background, allows chroma keying in video productions.

LED for miners, to increase visibility inside mines

Los Angeles Vincent Thomas Bridge illuminated with blue LEDs

LEDs are used in mining operations, as cap lamps to provide light for miners. Research has been done to improve LEDs for mining, to reduce glare and to increase illumination, reducing risk of injury to the miners.

LEDs are now used commonly in all market areas from commercial to home use: standard lighting, AV, stage, theatrical, architectural, and public installations, and wherever artificial light is used.

LEDs are increasingly finding uses in medical and educational applications, for example as mood enhancement, and new technologies such as AmBX, exploiting LED versatility. NASA has even sponsored research for the use of LEDs to promote health for astronauts.

Data Communication and other Signalling

Light can be used to transmit data and analog signals. For example, lighting white LEDs can be used in systems assisting people to navigate in closed spaces while searching necessary rooms or objects.

Assistive listening devices in many theaters and similar spaces use arrays of infrared LEDs to send sound to listeners' receivers. Light-emitting diodes (as well as semiconductor lasers) are used to send data over many types of fiber optic cable, from digital audio over TOSLINK cables to the very high bandwidth fiber links that form the Internet backbone. For some time, computers were commonly equipped with IrDA interfaces, which allowed them to send and receive data to nearby machines via infrared.

Because LEDs can cycle on and off millions of times per second, very high data bandwidth can be achieved.

Sustainable Lighting

Efficient lighting is needed for sustainable architecture. In 2009, US Department of Energy testing results on LED lamps showed an average efficacy of 35 lm/W, below that of typical CFLs, and as low as 9 lm/W, worse than standard incandescent bulbs. A typical 13-watt LED lamp emitted 450 to 650 lumens, which is equivalent to a standard 40-watt incandescent bulb.

However, as of 2011, there are LED bulbs available as efficient as 150 lm/W and even inexpensive low-end models typically exceed 50 lm/W, so that a 6-watt LED could achieve the same results as a standard 40-watt incandescent bulb. The latter has an expected lifespan of 1,000 hours, whereas an LED can continue to operate with reduced efficiency for more than 50,000 hours.

See the chart below for a comparison of common light types:

	LED	CFL	Incandescent
Lightbulb Projected Lifespan	50,000 hours	10,000 hours	1,200 hours
Watts Per Bulb (equiv. 60 watts)	10	14	60
Cost Per Bulb	$2.00	$2.00	$1.25
kWh of Electricity Used Over 50,000 Hours	500	700	3000
Cost of Electricity (@ 0.10 per kWh)	$50	$70	$300
Bulbs Needed for 50,000 Hours of Use	1	5	42
Equivalent 50,000 Hours Bulb Expense	$7.00	$10.00	$52.50
TOTAL Cost for 50,000 Hours	$57.00	$80.00	$352.50

Energy Consumption

In the US, one kilowatt-hour (3.6 MJ) of electricity currently causes an average 1.34 pounds (610 g) of CO_2 emission. Assuming the average light bulb is on for 10 hours a day, a 40-watt bulb causes 196 pounds (89 kg) of CO_2 emission per year. The 6-watt LED equivalent only causes 30 pounds (14 kg) of CO_2 over the same time span. A building's carbon footprint from lighting can, therefore, be reduced by 85% by exchanging all incandescent bulbs for new LEDs—if a building previously used only incandescent bulbs.

In practice, most buildings that use a lot of lighting use fluorescent lighting, which has 22% luminous efficiency compared with 5% for filaments, so changing to LED lighting would still give a 34% reduction in electrical power use and carbon emissions.

The reduction in carbon emissions depends on the source of electricity. Nuclear power in the United States produced 19.2% of electricity in 2011, so reducing electricity consumption in the U.S. reduces carbon emissions more than in France (75% nuclear electricity) or Norway (almost entirely hydroelectric).

Replacing lights that spend the most time lit results in the most savings, so LED lights in infrequently used locations bring a smaller return on investment.

Light Sources for Machine Vision Systems

Machine vision systems often require bright and homogeneous illumination, so features of interest are easier to process. LEDs are often used for this purpose, and this is likely to remain one of their major uses until the price drops low enough to make signaling and illumination uses more widespread. Barcode scanners are the most common example of machine vision, and many low-cost products use red LEDs instead of lasers. Optical computer mice are an example of LEDs in machine vision, as it is used to provide an even light source on the surface for the miniature camera within the mouse. LEDs constitute a nearly ideal light source for machine vision systems for several reasons:

- The size of the illuminated field is usually comparatively small and machine vision systems are often quite expensive, so the cost of the light source is usually a minor concern. However, it might not be easy to replace a broken light source placed within complex machinery, and here the long service life of LEDs is a benefit.
- LED elements tend to be small and can be placed with high density over flat or even-shaped substrates (PCBs etc.) so that bright and homogeneous sources that direct light from tightly controlled directions on inspected parts can be designed. This can often be obtained with small, low-cost lenses and diffusers, helping to achieve high light densities with control over lighting levels and homogeneity. LED sources can be shaped in several configurations (spot lights for reflective illumination; ring lights for coaxial illumination; backlights for contour illumination; linear assemblies; flat, large format panels; dome sources for diffused, omnidirectional illumination).
- LEDs can be easily strobed (in the microsecond range and below) and synchronized with imaging. High-power LEDs are available allowing well-lit images even with very short light pulses. This is often used to obtain crisp and sharp “still” images of quickly moving parts.
- LEDs come in several different colors and wavelengths, allowing easy use of the best color for each need, where different color may provide better visibility of features of interest. Having a precisely known spectrum lets tightly matched filters be used to separate informative bandwidth or to reduce disturbing effects of ambient light. LEDs usually operate at comparatively low working temperatures, simplifying heat management, and dissipation. This allows using plastic lenses, filters, and diffusers. Waterproof units can also easily be designed, allowing use in harsh or wet environments (food, beverage, oil industries).

LED digital display that can display four digits and points

Traffic light using LED

LED lights reacting dynamically to video feed via AmBX

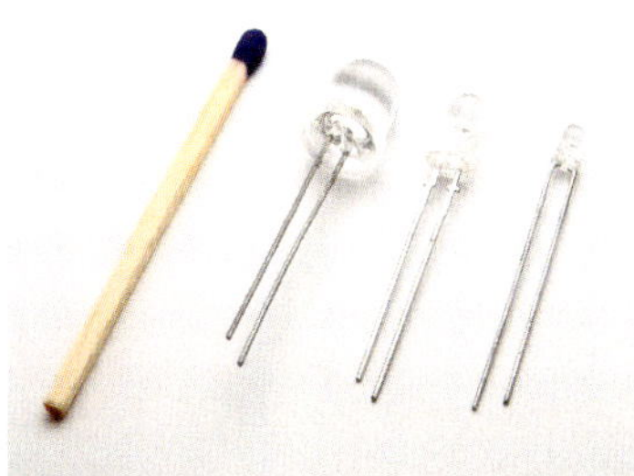

Different sized LEDs. 8 mm, 5 mm and 3 mm, with a wooden match-stick for scale.

Other Applications

LED wallpaper by Meystyle

The light from LEDs can be modulated very quickly so they are used extensively in optical fiber and free space optics communications. This includes remote controls, such as for TVs, VCRs, and LED Computers, where infrared LEDs are often used. Opto-isolators use an LED combined with a photodiode or phototransistor to provide a signal path with electrical isolation between two circuits. This is especially useful in medical equipment where the signals from a low-voltage sensor circuit (usually battery-powered) in contact with a living organism must be electrically isolated from any possible electrical failure in a recording or monitoring device operating at potentially dangerous voltages. An optoisolator also lets information be transferred between circuits that don't share a common ground potential.

Many sensor systems rely on light as the signal source. LEDs are often ideal as a light source due to the requirements of the sensors. LEDs are used as motion sensors, for example in optical computer mice. The Nintendo Wii's sensor bar uses infrared LEDs. Pulse oximeters use them for measuring oxygen saturation. Some flatbed scanners use arrays of RGB LEDs rather than the typical cold-cathode fluorescent lamp as the light source. Having independent control of three illuminated colors allows the scanner to calibrate itself for more accurate color balance, and there is no need for warm-up. Further, its sensors only need be monochromatic, since at any one time the page being scanned is only lit by one color of light. Since LEDs can also be used as photodiodes, they can be used for both photo emission and detection. This could be used, for example, in a touchscreen that registers reflected light from a finger or stylus. Many materials and biological systems are sensitive to, or dependent on, light. Grow lights use LEDs to increase photosynthesis in plants, and bacteria and viruses can be removed from water and other substances using UV LEDs for sterilization.

LEDs have also been used as a medium-quality voltage reference in electronic circuits. The forward voltage drop (e.g. about 1.7 V for a normal red LED) can be used instead of a Zener diode in low-voltage regulators. Red LEDs have the flattest I/V curve above the knee. Nitride-based LEDs have a fairly steep I/V curve and are useless for this purpose. Although LED forward voltage is far more current-dependent than a Zener diode, Zener diodes with breakdown voltages below 3 V are not widely available.

The progressive miniaturization of low-voltage lighting technology, such as LEDs and OLEDs, suitable to incorporate into low-thickness materials has fostered experimentation in combining light sources and wall covering surfaces for interior walls. The new possibilities offered by these developments have prompted some designers and companies, such as Meystyle, Ingo Maurer, Lomox and Samsung, to research and develop proprietary LED wallpaper technologies, some of which are currently available for commercial purchase. Other solutions mainly exist as prototypes or are in the process of being further refined.

Gunn Diode-Transferred Electron Device

A Gunn diode, also known as a transferred electron device (TED), is a form of diode, a two-terminal passive semiconductor electronic component, with negative resistance, used in high-frequency electronics. It is based on the "Gunn effect" discovered in 1962 by physicist J. B. Gunn. Its largest use is in electronic oscillators to generate microwaves, in applications such as radar speed guns, microwave relay data link transmitters, and automatic door openers.

Its internal construction is unlike other diodes in that it consists only of N-doped semiconductor material, whereas most diodes consist of both P and N-doped regions. It therefore does not conduct in only one direction and cannot rectify alternating current like other diodes, which is why some sources do not use the term *diode* but prefer TED. In the Gunn diode, three regions exist: two of those are heavily N-doped on each terminal, with a thin layer of lightly n-doped material between. When a voltage is applied to the device, the electrical gradient will be largest across the thin middle layer. If the voltage is increased, the current through the layer will first increase, but eventually, at higher field values, the conductive properties of the middle layer are altered, increasing its resistivity, and causing the current to fall. This means a Gunn diode has a region of negative differential resistance in its current-voltage characteristic curve, in which an increase of applied voltage, causes a decrease in current. This property allows it to amplify, functioning as a radio frequency amplifier, or to become unstable and oscillate when it is biased with a DC voltage.

A Russian-made Gunn diode

First observed in 1963. A microwave output was generated when a dc field (>critical threshold value) of several 1000 V/cm was applied across a short (very thin active region) n-type GaAs as InP, CdTe. It utilize a voltage-controlled differential negative resistance (DNR). The phenomenon that gives rise to DNR is the decrease in electron mobility with electric field. This decrease is caused by transfer of conduction of electrons electrons from a high mobility low-energy state to a low mobility, high-energy state under the action of high x (> few kV/cm). This transfer is made possible by the peculiar band structures of semiconductors like GaAs and InP, CdTe used as local oscillator and power amplifier (1 to 100 GHz).

Negative Differential Resistance

At low ξ, drift velocity of electrons (Vn) is linearly proportional to ξ.

Force on each electron = -q ξ and will accelerate opposite to ξ during the time between collision (ξ c). therefore, additional v component will be superimposed upon the thermal (random) motion of electrons (drift velocity) and net displacement of electrons in direction opposite to ξ.

Momentum applied to electron during free flight between collisions (force x time) = momentum gained by electron in same period.

$$-q\xi\tau_c = m_n v_n$$

Where τ_c is mean free time.

This implies that $v_n = -\left[\frac{q\tau_c}{m_n}\right]\xi$

Where $u_n = \left[\frac{q\tau_c}{m_n}\right]$ is the carrier transport parameters implies that how strongly motion of an electron is influenced by applied ξ.

$$v_n = -\mu_n \xi$$

Here we assume that tc is independent of ξ and is reasonable as long as vd is small compared to thermal velocity (vt) of carriers. As $v_d > v_t$, the field dependence on ξ will begin to depart from linear relationship.

For n-type GaAs, vd reaches maximum and then shows a decrease with increase in ξ. This phenomenon is due to energy band structure of GaAs. The band structure of GaAs allows the transfer of conduction electrons from a high mobility, energy minimum (called a valley; valley 1) to low mobility, higher energy satellite valleys.

Gunn Diode Oscillator

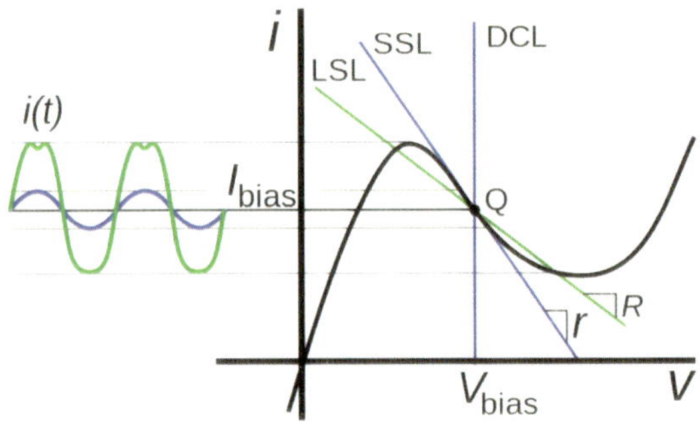

Gunn diode oscillator load lines. *DCL*: DC load line, which sets the Q point. *SSL*: negative resistance during startup while amplitude is small. Since $r < R$ poles are in RHP and amplitude of oscillations increases. *LSL*: large-signal load line. When the current swing approaches the edges of the negative resistance region *(green)*, the sine wave peaks are distorted ("clipped") and r decreases until it equals R.

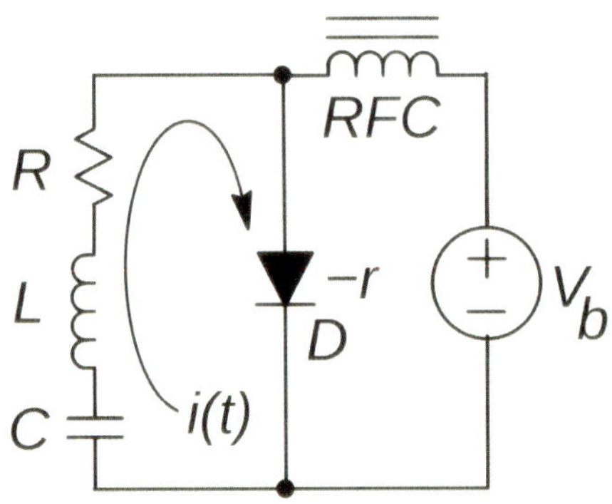

Gunn diode oscillator circuit

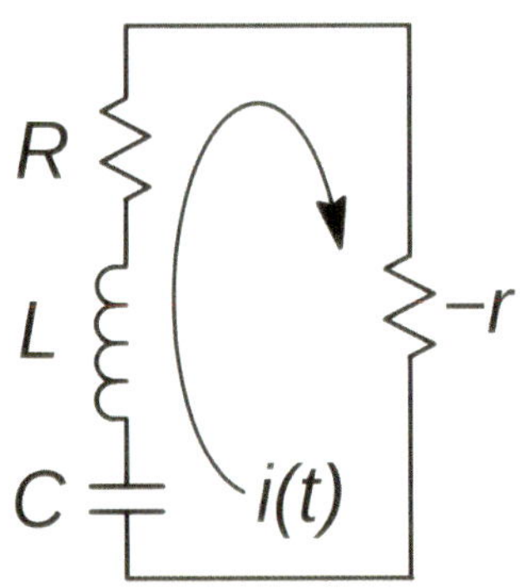

AC equivalent circuit

The common Gunn diode oscillator *(circuit diagrams)* illustrates how negative resistance oscillators work. The diode *D* has voltage controlled ("N" type) negative resistance and the voltage source V_b biases it into its negative resistance region where its differential resistance is $dv/di = -r$. The choke *RFC* prevents AC current from flowing through the bias source. R is the equivalent resistance due to damping and losses in the series tuned circuit LC, plus any load resistance. Analyzing the AC circuit with Kirchhoff's Voltage Law gives a differential equation for $i(t)$, the AC current

$$\frac{d^2i}{dt^2} + \frac{R-r}{L}\frac{di}{dt} + \frac{1}{LC}i = 0$$

Solving this equation gives a solution of the form

$$i(t) = i_0 e^{\alpha t} \cos(\omega t + \phi) \quad \text{where} \quad \alpha = \frac{r-R}{2L} \quad \omega = \sqrt{\frac{1}{LC} - \left(\frac{r-R}{2L}\right)^2}$$

This shows that the current through the circuit, $i(t)$, varies with time about the DC Q point, I_{bias}. When started from a nonzero initial current $i(t) = i_0$ the current oscillates sinusoidally at the resonant frequency ω of the tuned circuit, with amplitude either constant, increasing, or decreasing exponentially, depending on the value of α. Whether the circuit can sustain steady oscillations depends on the balance between R and r, the positive and negative resistance in the circuit:

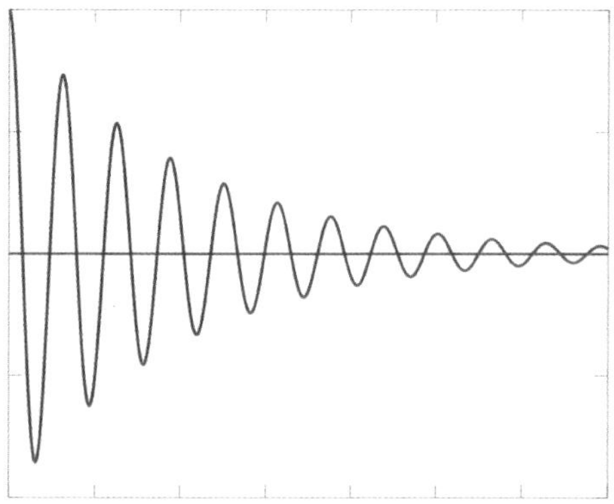

$r < R \Rightarrow \alpha < 0$: (poles in left half plane) If the diode's negative resistance is less than the positive resistance of the tuned circuit, the damping is positive. Any oscillations in the circuit will lose energy as heat in the resistance R and die away exponentially to zero, as in an ordinary tuned circuit. So the circuit does not oscillate.

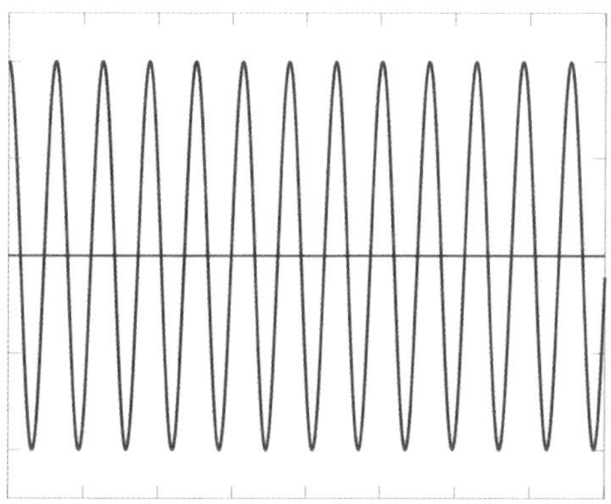

$r = R \Rightarrow \quad = $: (poles on $j\omega$ axis) If the positive and negative resistances are equal, the net resistance is zero, so the damping is zero. The diode adds just enough energy to compensate for energy lost in the tuned circuit and load, so oscillations in the circuit, once started, will continue at a constant amplitude. This is the condition during steady-state operation of the oscillator.

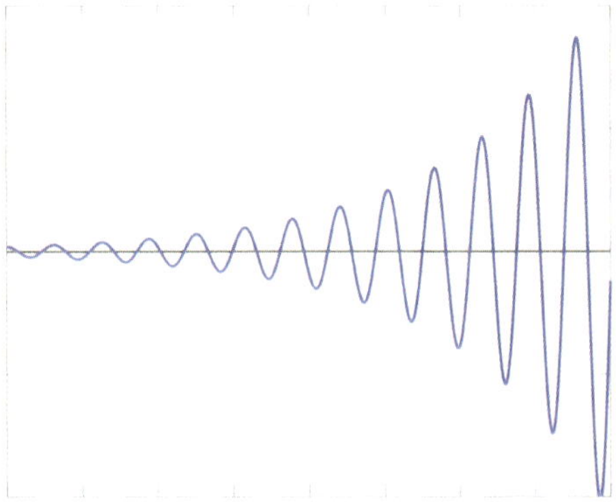

$r > R \Rightarrow \alpha > 0$: (poles in right half plane) If the negative resistance is greater than the positive resistance, damping is negative, so oscillations will grow exponentially in energy and amplitude. This is the condition during startup.

Practical oscillators are designed in region (3) above, with net negative resistance, to get oscillations started. A widely used rule of thumb is to make $R = r/3$. When the power is turned on, electrical noise in the circuit provides a signal i_0 to start spontaneous oscillations, which grow exponentially. However, the oscillations cannot grow forever; the nonlinearity of the diode eventually limits the amplitude.

At large amplitudes the circuit is nonlinear, so the linear analysis above does not strictly apply and differential resistance is undefined; but the circuit can be understood by considering r to be the "average" resistance over the cycle. As the amplitude of the sine wave exceeds the width of the negative resistance region and the voltage swing extends into regions of the curve with positive differential resistance, the average negative differential resistance r becomes smaller, and thus the total resistance $R - r$ and the damping α becomes less negative and eventually turns positive. Therefore, the oscillations will stabilize at the amplitude at which the damping becomes zero, which is when $r = R$.

Gunn diodes have negative resistance in the range −5 to −25 ohms. In oscillators where R is close to r; just small enough to allow the oscillator to start, the voltage swing will be mostly limited to the linear portion of the I–V curve, the output waveform will be nearly sinusoidal and the frequency will be most stable. In circuits in which R is far below r, the swing extends further into the nonlinear part of the curve, the clipping distortion of the output sine wave is more severe, and the frequency will be increasingly dependent on the supply voltage.

References

- Lowe, Doug (2013). "Electronics Components: Diodes". Electronics All-In-One Desk Reference For Dummies. John Wiley & Sons. Retrieved January 4, 2013
- Guarnieri, M. (2011). "Trailblazers in Solid-State Electronics". IEEE Ind. Electron. M. 5 (4): 46–47. doi:10.1109/MIE.2011.943016
- Narendran, N.; Y. Gu (2005). "Life of LED-based white light sources". IEEE/OSA Journal of Display Technology. 1 (1): 167–171. Bibcode:2005JDisT...1..167N. doi:10.1109/JDT.2005.852510
- Crecraft, David; Stephen Gergely (2002). Analog Electronics: Circuits, Systems and Signal Processing. Butterworth-Heinemann. p. 110. ISBN 0-7506-5095-8

- Diffenderfer, Robert (2005). Electronic Devices: Systems and Applications. Thomas Delmar Learning. pp. 95–100. ISBN 1401835147. Retrieved July 22, 2014
- Destriau, G. (1936). "Recherches sur les scintillations des sulfures de zinc aux rayons". Journal de Chemie Physique. 33: 587–625
- Deliyannis, T.; Yichuang Sun; J.K. Fidler (1998). Continuous-Time Active Filter Design. CRC Press. pp. 82–84. ISBN 0849325730. Archived from the original on 2017-12-21
- Comer, Donald T. (1996). "Zener Zap Anti-Fuse Trim in VLSI Circuits". VLSI Design. 5: 89. doi:10.1155/1996/23706
- Rakesh Kumar Garg, Ashish Dixit, Pavan Yadav, Basic Electronics, p. 150, Firewall Media, 2008 ISBN 8131803023
- H. Komizo, Y. Ito, H. Ashida, M. Shinoda: A 0.5-W CW IMPATT diode amplifier for high-capacity 11-GHz FM radio-relay equipment. 14-20 (Feb 1973). IEEE Journal Volume: 8. Issue: 1. ISSN 0018-9200
- Fundamentals of Semiconductors: Physics and Materials Properties, Peter Y. Yu, Manuel Cardona, pp.227-228, Springer, New York 2005, ISBN 3-540-25470-6
- Nader, Ronnie. "THE PEGASUS INCIDENT: THE LOSS OF THE FIRST ECUADORIAN SATELLITE AND ITS RECOVERY". academia.edu. Retrieved 4 April 2018
- Berlin, Leslie (2005). The Man Behind the Microchip: Robert Noyce and the Invention of Silicon Valley. Oxford University Press. ISBN 0-19-516343-5

5

Transistors: Types and Technology

A transistor is a semiconductor device that is used for the purpose of amplification or switching of electrical power and electronic signals. It is made of a semiconductor material and has three terminals that connect to an external circuit. The aim of this chapter is to explore the different types of transistors, such as bipolar junction transistors and field effect transistors. These topics are crucial for a complete understanding of the subject.

A transistor is a device that regulates current or voltage flow and acts as a switch or gate for electronic signals. Transistors consist of three layers of a semiconductor material, each capable of carrying a current.

The transistor was invented by three scientists at the Bell Laboratories in 1947, and it rapidly replaced the vacuum tube as an electronic signal regulator. A transistor regulates current or voltage flow and acts as a switch or gate for electronic signals. A transistor consists of three layers of a semiconductor material, each capable of carrying a current. A semiconductor is a material such as germanium and silicon that conducts electricity in a "semi-enthusiastic" way. It's somewhere between a real conductor such as copper and an insulator (like the plastic wrapped around wires).

The semiconductor material is given special properties by a chemical process called *doping*. The doping results in a material that either adds extra electrons to the material (which is then called *N-type* for the extra negative charge carriers) or creates "holes" in the material's crystal structure (which is then called *P-type* because it results in more positive charge carriers). The transistor's three-layer structure contains an N-type semiconductor layer sandwiched between P-type layers (a PNP configuration) or a P-type layer between N-type layers (an NPN configuration).

A small change in the current or voltage at the inner semiconductor layer (which acts as the control electrode) produces a large, rapid change in the current passing through the entire component. The component can thus act as a switch, opening and closing an electronic gate many times per second. Today's computers use circuitry made with complementary metal oxide semiconductor (CMOS) technology. CMOS uses two complementary transistors per gate (one with N-type material; the other with P-type material). When one transistor is maintaining a logic state, it requires almost no power.

Transistors are the basic elements in integrated circuits (IC), which consist of very large numbers of transistors interconnected with circuitry and baked into a single silicon microchip.

Symbols, Pins and Construction

Transistors are fundamentally three-terminal devices. On a bi-polar junction transistor (BJT), those pins are labeled collector (C), base (B), and emitter (E). The circuit symbols for both the NPN and PNP BJT are below:

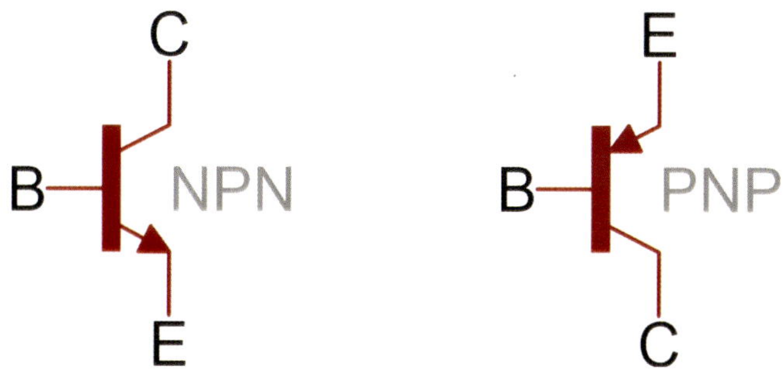

The only difference between an NPN and PNP is the direction of the arrow on the emitter. The arrow on an NPN points out, and on the PNP it points in. A useful mnemonic for remembering which is which is:
NPN: Not Pointing iN

Backwards logic, but it works.

Transistor Construction

Transistors rely on semiconductors to work their magic. A semiconductor is a material that's not quite a pure conductor (like copper wire) but also not an insulator (like air). The conductivity of a semiconductor – how easily it allows electrons to flow – depends on variables like temperature or the presence of more or less electrons.

A Transistor as Two Diodes

Transistors are kind of like an extension of another semiconductor component: diodes. In a way transistors are just two diodes with their cathodes (or anodes) tied together:

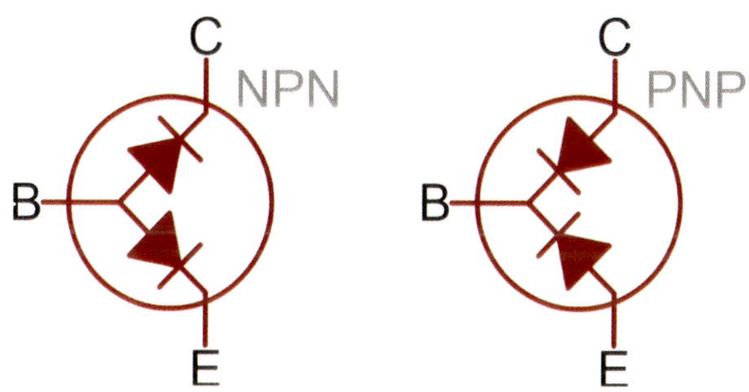

The diode connecting base to emitter is the important one here; it matches the direction of the arrow on the schematic symbol, and shows you which way current is intended to flow through the transistor.

The diode representation is a good place to start, but it's far from accurate. Don't base your understanding of a transistor's operation on that model (and definitely don't try to replicate it on a breadboard, it won't work). There's a whole lot of weird quantum physics level stuff controlling the interactions between the three terminals.

(This model *is* useful if you need to test a transistor. Using the diode (or resistance) test function on a multimeter, you can measure across the BE and BC terminals to check for the presence of those "diodes".)

Transistor Structure and Operation

Transistors are built by stacking three different layers of semiconductor material together. Some of those layers have extra electrons added to them (a process called "doping"), and others have electrons removed (doped with "holes" – the absence of electrons). A semiconductor material with *extra* electrons is called an n-type (*n* for negative because electrons have a negative charge) and a material with electrons removed is called a p-type (for positive). Transistors are created by either stacking an *n* on top of a *p* on top of an *n*, or *p* over *n* over *p*.

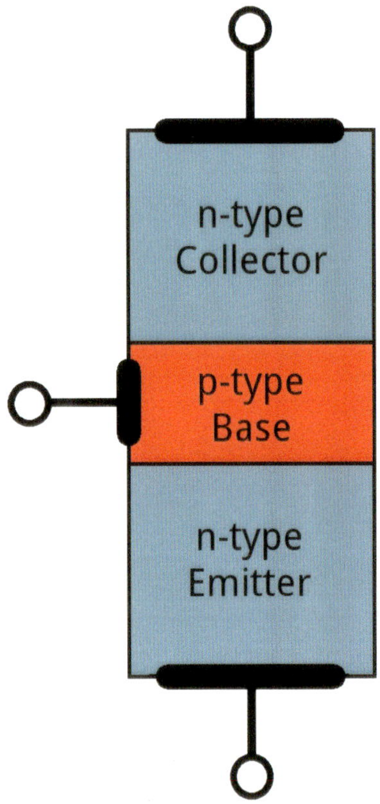

Simplified diagram of the structure of an NPN.

With some hand waving, we can say electrons can easily flow from *n* regions to *p* regions, as long as they have a little force (voltage) to push them. But flowing from a *p* region to an *n* region is really hard (requires a *lot* of voltage). But the special thing about a transistor – the part that makes our two-diode model obsolete – is the fact that electrons *can* easily flow from the p-type base to the n-type collector as long as the base-emitter junction is forward biased (meaning the base is at a higher voltage than the emitter).

A PNP works in a same but opposite fashion. The base still controls current flow, but that current flows in the opposite direction – from emitter to collector. Instead of electrons, the emitter emits "holes" (a conceptual absence of electrons) which are collected by the collector.

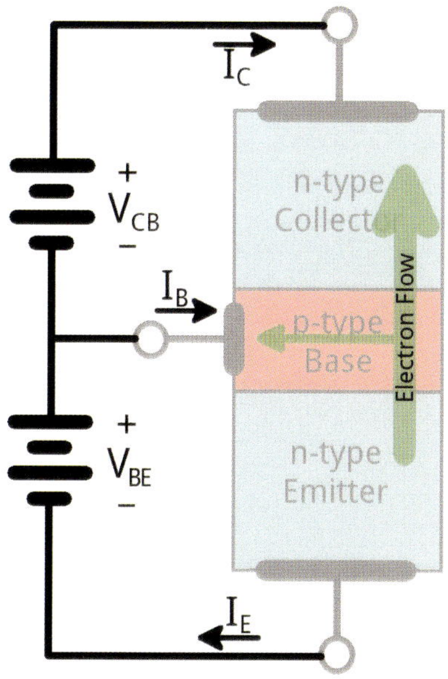

The NPN transistor is designed to pass electrons from the emitter to the collector (so conventional current flows from collector to emitter).

The emitter "emits" electrons into the base, which controls the number of electrons the emitter emits. Most of the electrons emitted are "collected" by the collector, which sends them along to the next part of the circuit.

The transistor is kind of like an electron valve. The base pin is like a handle you might adjust to allow more or less electrons to flow from emitter to collector. Let's investigate this analogy further...

Extending the Water Analogy

If you've been reading a lot of electricity concept, you're probably used to water analogies. We say that current is analogous to the flow rate of water, voltage is the pressure pushing that water through a pipe, and resistance is the width of the pipe.

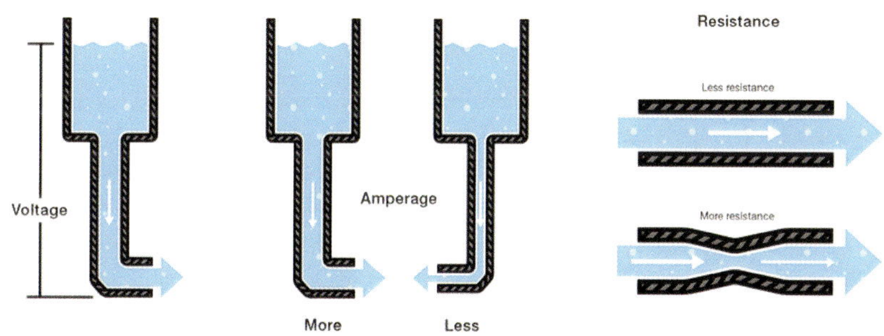

Unsurprisingly, the water analogy can be extended to transistors as well: a transistor is like a water valve – a mechanism we can use to control the flow rate.

There are three states we can use a valve in, each of which has a different effect on the flow rate in a system.

1) On – Short Circuit

A valve can be completely opened, allowing water to flow freely – passing through as if the valve wasn't even present.

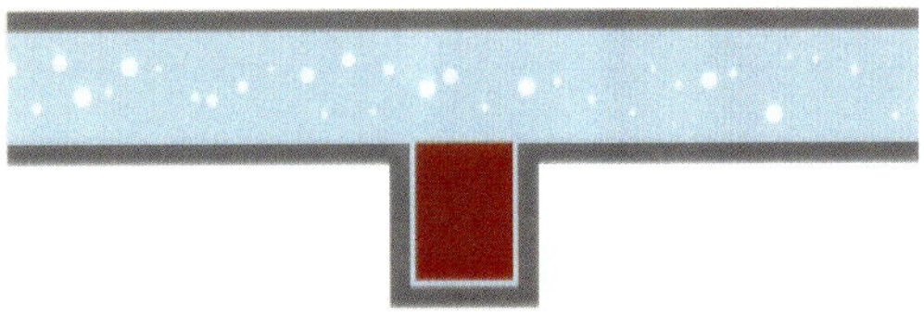

Transistor On

Likewise, under the right circumstances, a transistor can look like a short circuit between the collector and emitter pins. Current is free to flow through the collector, and out the emitter.

2) Off – Open Circuit

When it's closed, a valve can completely stop the flow of water.

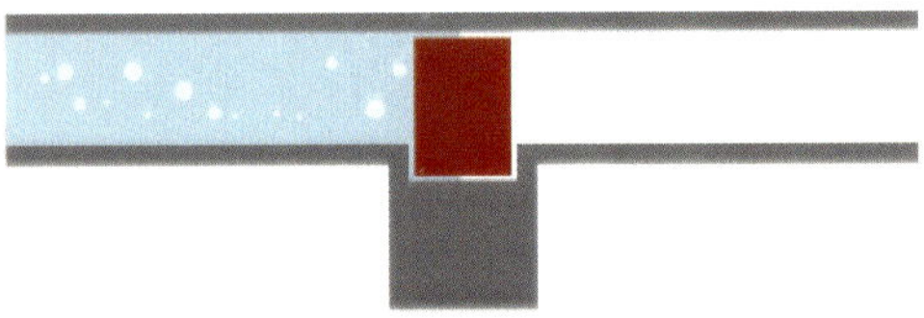

Transistor Off

In the same way, a transistor can be used to create an open circuit between the collector and emitter pins.

3) Linear Flow Control

With some precise tuning, a valve can be adjusted to finely control the flow rate to some point between fully open and closed.

Flow Control

A transistor can do the same thing – linearly controlling the current through a circuit at some point between fully off (an open circuit) and fully on (a short circuit).

From our water analogy, the width of a pipe is similar to the resistance in a circuit. If a valve can finely adjust the width of a pipe, then a transistor can finely adjust the resistance between collector and emitter. So, in a way, a transistor is like a variable, adjustable resistor.

Amplifying Power

There's another analogy we can wrench into this. Imagine if, with the slight turn of a valve, you could control the flow rate of the Hoover Dam's flow gates. The measly amount of force you might put into twisting that knob has the potential to create a force thousands of times stronger. We're stretching the analogy to its limits, but this idea carries over to transistors too. Transistors are special because they can amplify electrical signals, turning a low-power signal into a similar signal of much higher power.

Operation Modes

Unlike resistors, which enforce a linear relationship between voltage and current, transistors are non-linear devices. They have four distinct modes of operation, which describe the current flowing through them. (When we talk about current flow through a transistor, we usually mean current flowing from collector to emitter of an NPN.)

The four transistor operation modes are:

- Saturation – The transistor acts like a short circuit. Current freely flows from collector to emitter.
- Cut-off – The transistor acts like an open circuit. No current flows from collector to emitter.
- Active – The current from collector to emitter is proportional to the current flowing into the base.
- Reverse-Active – Like active mode, the current is proportional to the base current, but it flows in reverse. Current flows from emitter to collector (not, exactly, the purpose transistors were designed for).

To determine which mode a transistor is in, we need to look at the voltages on each of the three pins, and how they relate to each other. The voltages from base to emitter (V_{BE}), and the from base to collector (V_{BC}) set the transistor's mode:

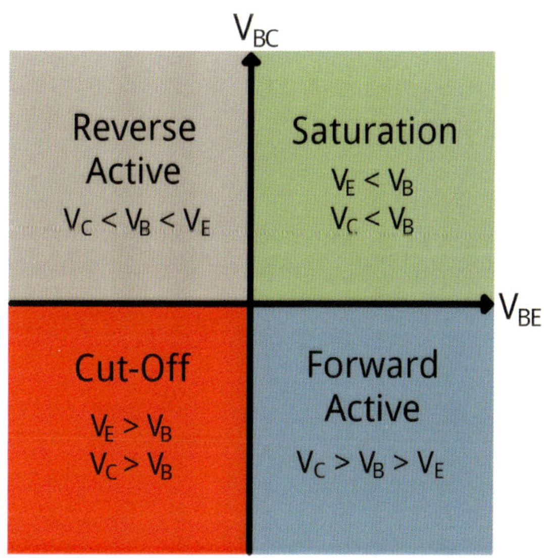

The simplified quadrant graph above shows how positive and negative voltages at those terminals affect the mode. In reality it's a bit more complicated than that.

Let's look at all four transistor modes individually; we'll investigate how to put the device into that mode, and what effect it has on current flow.

Note: The majority of this page focuses on NPN transistors. To understand how a PNP transistor works, simply flip the polarity or > and < signs.

Saturation Mode

Saturation is the on mode of a transistor. A transistor in saturation mode acts like a short circuit between collector and emitter.

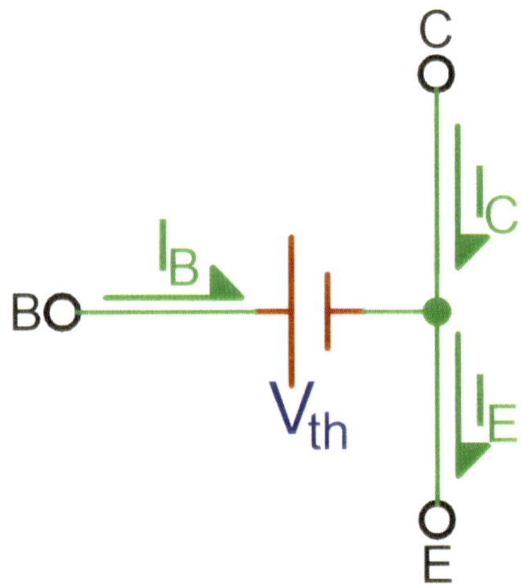

In saturation mode both of the "diodes" in the transistor are forward biased. That means V_{BE} must be greater than 0, *and* so must V_{BC}. In other words, V_B must be higher than both V_E and V_C.

$$V_B > V_C$$
$$V_B > V_E$$

Because the junction from base to emitter looks just like a diode, in reality, V_{BE} must be greater than a threshold voltage to enter saturation. There are many abbreviations for this voltage drop – V_{th}, V_γ, and V_d are a few – and the actual value varies between transistors (and even further by temperature). For a lot of transistors (at room temperature) we can estimate this drop to be about 0.6V.

Another reality bummer: there won't be perfect conduction between emitter and collector. A small voltage drop will form between those nodes. Transistor datasheets will define this voltage as CE saturation voltage $V_{CE(sat)}$ – a voltage from collector to emitter required for saturation. This value is usually around 0.05-0.2V. This value means that V_C must be slightly greater than V_E (but both still less than V_B) to get the transistor in saturation mode.

Cutoff Mode

Cutoff mode is the opposite of saturation. A transistor in cutoff mode is off – there is no collector current, and therefore no emitter current. It almost looks like an open circuit.

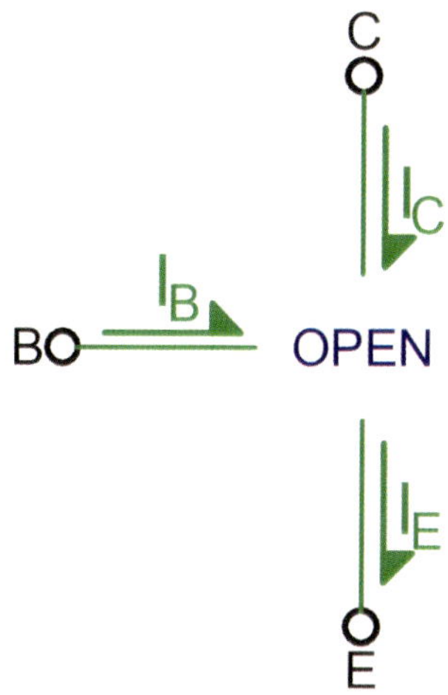

To get a transistor into cutoff mode, the base voltage must be less than both the emitter and collector voltages. V_{BC} and V_{BE} must both be negative.

$V_C > V_B$
$V_E > V_B$

In reality, V_{BE} can be anywhere between 0V and V_{th} (~0.6V) to achieve cutoff mode.

Active Mode

To operate in active mode, a transistor's V_{BE} must be greater than zero and V_{BC} must be negative. Thus, the base voltage must be less than the collector, but greater than the emitter. That also means the collector must be greater than the emitter.

$V_C > V_B > V_E$

In reality, we need a non-zero forward voltage drop (abbreviated either V_{th}, V_γ, or V_d) from base to emitter (V_{BE}) to "turn on" the transistor. Usually this voltage is usually around 0.6V.

Amplifying in Active Mode

Active mode is the most powerful mode of the transistor because it turns the device into an amplifier. Current going into the base pin amplifies current going into the collector and out the emitter.

Our shorthand notation for the gain (amplification factor) of a transistor is β (you may also see it as β_F, or h_{FE}). β linearly relates the collector current (I_C) to the base current (I_B):

$$I_C = \beta I_B$$

The actual value of β varies by transistor. It's usually around 100, but can range from 50 to 200... even 2000, depending on which transistor you're using and how much current is running through it. If your transistor had a β of 100, for example, that'd mean an input current of 1mA into the base could produce 100mA current through the collector.

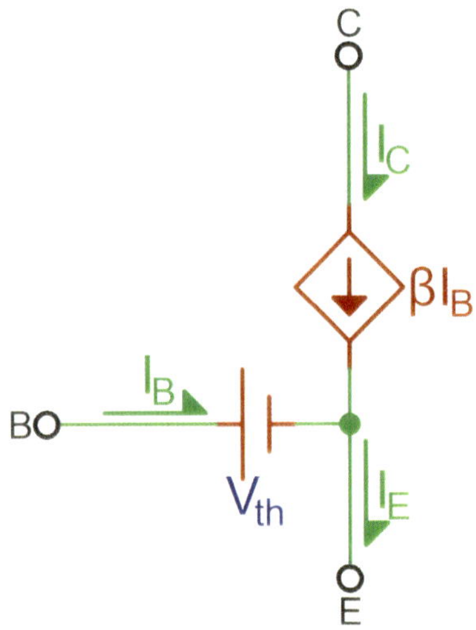

Active mode model. $V_{BE} = V_{th}$, *and* $I_C = \beta I_B$.

What about the emitter current, I_E? In active mode, the collector and base currents go *into* the device, and the I_E comes out. To relate the emitter current to collector current, we have another constant value: α. α is the common-base current gain, it relates those currents as such:

$$I_C = \alpha I_E$$

α is usually *very* close to, but less than, 1. That means I_C is very close to, but less than I_E in active mode.

You can use β to calculate α, or vice-versa:

$$\beta = \frac{\alpha}{(1-\alpha)}$$

$$\alpha = \frac{\beta}{\beta+1}$$

If β is 100, for example, that means α is 0.99. So, if I_C is 100mA, for example, then I_E is 101mA.

Reverse Active

Just as saturation is the opposite of cutoff, reverse active mode is the opposite of active mode. A transistor in reverse active mode conducts, even amplifies, but current flows in the opposite direction, from emitter to collector. The downside to reverse active mode is the β (β_R in this case) is *much* smaller.

To put a transistor in reverse active mode, the emitter voltage must be greater than the base, which must be greater than the collector ($V_{BE}<0$ and $V_{BC}>0$).

$$V_C < V_B < V_E$$

Reverse active mode isn't usually a state in which you want to drive a transistor. It's good to know it's there, but it's rarely designed into an application.

Relating to the PNP

PNP's work a lot like the NPN's – they have the same four modes – but everything is turned around. To find out which mode a PNP transistor is in, reverse all of the < and > signs.

For example, to put a PNP into saturation V_C and V_E must be higher than V_B. You pull the base low to turn the PNP on, and make it higher than the collector and emitter to turn it off. And, to put a PNP into active mode, V_E must be at a higher voltage than V_B, which must be higher than V_C.

Voltage relations	NPN Mode	PNP Mode
$V_E < V_B < V_C$	Active	Reverse
$V_E < V_B > V_C$	Saturation	Cutoff
$V_E > V_B < V_C$	Cutoff	Saturation
$V_E > V_B > V_C$	Reverse	Active

Another opposing characteristic of the NPNs and PNPs is the direction of current flow. In active and saturation modes, current in a PNP flows from emitter to collector. This means the emitter must generally be at a higher voltage than the collector.

Simplified Operation

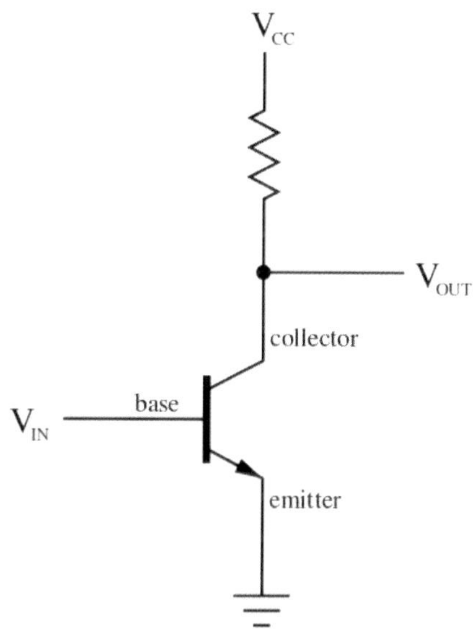

A simple circuit diagram to show the labels of a n–p–n bipolar transistor.

The essential usefulness of a transistor comes from its ability to use a small signal applied between one pair of its terminals to control a much larger signal at another pair of terminals. This property is called gain. It can produce a stronger output signal, a voltage or current, which is proportional to a weaker input signal; that is, it can act as an amplifier. Alternatively, the transistor can be used to turn current on or off in a circuit as an electrically controlled switch, where the amount of current is determined by other circuit elements.

There are two types of transistors, which have slight differences in how they are used in a circuit. A bipolar transistor has terminals labeled base, collector, and emitter. A small current at the base terminal (that is, flowing between the base and the emitter) can control or switch a much larger current between the collector and emitter terminals. For a field-effect transistor, the terminals are labeled gate, source, and drain, and a voltage at the gate can control a current between source and drain.

The image represents a typical bipolar transistor in a circuit. Charge will flow between emitter and collector terminals depending on the current in the base. Because internally the base and emitter connections behave like a semiconductor diode, a voltage drop develops between base and emitter while the base current exists. The amount of this voltage depends on the material the transistor is made from, and is referred to as V_{BE}.

Transistor as a Switch

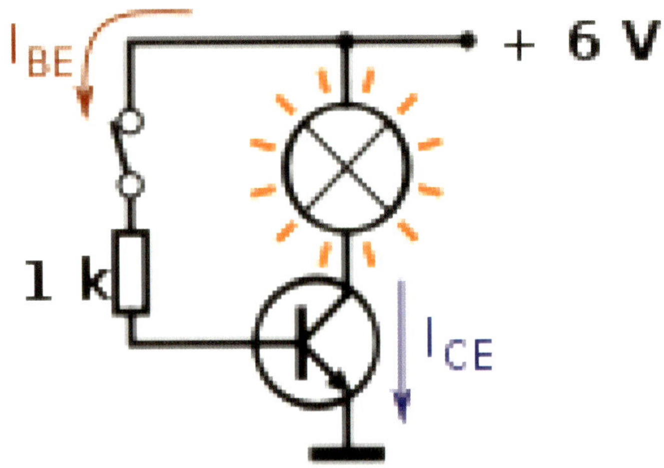

BJT used as an electronic switch, in grounded-emitter configuration.

Transistors are commonly used in digital circuits as electronic switches which can be either in an "on" or "off" state, both for high-power applications such as switched-mode power supplies and for low-power applications such as logic gates. Important parameters for this application include the current switched, the voltage handled, and the switching speed, characterised by the rise and fall times.

In a grounded-emitter transistor circuit, such as the light-switch circuit shown, as the base voltage rises, the emitter and collector currents rise exponentially. The collector voltage drops because of reduced resistance from collector to emitter. If the voltage difference between the collector and emitter were zero (or near zero), the collector current would be limited only by the load resistance (light bulb) and the supply voltage. This is called *saturation* because current is flowing from collector to emitter freely. When saturated, the switch is said to be *on*.

Providing sufficient base drive current is a key problem in the use of bipolar transistors as switches. The transistor provides current gain, allowing a relatively large current in the collector to be switched by a much smaller current into the base terminal. The ratio of these currents varies depending on the type of transistor, and even for a particular type, varies depending on the collector current. In the example light-switch circuit shown, the resistor is chosen to provide enough base current to ensure the transistor will be saturated.

In a switching circuit, the idea is to simulate, as near as possible, the ideal switch having the properties of open circuit when off, short circuit when on, and an instantaneous transition between the two states. Parameters are chosen such that the "off" output is limited to leakage currents too small to affect connected circuitry; the resistance of the transistor in the "on" state is too small to affect circuitry; and the transition between the two states is fast enough not to have a detrimental effect.

Transistor as an Amplifier

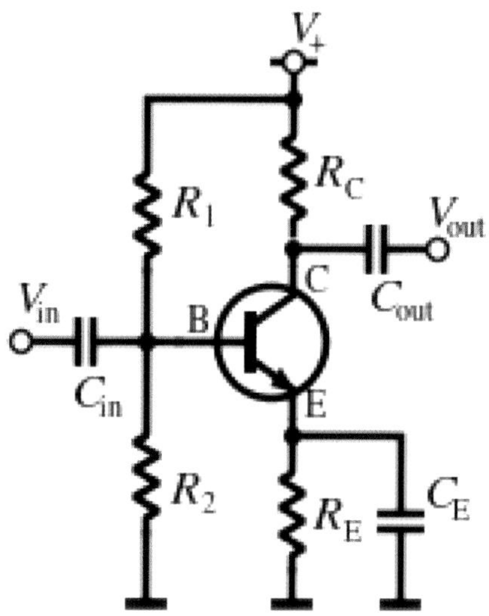

Amplifier circuit, common-emitter configuration with a voltage-divider bias circuit.

The common-emitter amplifier is designed so that a small change in voltage (V_{in}) changes the small current through the base of the transistor; the transistor's current amplification combined with the properties of the circuit means that small swings in V_{in} produce large changes in V_{out}.

Various configurations of single transistor amplifier are possible, with some providing current gain, some voltage gain, and some both.

From mobile phones to televisions, vast numbers of products include amplifiers for sound reproduction, radio transmission, and signal processing. The first discrete-transistor audio amplifiers barely supplied a few hundred milliwatts, but power and audio fidelity gradually increased as better transistors became available and amplifier architecture evolved.

Modern transistor audio amplifiers of up to a few hundred watts are common and relatively inexpensive.

Comparison with Vacuum Tubes

Before transistors were developed, vacuum (electron) tubes (or in the UK "thermionic valves" or just "valves") were the main active components in electronic equipment.

Advantages

The key advantages that have allowed transistors to replace vacuum tubes in most applications are

- no cathode heater (which produces the characteristic orange glow of tubes), reducing power consumption, eliminating delay as tube heaters warm up, and immune from cathode poisoning and depletion;
- very small size and weight, reducing equipment size;
- large numbers of extremely small transistors can be manufactured as a single integrated circuit;
- low operating voltages compatible with batteries of only a few cells;
- circuits with greater energy efficiency are usually possible. For low-power applications (e.g., voltage amplification) in particular, energy consumption can be very much less than for tubes;
- complementary devices available, providing design flexibility including complementary-symmetry circuits, not possible with vacuum tubes;
- very low sensitivity to mechanical shock and vibration, providing physical ruggedness and virtually eliminating shock-induced spurious signals (e.g., microphonics in audio applications);
- not susceptible to breakage of a glass envelope, leakage, outgassing, and other physical damage.

Limitations

Transistors have the following limitations:

- silicon transistors can age and fail;
- high-power, high-frequency operation, such as that used in over-the-air television broadcasting, is better achieved in vacuum tubes due to improved electron mobility in a vacuum;
- solid-state devices are susceptible to damage from very brief electrical and thermal events, including electrostatic discharge in handling; vacuum tubes are electrically much more rugged;
- sensitivity to radiation and cosmic rays (special radiation-hardened chips are used for spacecraft devices);
- vacuum tubes in audio applications create significant lower-harmonic distortion, the so-called tube sound, which some people prefer.

Types of Transistors

There are two types of transistors in present; they are bipolar junction transistor (BJT), field effect transistors (FET). A small current is flowing between the base and the emitter; base terminal can control a larger current flow between the collector and the emitter terminals. For a field-effect transistor, it also has the three terminals, they are gate, source, and drain, and a voltage at the gate can control a current between source and drain. The simple diagrams of BJT and FET are shown in figure below:

Bipolar Junction Transistor(BJT)

Field Effect Transistors(FET)

As you can see, transistors come in a variety of different sizes and shapes. One thing all of these transistors have in common is that they each have three leads.

Bipolar Junction Transistors

The invention of the bipolar transistor in 1948 ushered in a revolution in electronics. Technical feats previously requiring relatively large, mechanically fragile, power-hungry vacuum tubes were suddenly achievable with tiny, mechanically rugged, power-thrifty specks of crystalline silicon. This revolution made possible the design and manufacture of lightweight, inexpensive electronic devices that we now take for granted. Understanding how transistors function is of paramount importance to anyone interested in understanding modern electronics.

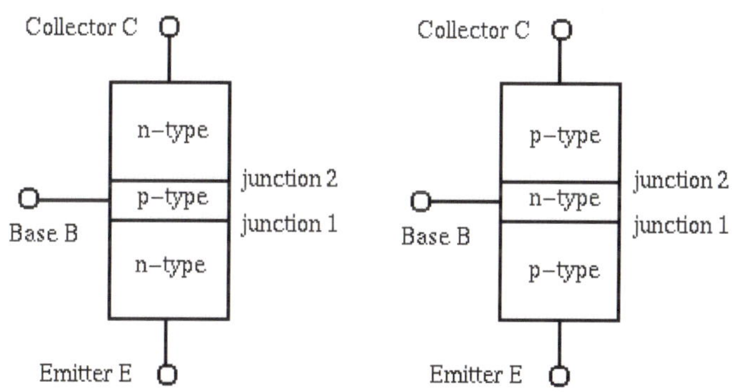

npn–transistor **pnp–transistor**

A bipolar transistor consists of a three-layer "sandwich" of doped (extrinsic) semiconductor materials, either P-N-P in Figure below(b) or N-P-N at (d). Each layer forming the transistor has a specific name, and each layer is provided with a wire contact for connection to a circuit. The schematic symbols are shown in Figure below(a) and (d).

The functional difference between a PNP transistor and an NPN transistor is the proper biasing (polarity) of the junctions when operating. For any given state of operation, the current directions and voltage polarities for each kind of transistor are exactly opposite each other.

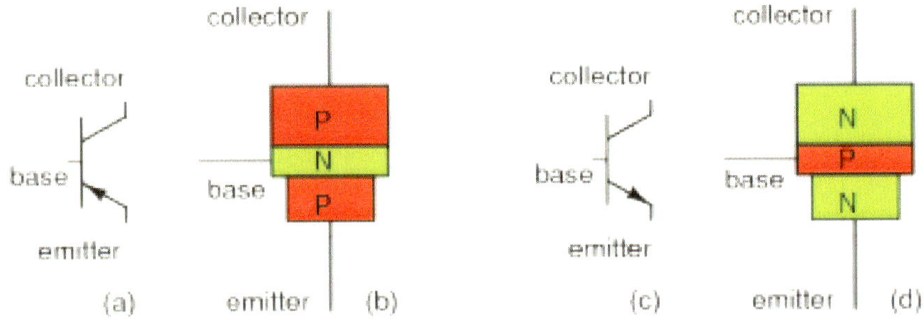

BJT transistor: (a) PNP schematic symbol, (b) physical layout (c) NPN symbol, (d) layout.

Bipolar transistors work as current-controlled current *regulators*. In other words, transistors restrict the amount of current passed according to a smaller, controlling current. The main current that is *controlled* goes from collector to emitter, or from emitter to collector, depending on the type of transistor it is (PNP or NPN, respectively). The small current that *controls* the main current goes from base to emitter, or from emitter to base, once again depending on the kind of transistor it is (PNP or NPN, respectively). According to the standards of semiconductor symbology, the arrow always points *against* the direction of electron flow. (Figure below)

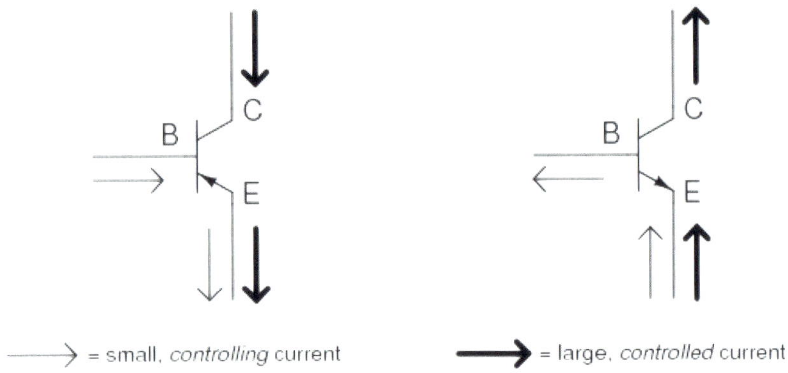

Small Base-Emitter current controls large Collector-Emitter current flowing against emitter arrow.

Bipolar transistors are called *bi*polar because the main flow of electrons through them takes place in *two* types of semiconductor material: P and N, as the main current goes from emitter to collector (or vice versa). In other words, two types of charge carriers—electrons and holes—comprise this main current through the transistor.

As you can see, the *controlling* current and the *controlled* current always mesh together through the emitter wire, and their electrons always flow *against* the direction of the transistor's arrow. This is the first and foremost rule in the use of transistors: all currents must be going in the proper directions for the device to work as a current regulator. The small, controlling current is usually referred to simply as the *base current* because it is the only current that goes through the base wire of the transistor. Conversely, the large, controlled current is referred to as the *collector current* because it is the only current that goes through the collector wire. The emitter current is the sum of the base and collector currents, in compliance with Kirchhoff's Current Law.

No current through the base of the transistor, shuts it off like an open switch and prevents current through the collector. A base current, turns the transistor on like a closed switch and allows a proportional amount of current through the collector. Collector current is primarily limited by the base current, regardless of the amount of voltage available to push it. The next section will explore in more detail the use of bipolar transistors as switching elements.

Function

BJTs come in two types, or polarities, known as PNP and NPN based on the doping types of the three main terminal regions. An NPN transistor comprises two semiconductor junctions that share a thin p-doped region, and a PNP transistor comprises two semiconductor junctions that share a thin n-doped region.

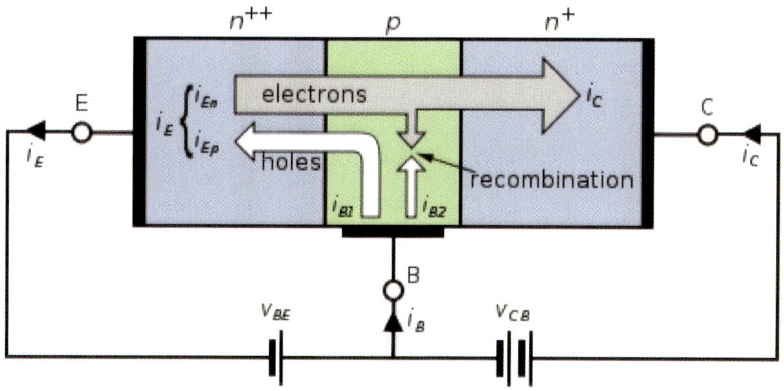

NPN BJT with forward-biased E–B junction and reverse-biased B–C junction

Charge flow in a BJT is due to diffusion of charge carriers across a junction between two regions of different charge concentrations. The regions of a BJT are called *emitter*, *collector*, and *base*. A discrete transistor has three leads for connection to these regions. Typically, the emitter region is heavily doped compared to the other two layers, whereas the majority charge carrier concentrations in base and collector layers are about the same (collector doping is typically ten times lighter than base doping). By design, most of the BJT collector current is due to the flow of charge carriers (electrons or holes) injected from a high-concentration emitter into the base where they are minority carriers that diffuse toward the collector, and so BJTs are classified as minority-carrier devices.

In typical operation, the base–emitter junction is forward-biased, which means that the p-doped side of the junction is at a more positive potential than the n-doped side, and the base–collector junction is reverse-biased. In an NPN transistor, when positive bias is applied to the base–emitter junction, the equilibrium is disturbed between the thermally generated carriers and the repelling electric field of the n-doped emitter depletion region. This allows thermally excited electrons to inject from the emitter into the base region. These electrons diffuse through the base from the region of high concentration near the emitter toward the region of low concentration near the collector. The electrons in the base are called minority carriers because the base is doped p-type, which makes holes the majority carrier in the base.

To minimize the fraction of carriers that recombine before reaching the collector–base junction, the transistor's base region must be thin enough that carriers can diffuse across it in much less time than the semiconductor's minority-carrier lifetime. In particular, the thickness of the base must be much less than the diffusion length of the electrons. The collector–base junction is reverse-biased, and so little electron injection occurs from the collector to the base, but electrons that diffuse through the base towards the collector are swept into the collector by the electric field in the depletion region of the collector–base junction. The thin *shared* base and asymmetric collector–emitter doping are what differentiates a bipolar transistor from two *separate* and oppositely biased diodes connected in series.

Voltage, Current, and Charge Control

The collector–emitter current can be viewed as being controlled by the base–emitter current (current control), or by the base–emitter voltage (voltage control). These views are related by the current–voltage relation of the base–emitter junction, which is the usual exponential current–voltage curve of a p–n junction (diode).

The physical explanation for collector current is the concentration of minority carriers in the base region. Due to low-level injection (in which there are much fewer excess carriers than normal majority carriers) the ambipolar transport rates (in which the excess majority and minority carriers flow at the same rate) is in effect determined by the excess minority carriers.

Detailed transistor models of transistor action, such as the Gummel–Poon model, account for the distribution of this charge explicitly to explain transistor behaviour more exactly. The charge-control view easily handles phototransistors, where minority carriers in the base region are created by the absorption of photons, and handles the dynamics of turn-off, or recovery time, which depends on charge in the base region recombining. However, because base charge is not a signal that is visible at the terminals, the current- and voltage-control views are generally used in circuit design and analysis.

In analog circuit design, the current-control view is sometimes used because it is approximately linear. That is, the collector current is approximately β_F times the base current. Some basic circuits can be designed by assuming that the emitter–base voltage is approximately constant and that collector current is β times the base current. However, to accurately and reliably design production BJT circuits, the voltage-control (for example, Ebers–Moll) model is required. The voltage-control model requires an exponential function to be taken into account, but when it is linearized such that the transistor can be modeled as a transconductance, as in the Ebers–Moll model, design for circuits such as differential amplifiers again becomes a mostly linear problem, so the voltage-control view is often preferred. For translinear circuits, in which the exponential I–V curve is key to the operation, the transistors are usually modeled as voltage-controlled current sources whose transconductance is proportional to their collector current. In general, transistor-level circuit design is performed using SPICE or a comparable analog-circuit simulator, so model complexity is usually not of much concern to the designer.

Turn-on, Turn-off, and Storage Delay

Most bipolar transistors, and especially power transistors, have long base-storage times when they are driven into saturation; the base storage limits turn-off time in switching applications. A Baker clamp can prevent the transistor from heavily saturating, which reduces the amount of charge stored in the base and thus improves switching time.

Transistor Parameters: Alpha (α) and Beta (β)

The proportion of electrons able to cross the base and reach the collector is a measure of the BJT efficiency. The heavy doping of the emitter region and light doping of the base region causes many more electrons to be injected from the emitter into the base than holes to be injected from the base into the emitter.

The common-emitter *current gain* is represented by β_F or the h-parameter h_{FE}; it is approximately the ratio of the DC collector current to the DC base current in forward-active region. It is typically greater than 50 for small-signal transistors, but can be smaller in transistors designed for high-power applications.

Another important parameter is the common-base *current gain*, α_F. The common-base current gain is approximately the gain of current from emitter to collector in the forward-active region. This ratio usually has a value close to unity; between 0.980 and 0.998. It is less than unity due to recombination of charge carriers as they cross the base region.

Alpha and beta are more precisely related by the following identities (NPN transistor):

$$\alpha_F = \frac{I_C}{I_E}, \qquad \beta_F = \frac{I_C}{I_B},$$

$$\alpha_F = \frac{\beta_F}{1+\beta_F} \Leftrightarrow \beta_F = \frac{\alpha_F}{1-\alpha_F}.$$

Theory and Modeling

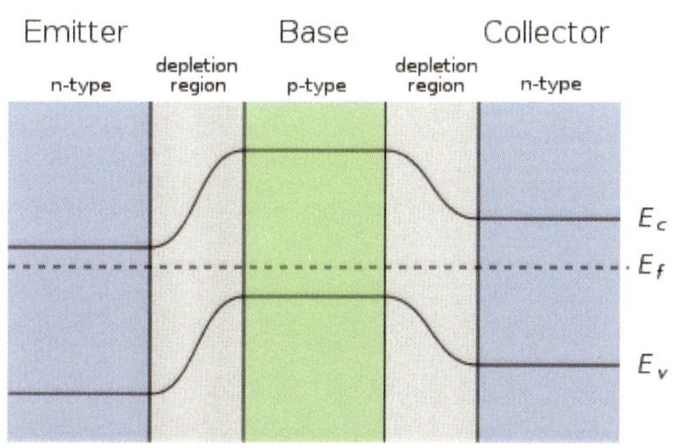

Band diagram for NPN transistor at equilibrium

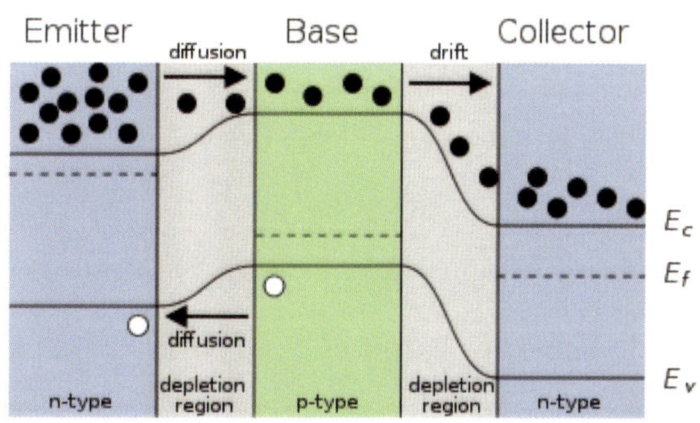

Band diagram for NPN transistor in active mode, showing injection of electrons from emitter to base, and their overshoot into the collector

Transistors can be thought of as two diodes (P–N junctions) sharing a common region that minority carriers can move through. A PNP BJT will function like two diodes that share an N-type cathode region, and the NPN like two diodes sharing a P-type anode region. Connecting two diodes with wires will not make a transistor, since minority carriers will not be able to get from one P–N junction to the other through the wire.

Both types of BJT function by letting a small current input to the base control an amplified output from the collector. The result is that the transistor makes a good switch that is controlled by its base input. The BJT also makes a good amplifier, since it can multiply a weak input signal to about 100 times its original strength. Networks of transistors are used to make powerful amplifiers with many different applications. Focus is on the NPN bipolar transistor. In the NPN transistor in what is called active mode, the base–emitter voltage V_{BE} and collector–base voltage V_{CB} are positive, forward biasing the emitter–base junction and reverse-biasing the collector–base junction. In the active mode of operation, electrons are injected from the forward biased n-type emitter region into the p-type base where they diffuse as minority carriers to the reverse-biased n-type collector and are swept away by the electric field in the reverse-biased collector–base junction.

Large-signal Models

In 1954, Jewell James Ebers and John L. Moll introduced their mathematical model of transistor currents.

Ebers–Moll Model

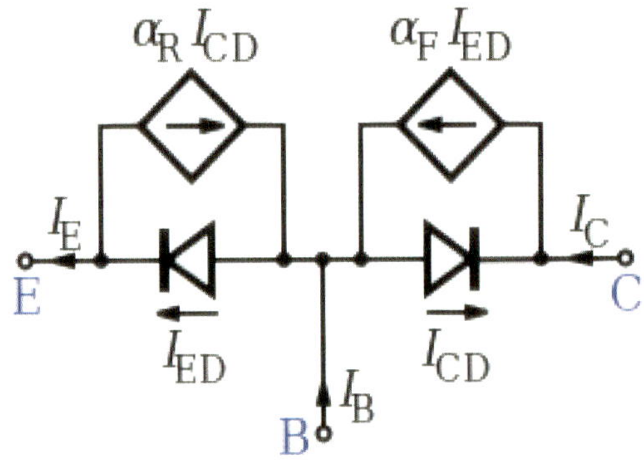

Ebers–Moll model for an NPN transistor. I_B, I_C and I_E are the base, collector and emitter currents; I_{CD} and I_{ED} are the collector and emitter diode currents; α_F and α_R are the forward and reverse common-base current gains

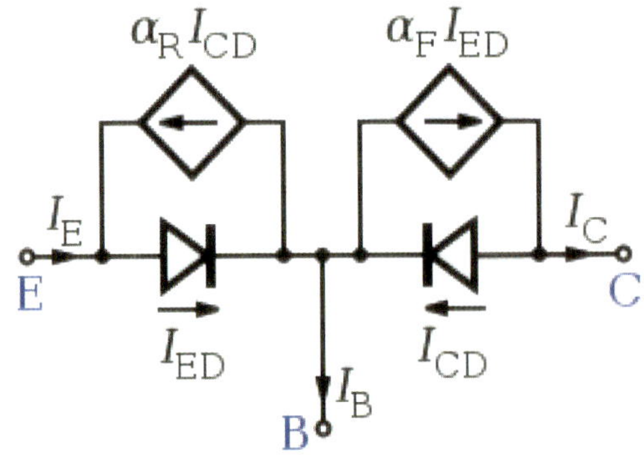

Ebers–Moll model for a PNP transistor

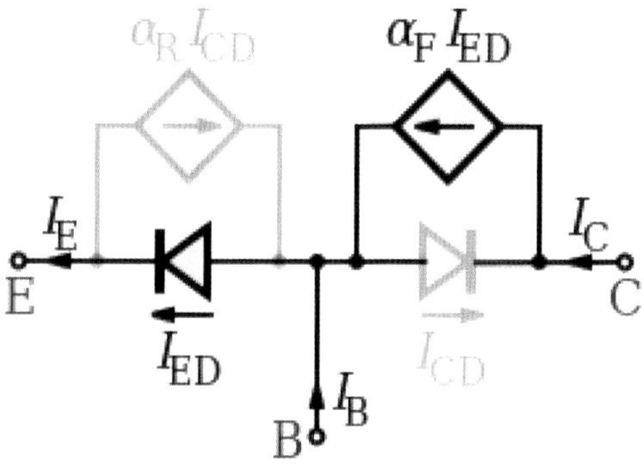

Approximated Ebers–Moll model for an NPN transistor in the forward active mode. The collector diode is reverse-biased so I_{CD} is virtually zero. Most of the emitter diode current (α_F is nearly 1) is drawn from the collector, providing the amplification of the base current

The DC emitter and collector currents in active mode are well modeled by an approximation to the Ebers–Moll model:

$$I_E = I_{ES}\left(e^{\frac{V_{BE}}{V_T}} - 1\right)$$

$$I_C = \alpha_F I_E$$

$$I_B = (1-\alpha_F) I_E$$

The base internal current is mainly by diffusion and

$$J_{n(base)} = \frac{1}{W} q D_n n_{bo} e^{\frac{V_{EB}}{V_T}}$$

where

- V_T is the thermal voltage kT/q (approximately 26 mV at 300 K ≈ room temperature).
- I_E is the emitter current
- I_C is the collector current
- α_F is the common base forward short-circuit current gain (0.98 to 0.998)
- I_{ES} is the reverse saturation current of the base–emitter diode (on the order of 10−15 to 10−12 amperes)
- V_{BE} is the base–emitter voltage
- D_n is the diffusion constant for electrons in the p-type base
- W is the base width

The α and forward β parameters are as described previously. A reverse β is sometimes included in the model.

The unapproximated Ebers–Moll equations used to describe the three currents in any operating region are given below. These equations are based on the transport model for a bipolar junction transistor.

$$i_C = I_S\left[\left(e^{\frac{V_{BE}}{V_T}} - e^{\frac{V_{BC}}{V_T}}\right) - \frac{1}{\beta_R}\left(e^{\frac{V_{BC}}{V_T}} - 1\right)\right]$$

$$i_B = I_S\left[\frac{1}{\beta_F}\left(e^{\frac{V_{BE}}{V_T}} - 1\right) + \frac{1}{\beta_R}\left(e^{\frac{V_{BC}}{V_T}} - 1\right)\right]$$

$$i_E = I_S\left[\left(e^{\frac{V_{BE}}{V_T}} - e^{\frac{V_{BC}}{V_T}}\right) + \frac{1}{\beta_F}\left(e^{\frac{V_{BE}}{V_T}} - 1\right)\right]$$

where

- i_C is the collector current
- i_B is the base current
- i_E is the emitter current
- β_F is the forward common emitter current gain (20 to 500)
- β_R is the reverse common emitter current gain (0 to 20)
- I_S is the reverse saturation current (on the order of 10−15 to 10−12 amperes)
- V_T is the thermal voltage (approximately 26 mV at 300 K ≈ room temperature).
- V_{BE} is the base–emitter voltage
- V_{BC} is the base–collector voltage

Base-width Modulation

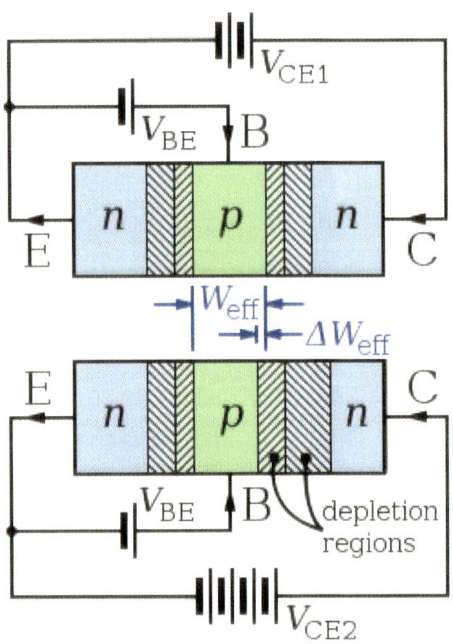

Top: NPN base width for low collector–base reverse bias; Bottom: narrower NPN base width for large collector–base reverse bias. Hashed regions are depleted regions

As the collector–base voltage ($V_{CB} = V_{CE} - V_{BE}$) varies, the collector–base depletion region varies in size. An increase in the collector–base voltage, for example, causes a greater reverse bias across the collector–base junction, increasing the collector–base depletion region width, and decreasing the width of the base. This variation in base width often is called the Early effect after its discoverer James M. Early.

Narrowing of the base width has two consequences:

- There is a lesser chance for recombination within the "smaller" base region.
- The charge gradient is increased across the base, and consequently, the current of minority carriers injected across the emitter junction increases.

Both factors increase the collector or "output" current of the transistor in response to an increase in the collector–base voltage.

In the forward-active region, the Early effect modifies the collector current (i_C) and the forward common emitter current gain (β_F) as given by:

$$i_C = I_S e^{\frac{v_{BE}}{V_T}} \left(1 + \frac{V_{CE}}{V_A}\right)$$

$$\beta_F = \beta_{F0} \left(1 + \frac{V_{CB}}{V_A}\right)$$

$$r_o = \frac{V_A}{I_C}$$

where:

- V_{CE} is the collector–emitter voltage
- V_A is the Early voltage (15 V to 150 V)
- β_{F0} is forward common-emitter current gain when $V_{CB} = 0$ V
- r_o is the output impedance
- I_C is the collector current

Punchthrough

When the base–collector voltage reaches a certain (device-specific) value, the base–collector depletion region boundary meets the base–emitter depletion region boundary. When in this state the transistor effectively has no base. The device thus loses all gain when in this state.

Gummel–Poon charge-control Model

The Gummel–Poon model is a detailed charge-controlled model of BJT dynamics, which has been adopted and elaborated by others to explain transistor dynamics in greater detail than the terminal-based models typically do. This model also includes the dependence of transistor β -values upon the direct current levels in the transistor, which are assumed current-independent in the Ebers–Moll model.

Small-signal Models

Hybrid-pi Model

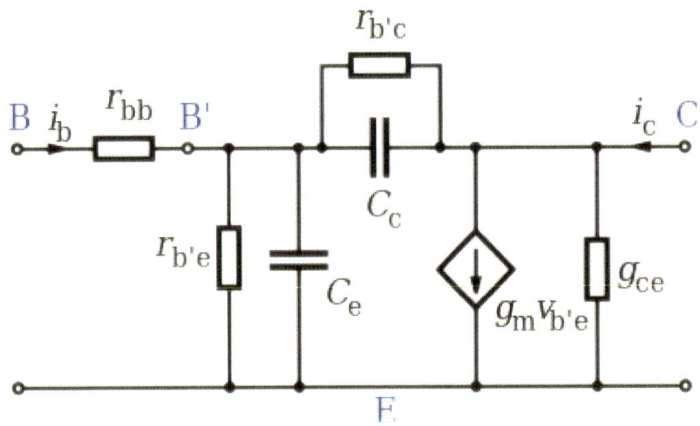

Hybrid-pi model

The hybrid-pi model is a popular circuit model used for analyzing the small signal behavior of bipolar junction and field effect transistors. Sometimes it is also called *Giacoletto model* because it was introduced by L.J. Giacoletto in 1969. The model can be quite accurate for low-frequency circuits and can easily be adapted for higher-frequency circuits with the addition of appropriate inter-electrode capacitances and other parasitic elements.

h-parameter Model

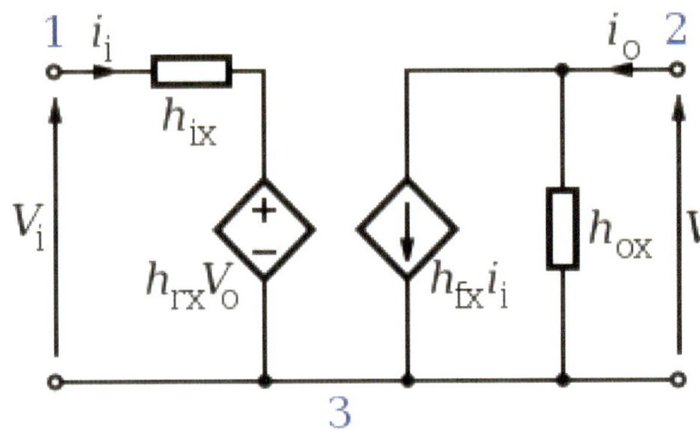

Generalized h-parameter model of an NPN BJT.
Replace x with e, b or c for CE, CB and CC topologies respectively.

Another model commonly used to analyze BJT circuits is the h-parameter model, closely related to the hybrid-pi model and the y-parameter two-port, but using input current and output voltage as independent variables, rather than input and output voltages. This two-port network is particularly suited to BJTs as it lends itself easily to the analysis of circuit behaviour, and may be used to develop further accurate models. As shown, the term, *x*, in the model represents a different BJT lead depending on the topology used. For common-emitter mode the various symbols take on the specific values as:

- Terminal 1, base
- Terminal 2, collector

- Terminal 3 (common), emitter; giving x to be e
- i_i, base current (i_b)
- i_o, collector current (i_c)
- V_{in}, base-to-emitter voltage (V_{BE})
- V_o, collector-to-emitter voltage (V_{CE})

and the h-parameters are given by:

- $h_{ix} = h_{ie}$, the input impedance of the transistor (corresponding to the base resistance r_{pi}).
- $h_{rx} = h_{re}$, represents the dependence of the transistor's I_B–V_{BE} curve on the value of V_{CE}. It is usually very small and is often neglected (assumed to be zero).
- $h_{fx} = h_{fe}$, the current-gain of the transistor. This parameter is often specified as h_{FE} or the DC current-gain (β_{DC}) in datasheets.
- $h_{ox} = 1/h_{oe}$, the output impedance of transistor. The parameter h_{oe} usually corresponds to the output admittance of the bipolar transistor and has to be inverted to convert it to an impedance.

As shown, the h-parameters have lower-case subscripts and hence signify AC conditions or analyses. For DC conditions they are specified in upper-case. For the CE topology, an approximate h-parameter model is commonly used which further simplifies the circuit analysis. For this the h_{oe} and h_{re} parameters are neglected (that is, they are set to infinity and zero, respectively). The h-parameter model as shown is suited to low-frequency, small-signal analysis. For high-frequency analyses the inter-electrode capacitances that are important at high frequencies must be added.

The h refers to its being an h-parameter, a set of parameters named for their origin in a *hybrid equivalent circuit* model. F is from *forward current amplification* also called the current gain. E refers to the transistor operating in a *common emitter* (CE) configuration. Capital letters used in the subscript indicate that h_{FE} refers to a direct current circuit.

Industry Models

The Gummel–Poon SPICE model is often used, but it suffers from several limitations. These have been addressed in various more advanced models: Mextram, VBIC, HICUM, Modella.

Bipolar Transistor Configurations

Depending on which of the three terminals is used as common terminal, there can be three possible configurations for the two-port network formed by a transistor:

- Common emitter (CE),
- Common base (CB),
- Common collector (CC).

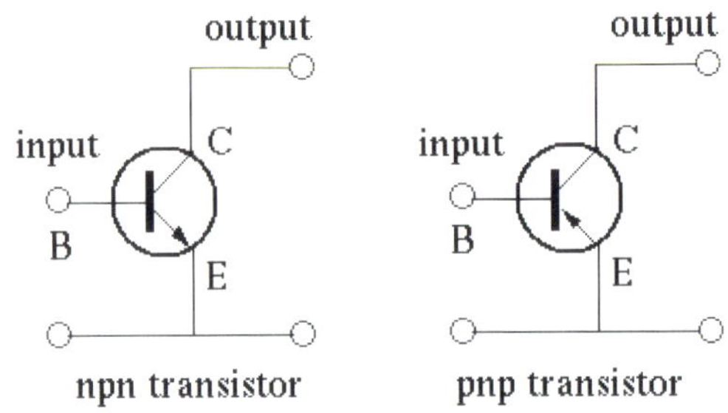

- Common-Base (CB) configuration

Two voltages V_{BE} and V_{CB} are applied respectively to the emitter *E* and collector *C*, with respect to the common base *B*, so that the BE junction is forward biased while the CB junction is reverse biased.

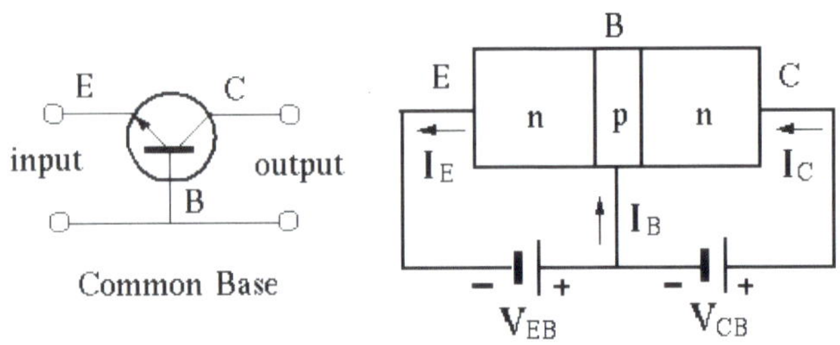

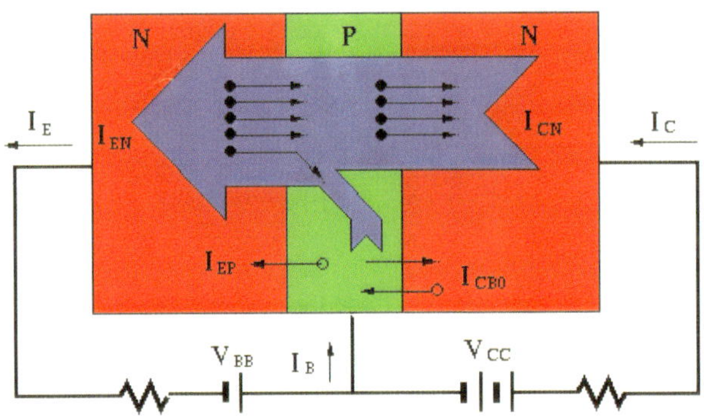

Note that the polarity of V_{BE} and direction of I_B associated with the PN-junction between E and B are the same as those associated with a diode, voltage polarity: positive on P, negative on N, current direction: from P to N, but V_{CB} and the direction of I_C associated with the PN-junction between the base and collector are defined oppositely.

The behavior of the NPN-transistor is determined by its two PN-junctions:

- The forward biased base-emitter (BE) PN-junction allows the free electrons in emitter to go through the PN-junction to arrive at the base, forming the emitter current I_E.
- As the P-type base is thin and lightly doped, only a small number of the electrons from the emitter are combined with the holes in base to form the base current I_B.
- Most of the electrons coming from the emitter become minority carriers in the P-type base, and they go through the reverse biased collector-base PN junction to arrive at the collector.

- The percentage of those electrons that arrive at the collector out of the electrons from the emitter is defined as α (e.g., $\alpha = 99\% = 0.99$, depending on the doping and geometry of the material). The total collector current I_C is therefore $I_C = \alpha I_E$.

The *current gain* or *current transfer ratio* is defined as the ratio between the emitter (input) current I_E and the collector (output) current I_C:

$$\frac{I_C}{I_E} \approx \alpha < 1, \quad \text{i.e.} \quad I_C = \alpha I_E$$

The base current I_B is:

$$I_\mathbf{B} = I_E - I_C \approx I_E - \alpha I_E = (1-\alpha) I_{E,} \quad \text{i.e.} \ I_E = \frac{1}{1-\alpha} I_\mathbf{B}$$

The CB configuration can be considered as a 2-port circuit. The input port is formed by the emitter and base, the output port is formed by the collector and base. The relationships between the current and voltage of both the input and output ports are described by the following input and output characteristics.

- Input characteristics:

  The input current I_E is a function of V_{CE} as well as the input voltage V_{BE}, hich is much more dominant:

$$I_E = f(V_{BE}, V_{CB}) \approx f(V_{BE}) = \frac{1}{1-\alpha} I_\mathbf{B} = \frac{1}{1-\alpha} I_0 (e^{V_{BE}/V_T} - 1)$$

where

$$I_\mathbf{B} = I_0 (e^{V_{BE}/V_T} - 1)$$

  This relationship between I_B and V_{BE} as the EB junction is very similar to the relationship of I_D and V_D of a diode. Also, we also note higher $V_{CB} > 0$ can slightly increase I_E.

- Output characteristics:

  The output current I_C is a function of the output voltage V_{CE} as well as the input current I_E, which is much more dominant:

$$I_C = f(I_E, V_{CB}) \approx f(I_E) = \alpha I_E$$

  Here the approximation is based on the assumption that $V_{CB} > 0.2V$ (in linear region). As $V_{CB} > 0$, i.e., the CB junction is reverse biased, the current I_C depends totally on I_E. When $I_E = 0$, $I_C = I_{CB0}$ is the current caused by the minority carriers crossing the PN-junction. This is similar to the diode current-voltage characteristics seen before, except both axes are reversed (the polarity of V_{CB} and the direction I_C are oppositely defined). When I_E is increased, $I_C = \alpha I_E$ is increased correspondingly. Higher V_{CB} can slightly increase I_C. As

$I_C = \alpha I_E < I_E$, CB configuration does not have current-amplification effect. However, if V_{BE} is held constant, I_E and therefore I_C will also be held constant, i.e., CB transistor circuit can be used as a current source.

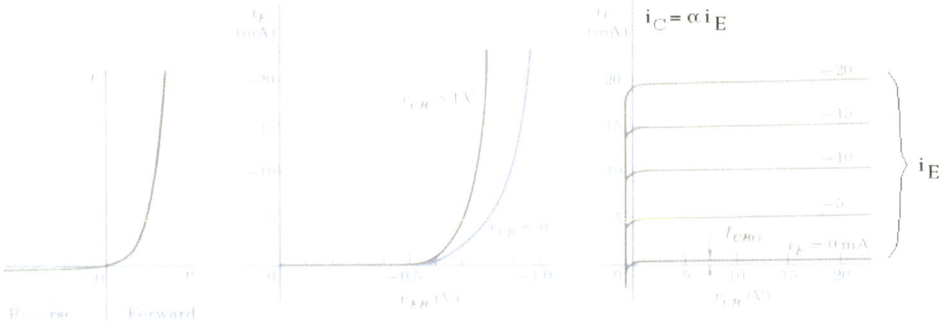

(a) Diode characteristic (b) Emitter characteristics (c) Collector characteristics

- Common-Emitter (CE) configuration

Two voltages V_{BE} and V_{CE} are applied respectively to the base B and collector C with respect to the common emitter E. As typically $V_{CE} > V_{BE}$, the BE junction is forward biased but the CB junction is reverse biased, same as the CB configuration. The voltages of CB and CE configurations are related by:

$$V_{CE} = V_{CB} + V_{BE}, \qquad or \qquad V_{CB} = V_{CE} - V_{BE}$$

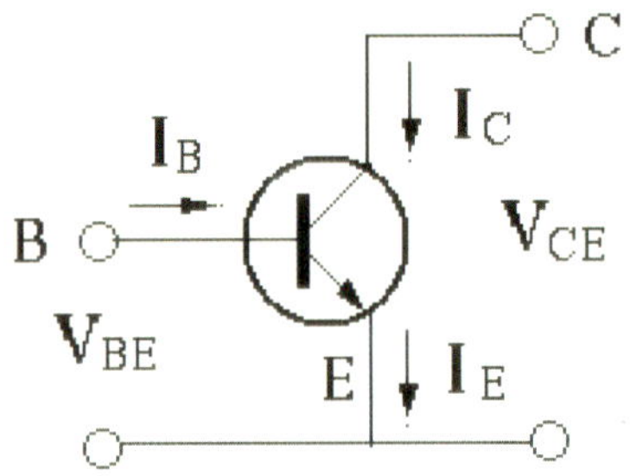

Common Emitter

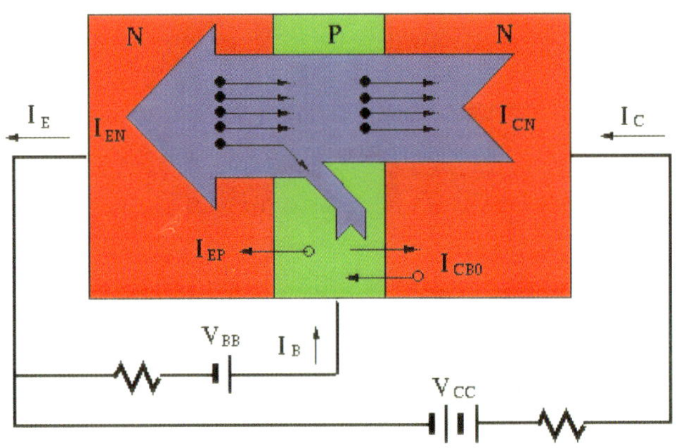

The base current I_B is treated as the input current, and the collector current I_C is treated as the output current:

$$I_C = \alpha I_E = \alpha(I_C + I_\mathbf{B})$$

Solving this equation for I_C, we get the relationship between the output I_C and the input I_B:

$$I_C = \frac{\alpha}{1-\alpha} I_\mathbf{B} = \beta I_\mathbf{B}$$

where we have defined the CE *current gain*, the ratio of the output current I_C and the input current I_B:

$$\beta = \frac{\alpha}{1-\alpha} \approx \frac{I_C}{I_\mathbf{B}}$$

The two parameters α and β are related by any of the following:

$$\beta = \frac{\alpha}{1-\alpha}, \quad \alpha = \frac{\beta}{1+\beta}, \quad 1+\beta = \frac{1}{1-\alpha}, \quad 1-\alpha = \frac{1}{1+\beta}$$

For example, if $\alpha = 0.99$, then $\beta = 0.99/(1-0.99) = 99$

The CE configuration can be considered as a 2-port circuit. The input port is formed by the base and emitter, the output port is formed by the collector and emitter. The relationships between the current and voltage of both the input and output ports are described by the following input and output characteristics.

- Input characteristics:

  Same as in the case of common-base configuration, the EB junction of the common-emitter configuration can also be considered as a forward biased diode, the current-voltage characteristics is similar to that of a diode:

$$I_\mathbf{B} = f(V_{BE}, V_{CE}) \approx f(V_{BE}) = I_0(e^{V_{BE}/V_T} - 1)$$

V_{CE} has little effect on I_B.

Output characteristics:

$$I_C = f(I_\mathbf{B}, V_{CE}) \approx f(I_\mathbf{B}) = \beta I_\mathbf{B} \qquad \text{(in linear region)}$$

Higher $V_{CE} > 0$ can slightly increase I_C.

The CB junction is reverse biased, the current $I_C = \beta I_B$ depends on the current I_B. When $I_B = 0$, $I_C = 0$, the current caused by the minority carriers crossing the PN-junctions. When I_B is increased, I_C is correspondingly increased by β fold.

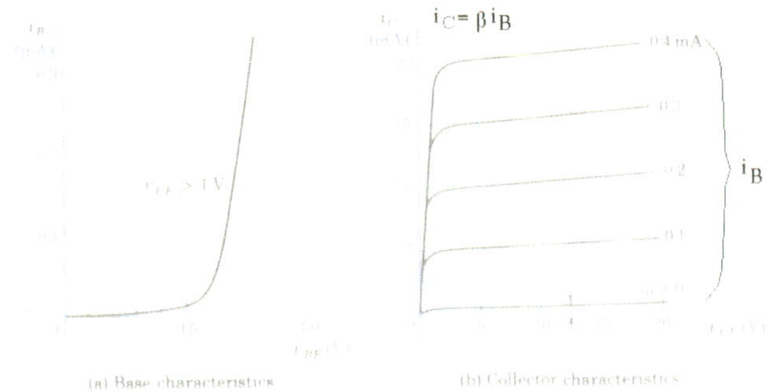

The relationship between the input and output currents of both CB and CE configurations is summarized below:

- CB

$$\begin{cases} I_{out} = I_C = \alpha I_E = \alpha I_{in}, & (\alpha : \text{CB cureent gain}) \\ I_B = I_E - I_C = I_E - \alpha I_E = (1-\alpha)\, I_E \end{cases}$$

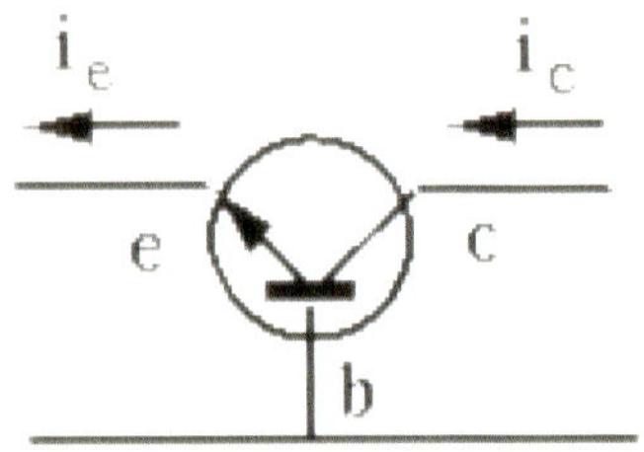

common base

- CE

$$\begin{cases} I_{out} = I_C = \beta I_B = \beta I_{in}, & (\beta : \text{CE current gain}) \\ I_E = I_C + I_B = (\beta+1) I_B \end{cases}$$

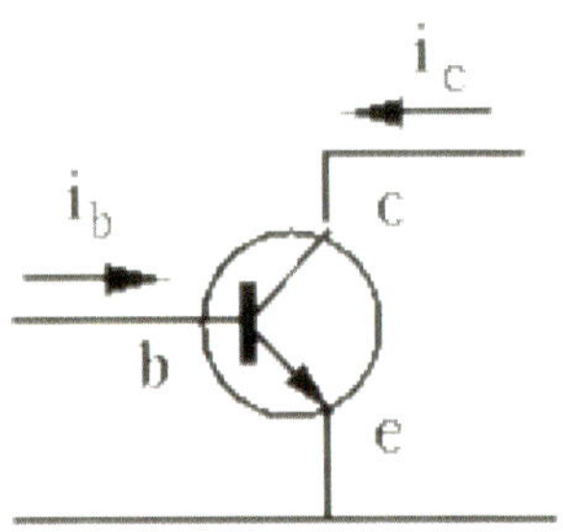

common emiter

The collector characteristics of the common-base (CB) and common-emitter (CE) configurations have the following differences:

- In CB circuit $I_C = \alpha I_E$ is slightly less than I_E, while in CE circuit $I_C = \beta I_B$ is much greater than I_B.

- In CB circuit, $I_C > 0$ when $V_{CB} = 0$; while in CE circuit $I_C = 0$ when $V_{CE} = 0$ (as $V_{CB} = -V_{BE}$ has the effect of suppressing I_C).
- Increased V_{CE} will slightly increase $\underline{\alpha}$ but more greatly increase $\beta = \alpha/(1-\alpha)$, thereby causing more significantly increased I_C.
- I_E in CB is a function of two variables V_{BE} and V_{CB}, but the former is much more significant then the latter. I_B in CE is a function of two variables V_{BE} and V_{CE}, but the former is much more significant then the latter.
- I_C in CB is a function of two variables V_{CB} and I_E. When V_{CB} is small, its slight increase will cause significant increase of I_C. But its further increase will not cause much change in I_C due to saturation (all available charge carriers travel at the saturation velocity to arrive at collector C), $I_C = \alpha I_E$ is mostly determined by I_E.
- I_C in CE is a function of two variables V_{CE} and I_B. When V_{CE} is small ($\approx 0.3V$), its slight increase will cause significant increase of I_C. But when $V_{CE} > 0.3V$, its further increase will not cause much change in I_C due to saturation (all available charge carriers travel at the saturation velocity to arrive at collector C), $I_C = \beta I_B$ is mostly determined by I_B.

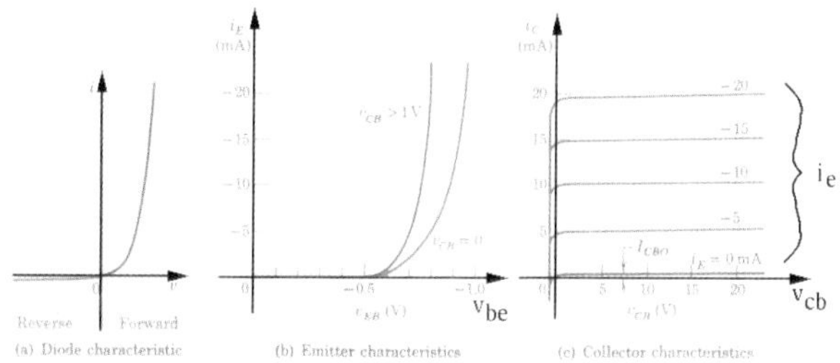

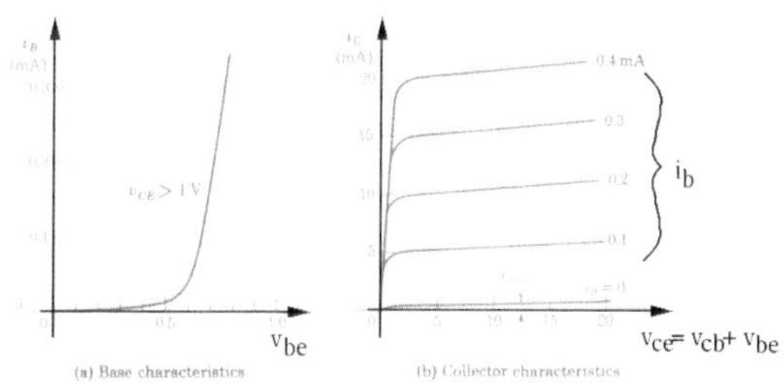

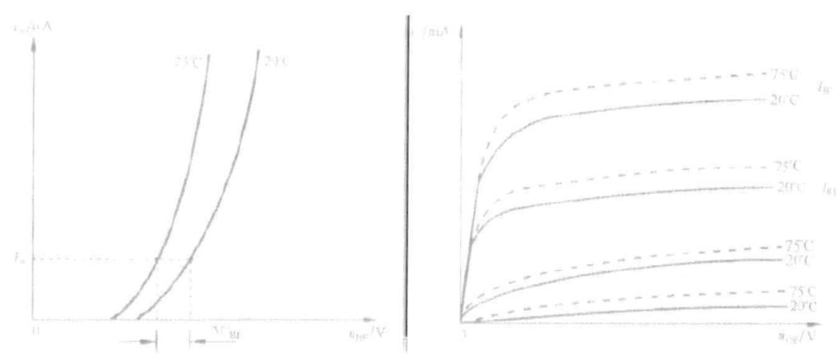

Various parameters of a transistor change as functions of temperature. For example, β increases along with temperature.

- Applications I: Switches

One of the most fundamental applications of a transistor is using it to control the flow of power to another part of the circuit – using it as an electric switch. Driving it in either cutoff or saturation mode, the transistor can create the binary on/off effect of a switch.

Transistor switches are critical circuit-building blocks; they're used to make logic gates, which go on to create microcontrollers, microprocessors, and other integrated circuits. Below are a few example circuits.

Transistor Switch

Let's look at the most fundamental transistor-switch circuit: an NPN switch. Here we use an NPN to control a high-power LED:

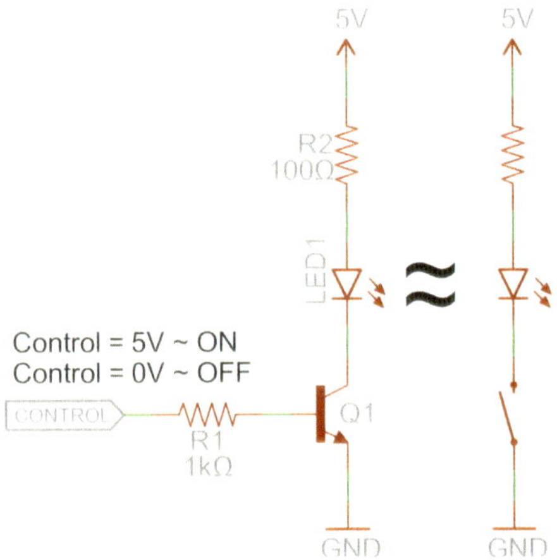

Our control input flows into the base, the output is tied to the collector, and the emitter is kept at a fixed voltage.

While a normal switch would require an actuator to be physically flipped, this switch is controlled by the voltage at the base pin. A microcontroller I/O pin, like those on an Arduino, can be programmed to go high or low to turn the LED on or off.

When the voltage at the base is greater than 0.6V (or whatever your transistor's V_{th} might be), the transistor starts saturating and looks like a short circuit between collector and emitter. When the voltage at the base is less than 0.6V the transistor is in cutoff mode – no current flows because it looks like an open circuit between C and E.

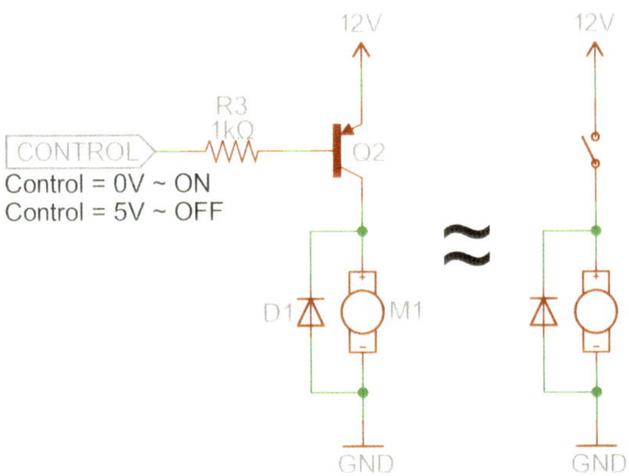

The circuit above is called a low-side switch, because the switch – our transistor – is on the low (ground) side of the circuit. Alternatively, we can use a PNP transistor to create a high-side switch:

Similar to the NPN circuit, the base is our input, and the emitter is tied to a constant voltage. This time however, the emitter is tied high, and the load is connected to the transistor on the ground side.

This circuit works just as well as the NPN-based switch, but there's one huge difference: to turn the load "on" the base must be low. This can cause complications, especially if the load's high voltage (V_{CC} in this picture) is higher than our control input's high voltage. For example, this circuit wouldn't work if you were trying to use a 5V-operating Arduino to switch on a 12V motor. In that case it'd be impossible to turn the switch off because V_B would always be less than V_E.

Base Resistors

You'll notice that each of those circuits uses a series resistor between the control input and the base of the transistor. Don't forget to add this resistor! A transistor without a resistor on the base is like an LED with no current-limiting resistor.

Recall that, in a way, a transistor is just a pair of interconnected diodes. We're forward-biasing the base-emitter diode to turn the load on. The diode only needs 0.6V to turn on, more voltage than that means more current. Some transistors may only be rated for a maximum of 10-100mA of current to flow through them. If you supply a current over the maximum rating, the transistor might blow up.

The series resistor between our control source and the base limits current into the base. The base-emitter node can get its happy voltage drop of 0.6V, and the resistor can drop the remaining voltage. The value of the resistor, and voltage across it, will set the current.

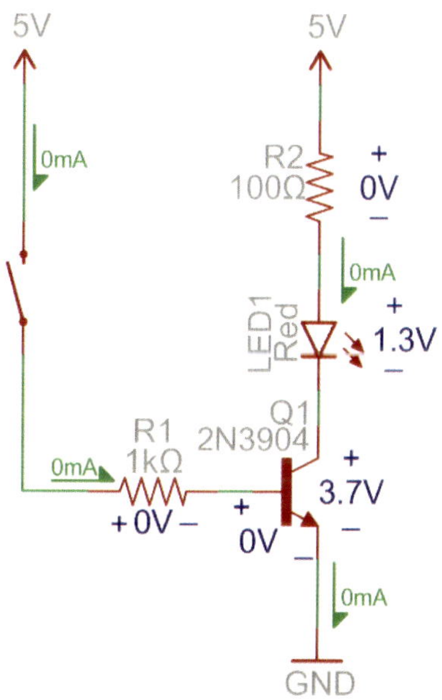

The resistor needs to be large enough to effectively *limit* the current, but small enough to feed the base *enough* current. 1mA to 10mA will usually be enough, but check your transistor's datasheet to make sure.

Digital Logic

Transistors can be combined to create all our fundamental logic gates: AND, OR, and NOT.

(Note: These days MOSFETS are more likely to be used to create logic gates than BJTs. MOSFETs are more power-efficient, which makes them the better choice.)

Inverter

Here's a transistor circuit that implements an inverter, or NOT gate:

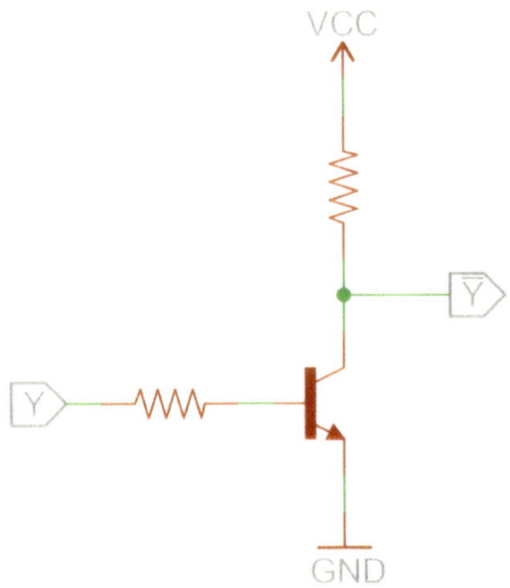

An inverter built out of transistors.

Here a high voltage into the base will turn the transistor on, which will effectively connect the collector to the emitter. Since the emitter is connected directly to ground, the collector will be as well (though it will be slightly higher, somewhere around $V_{CE(sat)}$ ~ 0.05-0.2V). If the input is low, on the other hand, the transistor looks like an open circuit, and the output is pulled up to VCC.

AND Gate

Here are a pair of transistors used to create a 2-input AND gate:

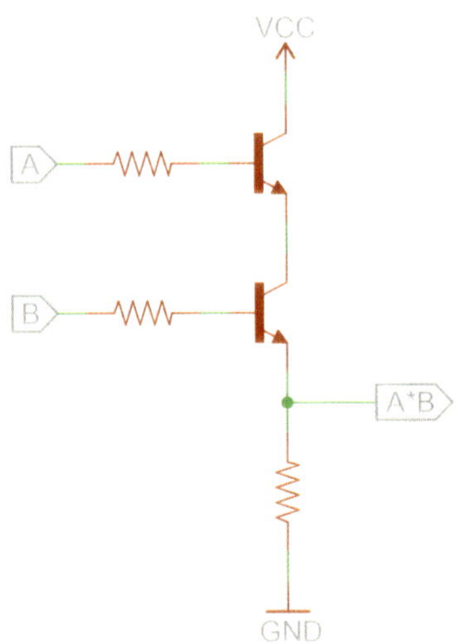

2-input AND gate built out of transistors.

If either transistor is turned off, then the output at the second transistor's collector will be pulled low. If both transistors are "on" (bases both high), then the output of the circuit is also high.

OR Gate

And, finally, here's a 2-input OR gate:

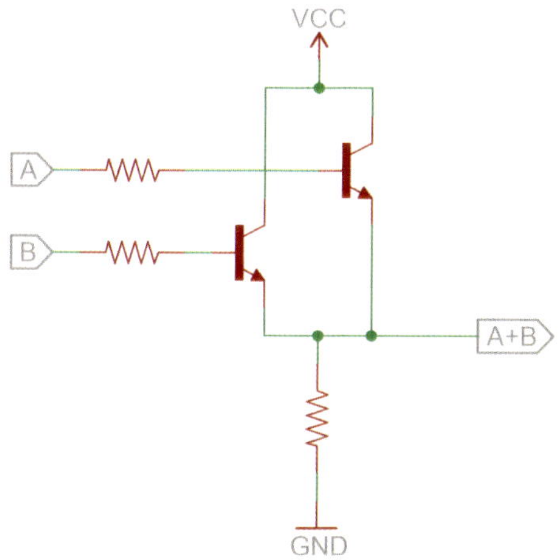

2-input OR gate built out of transistors.

In this circuit, if either (or both) A or B are high, that respective transistor will turn on, and pull the output high. If both transistors are off, then the output is pulled low through the resistor.

H-Bridge

An H-bridge is a transistor-based circuit capable of driving motors both clockwise and counter-clockwise. It's an incredibly popular circuit – the driving force behind countless robots that must be able to move both forward *and* backward.

Fundamentally, an H-bridge is a combination of four transistors with two inputs lines and two outputs:

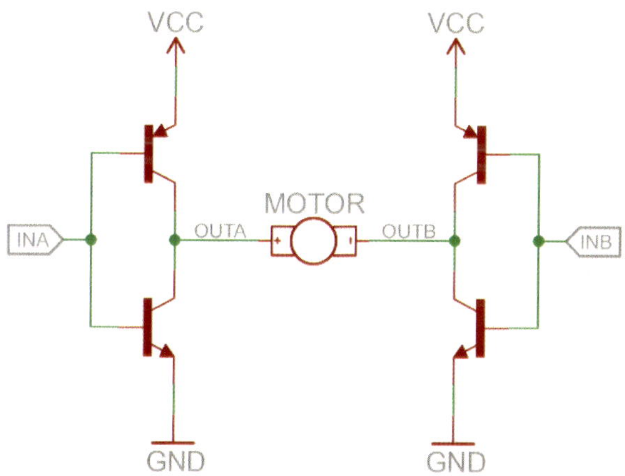

Can you guess why it's called an H bridge?

(Note: there's usually quite a bit more to a well-designed H-bridge including flyback diodes, base resistors and Schmidt triggers.)

If both inputs are the same voltage, the outputs to the motor will be the same voltage, and the motor won't be able to spin. But if the two inputs are opposite, the motor will spin in one direction or the other.

The H-bridge has a truth table that looks a little like this:

Input A	Input B	Output A	Output B	Motor Direction
0	0	1	1	Stopped (braking)
0	1	1	0	Clockwise
1	0	0	1	Counter-clockwise
1	1	0	0	Stopped (braking)

Oscillators

An oscillator is a circuit that produces a periodic signal that swings between a high and low voltage. Oscillators are used in all sorts of circuits: from simply blinking an LED to the producing a clock signal to drive a microcontroller. There are lots of ways to create an oscillator circuit including quartz crystals, op amps, and, of course, transistors.

Here's an example oscillating circuit, which we call an astable multivibrator. By using feedback we can use a pair of transistors to create two complementing, oscillating signals.

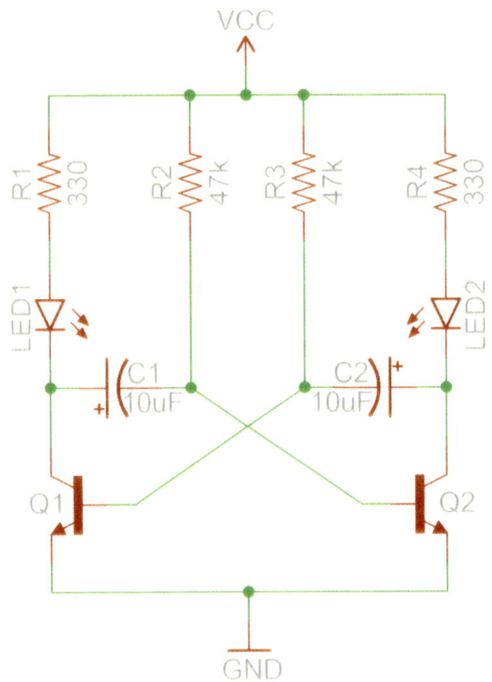

Aside from the two transistors, the capacitors are the real key to this circuit. The caps alternatively charge and discharge, which causes the two transistors to alternatively turn on and off.

Analyzing this circuit's operation is an excellent study in the operation of both caps and transistors. To begin, assume C1 is fully charged (storing a voltage of about V_{CC}), C2 is discharged, Q1 is on, and Q2 is off. Here's what happens after that:

- If Q1 is on, then C1's left plate (on the schematic) is connected to about 0V. This will allow C1 to discharge through Q1's collector.
- While C1 is discharging, C2 quickly charges through the lower value resistor – R4.
- Once C1 fully discharges, its right plate will be pulled up to about 0.6V, which will turn on Q2.
- At this point we've swapped states: C1 is discharged, C2 is charged, Q1 is off, and Q2 is on. Now we do the same dance the other way.
- Q2 being on allows C2 to discharge through Q2's collector.
- While Q1 is off, C1 can charge, relatively quickly through R1.
- Once C2 fully discharges, Q1 will be turn back on and we're back in the state we started in.

By picking specific values for C1, C2, R2, and R3 (and keeping R1 and R4 relatively low), we can set the speed of our multivibrator circuit:

$$f = \frac{1}{\ln(2) \cdot (R_2C_1 + R_3C_2)}$$

So, with the values for caps and resistors set to 10μF and 47kΩ respectively, our oscillator frequency is about 1.5 Hz. That means each LED will blink about 1.5 times per second.

- Applications II: Amplifiers

Some of the most powerful transistor applications involve amplification: turning a low power signal into one of higher power. Amplifiers can increase the voltage of a signal, taking something from the μV range and converting it to a more useful mV or V level. Or they can amplify current, useful for turning the μA of current produced by a photodiode into a current of much higher magnitude. There are even amplifiers that take a current in, and produce a higher voltage, or vice-versa (called transresistance and transconductance respectively).

Transistors are a key component to many amplifying circuits. There are a seemingly infinite variety of transistor amplifiers out there, but fortunately a lot of them are based on some of these more primitive circuits. Remember these circuits, and, hopefully, with a bit of pattern-matching, you can make sense of more complex amplifiers.

Common Configurations

Three of the most fundamental transistor amplifiers are: common emitter, common collector and common base. In each of the three configurations one of the three nodes is permanently tied to a common voltage (usually ground), and the other two nodes are either an input or output of the amplifier.

Common Emitter

Common emitter is one of the more popular transistor arrangements. In this circuit the emitter is tied to a voltage common to both the base and collector (usually ground). The base becomes the signal input, and the collector becomes the output.

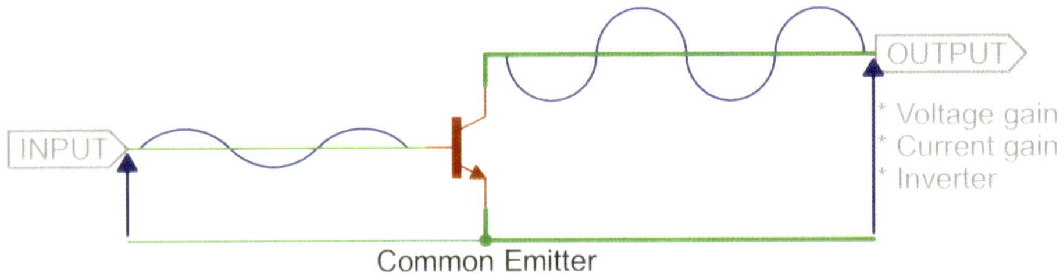

The common emitter circuit is popular because it's well-suited for voltage amplification, especially at low frequencies. They're great for amplifying audio signals, for example. If you have a small 1.5V peak-to-peak input signal, you could amplify that to a much higher voltage using a slightly more complicated circuit, like:

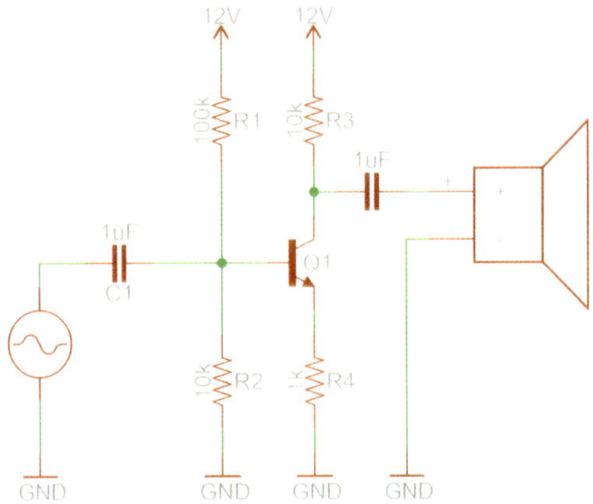

One quirk of the common emitter, though, is that it inverts the input signal.

Common Collector (Emitter Follower)

If we tie the collector pin to a common voltage, use the base as an input, and the emitter as an output, we have a common collector. This configuration is also known as an emitter follower.

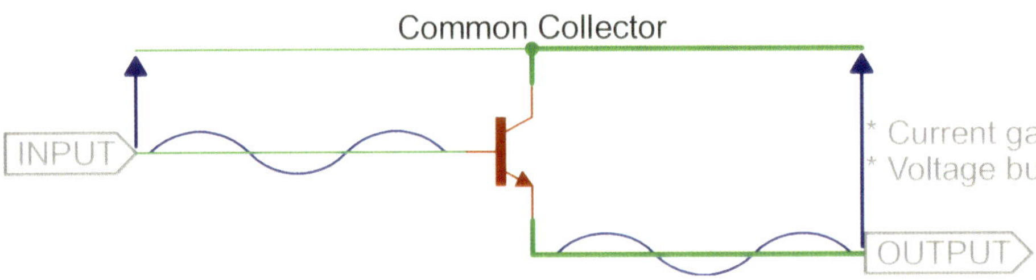

The common collector doesn't do any voltage amplification (in fact, the voltage out will be 0.6V lower than the voltage in). For that reason, this circuit is sometimes called a voltage follower.

This circuit does have great potential as a current amplifier. In addition to that, the high current gain combined with near unity voltage gain makes this circuit a great voltage buffer. A voltage buffer prevents a load circuit from undesirably interfering with the circuit driving it.

For example, if you wanted to deliver 1V to a load, you could go the easy way and use a voltage divider, or you could use an emitter follower.

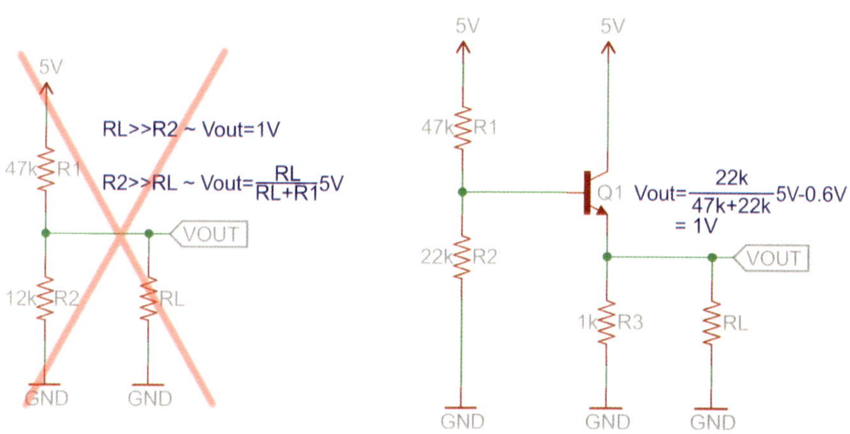

As the load gets larger (which, conversely, means the resistance is lower) the output of the voltage divider circuit drops. But the voltage output of the emitter follower remains steady, regardless of what the load is. Bigger loads can't "load down" an emitter follower, like they can circuits with larger output impedances.

Common Base

We'll talk about common base to provide some closure to this section, but this is the least popular of the three fundamental configurations. In a common base amplifier, the emitter is an input and the collector an output. The base is common to both.

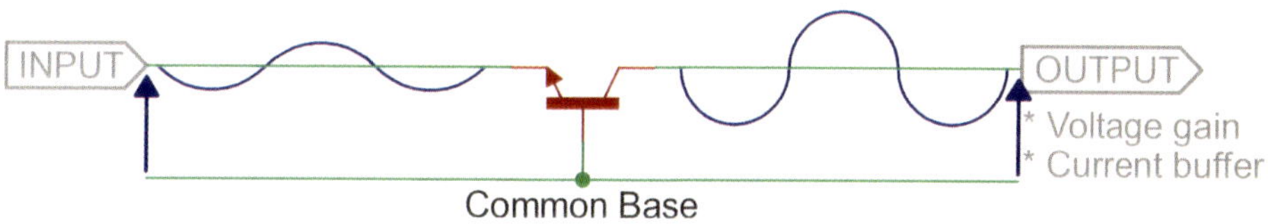

Common base is like the anti-emitter-follower. It's a decent voltage amplifier, and current in is about equal to current out (actually current in is slightly greater than current out).

The common base circuit works best as a current buffer. It can take an input current at a low input impedance, and deliver nearly that same current to a higher impedance output.

These three amplifier configurations are at the heart of many more complicated transistor amplifiers. They each have applications where they shine, whether they're amplifying current, voltage, or buffering.

	Common Emitter	Common Collector	Common Base
Voltage Gain	Medium	Low	High
Current Gain	Medium	High	Low
Input Impedance	Medium	High	Low
Output Impedance	Medium	Low	High

Multistage Amplifiers

We could go on and on about the great variety of transistor amplifiers out there. Here are a few quick examples to show off what happens when you combine the single-stage amplifiers above:

Darlington

The Darlington amplifier runs one common collector into another to create a high current gain amplifier.

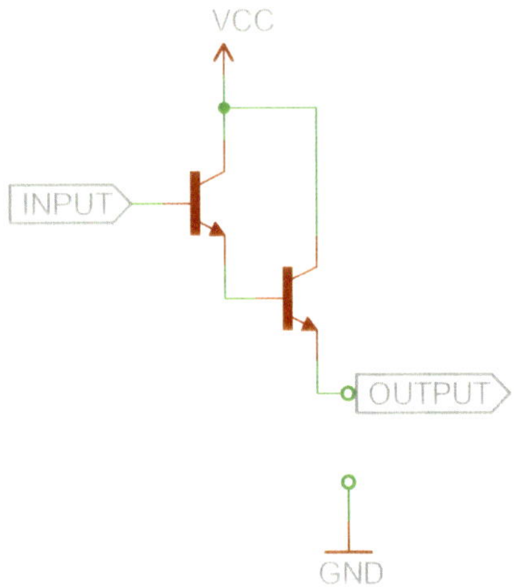

Voltage out is *about* the same as voltage in (minus about 1.2V-1.4V), but the current gain is the product of *two* transistor gains. That's β^2 – upwards of 10,000.

The Darlington pair is a great tool if you need to drive a large load with a very small input current.

Differential Amplifier

A differential amplifier subtracts two input signals and amplifies that difference. It's a critical part of feedback circuits, where the input is compared against the output, to produce a future output.

Here's the foundation of the differential amp:

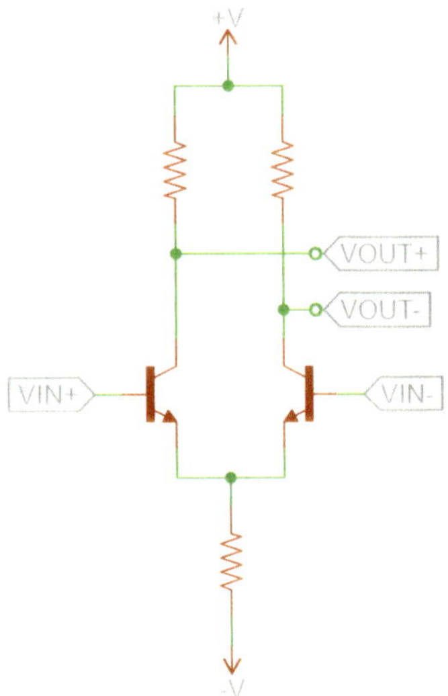

This circuit is also called a long tailed pair. It's a pair of common-emitter circuits that are compared against each other to produce a differential output. Two inputs are applied to the bases of the transistors; the output is a differential voltage across the two collectors.

Push-Pull Amplifier

A push-pull amplifier is a useful "final stage" in many multi-stage amplifiers. It's an energy efficient power amplifier, often used to drive loudspeakers.

The fundamental push-pull amp uses an NPN and PNP transistor, both configured as common collectors:

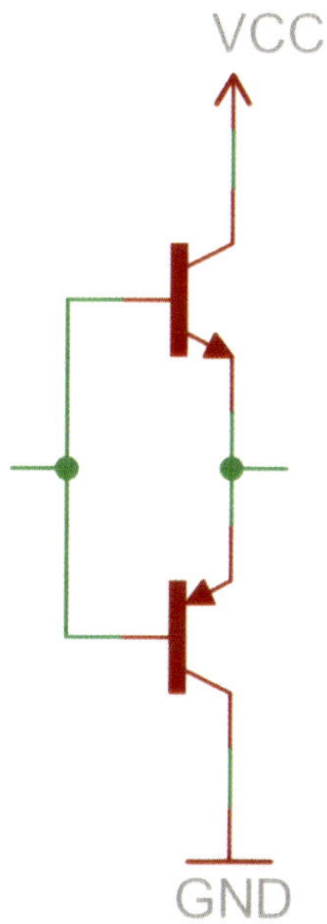

The push-pull amp doesn't really amplify voltage (voltage out will be slightly less than that in), but it does amplify current. It's especially useful in bi-polar circuits (those with positive and negative supplies), because it can both "push" current into the load from the positive supply, and "pull" current out and sink it into the negative supply.

If you have a bi-polar supply (or even if you don't), the push-pull is a great final stage to an amplifier, acting as a buffer for the load.

Putting Them Together (An Operational Amplifier)

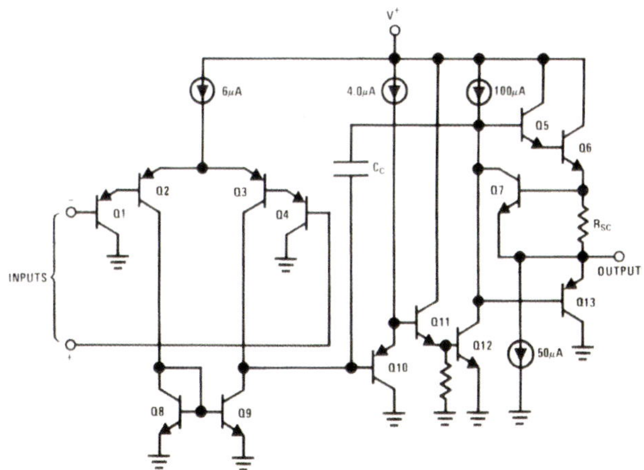

The internals of an LM358 operational amplifier. Recognize some amplifiers?

Let's look at a classic example of a multi-stage transistor circuit: an Op Amp. Being able to recognize common transistor circuits, and understanding their purpose can get you a long way! Here is the circuit inside an LM3558, a really simple op amp:

There's certainly more complexity here than you may be prepared to digest, however you might see some familiar topologies:

- Q1, Q2, Q3, and Q4 form the input stage. Looks a lot like an common collector (Q1 and Q4) into a differential amplifier, right? It just looks upside down, because it's using PNP's. These transistors help to form the input differential stage of the amplifier.
- Q11 and Q12 are part of the second stage. Q11 is a common collector and Q12 is a common emitter. This pair of transistors will buffer the signal from Q3's collector, and provide a high gain as the signal goes to the final stage.
- Q6 and Q13 are part of the final stage, and they should look familiar as well (especially if you ignore R_{SC}) – it's a push-pull! This stage buffers the output, allowing it to drive larger loads.
- There are a variety of other common configurations in there that we haven't talked about. Q8 and Q9 are configured as a current mirror, which simply copies the amount of current through one transistor into the other.

After this crash course in transistors, we wouldn't expect you to understand what's going on in this circuit, but if you can begin to identify common transistor circuits you're on the right track!

N-P-N Bipolar Junction Transistor

NPN Transistor

When a single p-type semiconductor layer is sandwiched between two n-type semiconductor layers, an npn transistor is formed.

NPN Transistor Symbol

The circuit symbol and diode analogy of npn transistor is shown in the below figure.

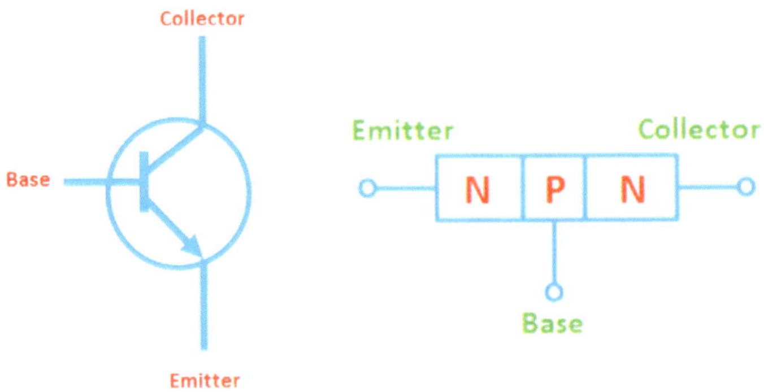

NPN Transistor symbol

In the above figure, it is shown that the electric current always flows from p-region to n-region.

A Bipolar NPN Transistor Configuration

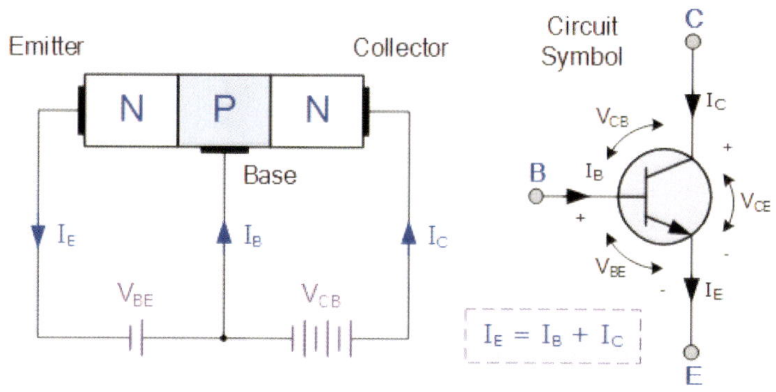

(Note: Arrow defines the emitter and conventional current flow, "out" for a Bipolar NPN Transistor.)

The construction and terminal voltages for a bipolar NPN transistor are shown above. The voltage between the Base and Emitter (V_{BE}), is positive at the Base and negative at the Emitter because for an NPN transistor, the Base terminal is always positive with respect to the Emitter. Also the Collector supply voltage is positive with respect to the Emitter (V_{CE}). So for a bipolar NPN transistor to conduct the Collector is always more positive with respect to both the Base and the Emitter.

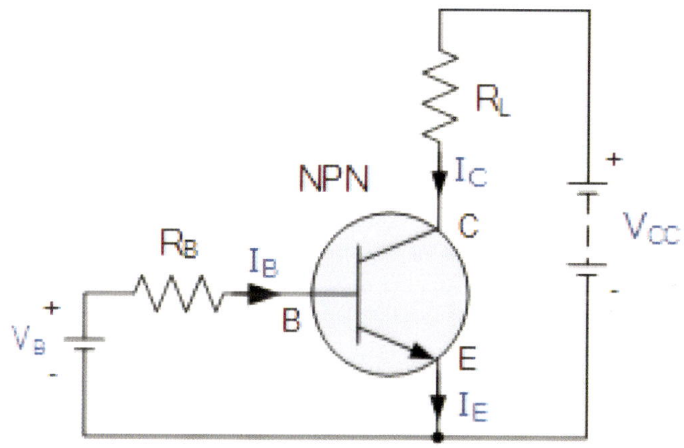

NPN Transistor Connection

Then the voltage sources are connected to an NPN transistor as shown. The Collector is connected to the supply voltage V_{CC} via the load resistor, RL which also acts to limit the maximum current flowing through the device. The Base supply voltage V_B is connected to the Base resistor R_B, which again is used to limit the maximum Base current.

So in a NPN Transistor it is the movement of negative current carriers (electrons) through the Base region that constitutes transistor action, since these mobile electrons provide the link between the Collector and Emitter circuits. This link between the input and output circuits is the main feature of transistor action because the transistors amplifying properties come from the consequent control which the Base exerts upon the Collector to Emitter current.

Then we can see that the transistor is a current operated device (Beta model) and that a large current (Ic) flows freely through the device between the collector and the emitter terminals when the transistor is switched "fully-ON". However, this only happens when a small biasing current (Ib) is flowing into the base terminal of the transistor at the same time thus allowing the Base to act as a sort of current control input.

The transistor current in a bipolar NPN transistor is the ratio of these two currents (Ic/Ib), called the *DC Current Gain* of the device and is given the symbol of hfe or nowadays Beta, (β).

The value of β can be large up to 200 for standard transistors, and it is this large ratio between Ic and Ib that makes the bipolar NPN transistor a useful amplifying device when used in its active region as Ib provides the input and Ic provides the output. Note that Beta has no units as it is a ratio.

Also, the current gain of the transistor from the Collector terminal to the Emitter terminal, Ic/Ie, is called Alpha, (α), and is a function of the transistor itself (electrons diffusing across the junction). As the emitter current Ie is the sum of a very small base current plus a very large collector current, the value of alpha (α), is very close to unity, and for a typical low-power signal transistor this value ranges from about 0.950 to 0.999

α and β Relationship in a NPN Transistor

$$\text{DC Current Gain} = \frac{\text{Output Current}}{\text{Input Current}} = \frac{I_C}{I_B}$$

$$I_E = I_B + I_C \ \ldots\ldots (\text{KCL}) \quad \text{and} \quad \frac{I_C}{I_E} = \alpha$$

$$\text{Thus}: \ I_B = I_E - I_C$$

$$I_B = I_E - \alpha I_E$$

$$I_B = I_E(1 - \alpha)$$

$$\therefore \beta = \frac{I_C}{I_B} = \frac{I_C}{I_E(1-a)} = \frac{\alpha}{1-\alpha}$$

By combining the two parameters α and β we can produce two mathematical expressions that gives the relationship between the different currents flowing in the transistor.

$$\alpha = \frac{\beta}{\beta+1} \quad \text{or} \quad \alpha = \beta(1-\alpha)$$

$$\beta = \frac{\alpha}{1-\alpha} \quad \text{or} \quad \beta = \alpha = (1+\alpha)$$

$$\text{If } \alpha = 0.99 \qquad \beta = \frac{0.99}{0.01} = 99$$

The values of Beta vary from about 20 for high current power transistors to well over 1000 for high frequency low power type bipolar transistors. The value of Beta for most standard NPN transistors can be found in the manufactures data sheets but generally range between 50 – 200.

The equation above for Beta can also be re-arranged to make Ic as the subject, and with a zero base current (Ib = 0) the resultant collector current Ic will also be zero, (β*0). Also when the base current is high the corresponding collector current will also be high resulting in the base current controlling the collector current. One of the most important properties of the Bipolar Junction Transistor is that a small base current can control a much larger collector current. Consider the following example.

- NPN Transistor Example

A bipolar NPN transistor has a DC current gain, (Beta) value of 200. Calculate the base current Ib required to switch a resistive load of 4mA.

$$I_B = \frac{I_C}{\beta} = \frac{4\times10^{-3}}{200} = 20\mu A$$

Therefore, $\beta = 200$, Ic = 4mA and Ib = 20μA.

One other point to remember about Bipolar NPN Transistors. The collector voltage, (Vc) must be greater and positive with respect to the emitter voltage, (Ve) to allow current to flow through the transistor between the collector-emitter junctions. Also, there is a voltage drop between the Base and the Emitter terminal of about 0.7V (one diode volt drop) for silicon devices as the input characteristics of an NPN Transistor are of a forward biased diode.

Then the base voltage, (Vbe) of a NPN transistor must be greater than this 0.7V otherwise the transistor will not conduct with the base current given as.

$$I_B = \frac{V_B - V_{BE}}{R_B}$$

Where: Ib is the base current, Vb is the base bias voltage, Vbe is the base-emitter volt drop (0.7v) and Rb is the base input resistor. Increasing Ib, Vbe slowly increases to 0.7V but Ic rises exponentially.

- NPN Transistor Example

An NPN Transistor has a DC base bias voltage, Vb of 10v and an input base resistor, Rb of 100kΩ. What will be the value of the base current into the transistor.

$$I_B = \frac{V_B - V_{BE}}{R_B} = \frac{10 - 0.7}{100\,k\,\Omega} = 93\mu A$$

Therefore, Ib = 93μA.

The Common Emitter Configuration

As well as being used as a semiconductor switch to turn load currents "ON" or "OFF" by controlling the Base signal to the transistor in either its saturation or cut-off regions, Bipolar NPN Transistors can also be used in its active region to produce a circuit which will amplify any small AC signal applied to its Base terminal with the Emitter grounded.

If a suitable DC "biasing" voltage is firstly applied to the transistors Base terminal thus allowing it to always operate within its linear active region, an inverting amplifier circuit called a single stage common emitter amplifier is produced.

One such *Common Emitter Amplifier* configuration of an NPN transistor is called a Class A Amplifier. A "Class A Amplifier" operation is one where the transistors Base terminal is biased in such a way as to forward bias the Base-emitter junction.

The result is that the transistor is always operating halfway between its cut-off and saturation regions, thereby allowing the transistor amplifier to accurately reproduce the positive and negative halves of any AC input signal superimposed upon this DC biasing voltage.

Without this "Bias Voltage" only one half of the input waveform would be amplified. This common emitter amplifier configuration using an NPN transistor has many applications but is commonly used in audio circuits such as pre-amplifier and power amplifier stages.

With reference to the common emitter configuration shown below, a family of curves known as the Output Characteristics Curves, relates the output collector current, (Ic) to the collector voltage, (Vce) when different values of Base current, (Ib). Output characteristics curves are applied to the transistor for transistors with the same β value.

A DC "Load Line" can also be drawn onto the output characteristics curves to show all the possible operating points when different values of base current are applied. It is necessary to set the initial value of Vce correctly to allow the output voltage to vary both up and down when amplifying AC input signals and this is called setting the operating point or Quiescent Point, Q-point for short and this is shown below.

Single Stage Common Emitter Amplifier Circuit

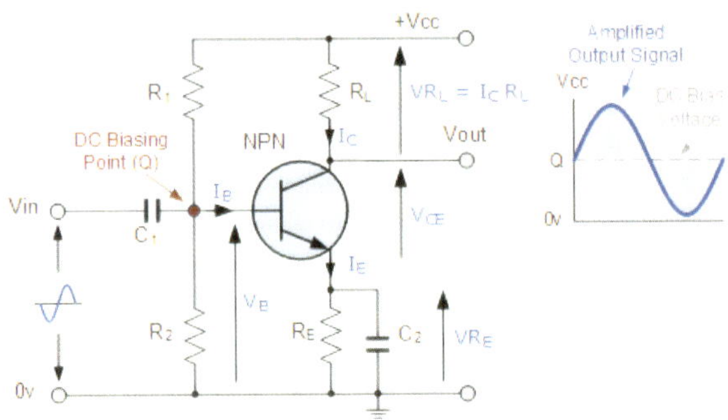

Output Characteristics Curves of a Typical Bipolar Transistor

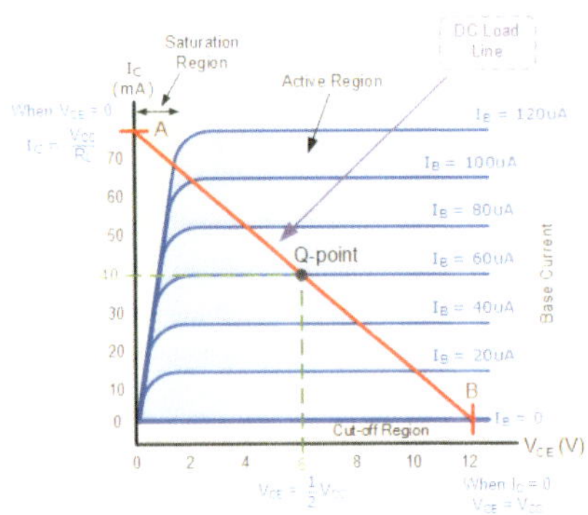

The most important factor to notice is the effect of Vce upon the collector current Ic when Vce is greater than about 1.0 volts. We can see that Ic is largely unaffected by changes in Vce above this value and instead it is almost entirely controlled by the base current, Ib. When this happens we can say then that the output circuit represents that of a "Constant Current Source".

It can also be seen from the common emitter circuit above that the emitter current Ie is the sum of the collector current, Ic and the base current, Ib, added together so we can also say that Ie = Ic + Ib for the common emitter (CE) configuration.

By using the output characteristics curves in our example above and also Ohm´s Law, the current flowing through the load resistor, (R_L), is equal to the collector current, Ic entering the transistor which in turn corresponds to the supply voltage, (Vcc) minus the voltage drop between the collector and the emitter terminals, (Vce) and is given as:

$$\text{Collector Current, } I_C = \frac{V_{CC} - V_{CE}}{R_L}$$

Also, a straight line representing the Dynamic Load Line of the transistor can be drawn directly onto the graph of curves above from the point of "Saturation" (A) when Vce = 0 to the point of "Cut-off" (B) when Ic = 0 thus giving us the "Operating" or Q-point of the transistor. These two points are joined together by a straight line and any position along this straight line represents the "Active Region" of the transistor. The actual position of the load line on the characteristics curves can be calculated as follows:

$$\text{When} : (V_{CE} = 0) \quad I_C = \frac{V_{CC} - 0}{R_L}, \ I_C = \frac{V_{CC}}{R_L}$$

$$\text{When} : (I_C = 0) \quad 0 = \frac{V_{CC} - V_{CE}}{R_L}, \ V_{CC} = V_{CE}$$

Then, the collector or output characteristics curves for Common Emitter NPN Transistors can be used to predict the Collector current, Ic, when given Vce and the Base current, Ib. A Load Line can also be constructed onto the curves to determine a suitable Operating or Q-point which can be set by adjustment of the base current. The slope of this load line is equal to the reciprocal of the load resistance which is given as: $-1/R_L$

Then we can define a NPN Transistor as being normally "OFF" but a small input current and a small positive voltage at its Base (B) relative to its Emitter (E) will turn it "ON" allowing a much large Collector-Emitter current to flow. NPN transistors conduct when Vc is much greater than Ve.

In the next topic about Bipolar Transistors, we will look at the opposite or complementary form of the *NPN Transistor* called the PNP Transistor and show that the PNP Transistor has very similar characteristics to the bipolar NPN transistor except that the polarities (or biasing) of the current and voltage directions are reversed.

P-N-P Bipolar Junction Transistor

PNP Transistor

The transistor in which one n-type material is doped with two p-type materials such type of transistor is known as PNP transistor. It is a current controlled device. The small amount of base current controlled both the emitter and collector current. The PNP transistor has two crystal diodes connected back to back. The left side of the diode in known as the emitter-base diode and the right side of the diode is known as the collector-base diode.

The hole is the majority carriers of the PNP transistors which constitute the current in it. The current inside the transistor is constituted because of the changing position of holes and in the leads of the transistor it is because of the flow of the electrons. The PNP transistor turns on when a small current flows through the base. The direction of current in PNP transistor is from the emitter to collector.

The letter of the PNP transistor indicates the voltage requires by the emitter, collector and the base of the transistor. The base of the PNP transistor has always been negative with respect to the emitter and collector. In PNP transistor, the electrons are taken from the base terminal. The current which enters into the base is amplified into the collector ends.

Symbol of PNP Transistor

The symbol of PNP transistor is shown in the figure below. The inward arrow shows that the direction of current in PNP transistor is from the emitter to collector.

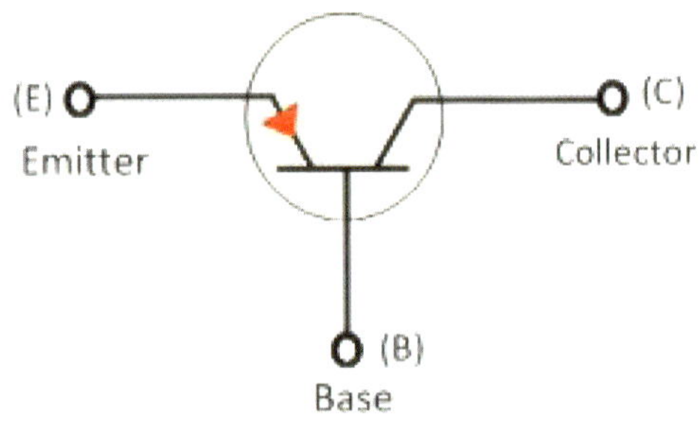

A PNP Transistor Configuration

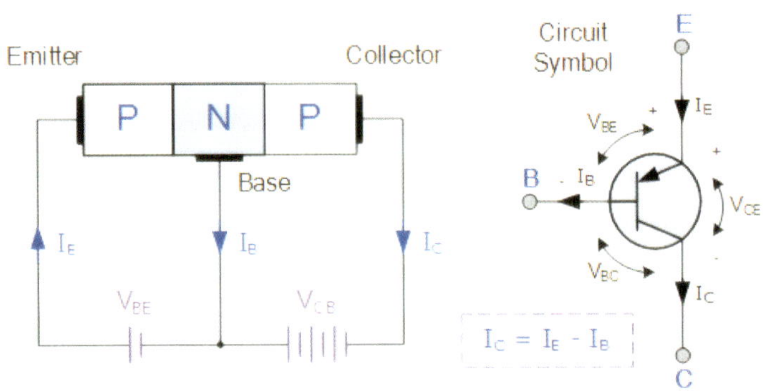

Arrow defines the emitter and conventional current flow, "in" for a PNP transistor.

The construction and terminal voltages for an NPN transistor are shown above. The PNP Transistor has very similar characteristics to their NPN bipolar cousins, except that the polarities (or biasing) of the current and voltage directions are reversed for any one of the possible three configurations, Common Base, Common Emitter and Common Collector.

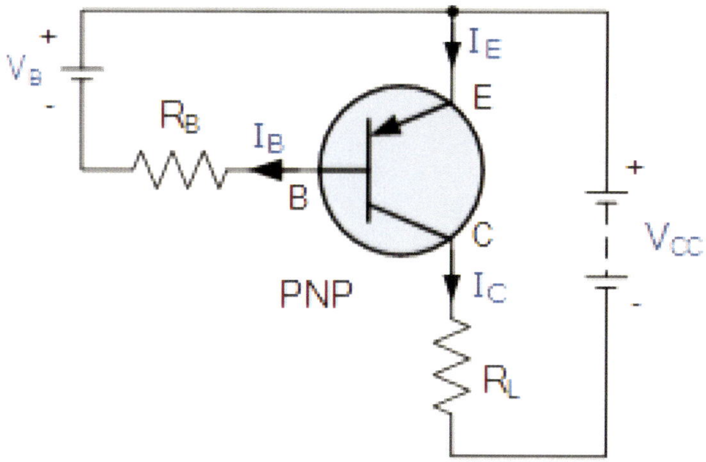

PNP Transistor Connection

The voltage between the Base and Emitter (V_{BE}), is now negative at the Base and positive at the Emitter because for a PNP transistor, the Base terminal is always biased negative with respect to the Emitter.

Also the Emitter supply voltage is positive with respect to the Collector (V_{CE}). So for a PNP transistor to conduct the Emitter is always more positive with respect to both the Base and the Collector.

The voltage sources are connected to a PNP transistor are as shown. This time the Emitter is connected to the supply voltage V_{CC} with the load resistor, RL which limits the maximum current flowing through the device connected to the Collector terminal. The Base voltage V_B which is biased negative with respect to the Emitter and is connected to the Base resistor R_B, which again is used to limit the maximum Base current.

To cause the Base current to flow in a PNP transistor the Base needs to be more negative than the Emitter (current must leave the base) by approx 0.7 volts for a silicon device or 0.3 volts for a germanium device with the formulas used to calculate the Base resistor, Base current or Collector current are the same as those used for an equivalent NPN transistor and is given as.

$$I_C = I_E - I_B$$

$$I_C = \beta . I_B \quad I_B - \frac{I_C}{\beta}$$

We can see that the fundamental differences between a NPN Transistor and a PNP Transistor is the proper biasing of the transistors junctions as the current directions and voltage polarities are always opposite to each other. So for the circuit above: Ic = Ie – Ib as current must leave the Base.

Generally, the PNP transistor can replace NPN transistors in most electronic circuits, the only difference is the polarities of the voltages, and the directions of the current flow. PNP transistors can also be used as switching devices and an example of a PNP transistor switch is shown below.

A PNP Transistor Circuit

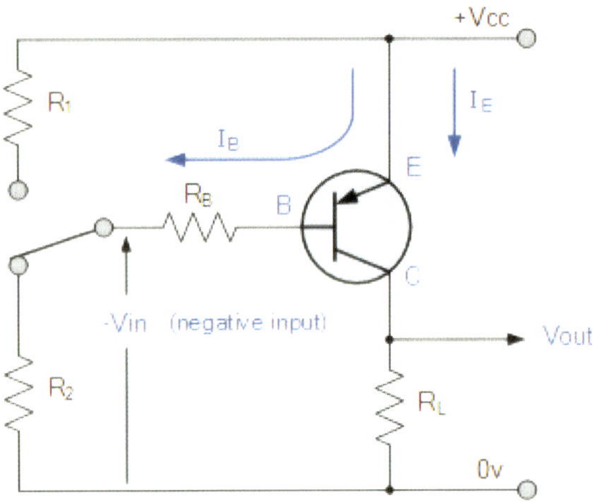

The Output Characteristics Curves for a PNP transistor look very similar to those for an equivalent NPN transistor except that they are rotated by 180° to take account of the reverse polarity voltages and currents, (that is for a PNP transistor, electron current flows out of the base and collector towards the battery). The same dynamic load line can be drawn onto the I-V curves to find the PNP transistors operating points.

Transistor Matching

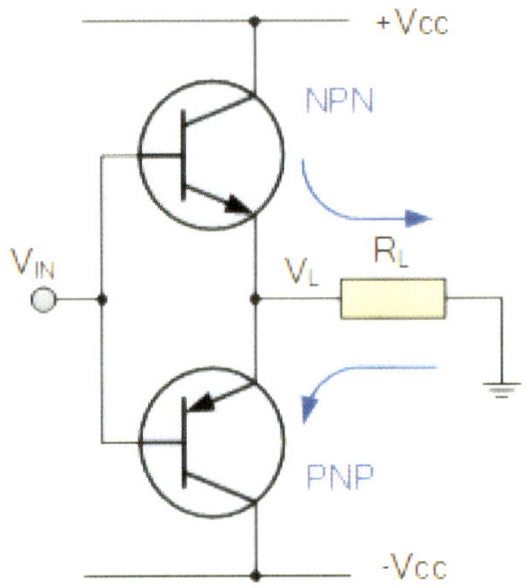

Complementary Transistors

You may think what is the point of having a PNP Transistor, when there are plenty of NPN Transistors available that can be used as an amplifier or solid-state switch?. Well, having two different types of transistors "PNP" and "NPN", can be a great advantage when designing power amplifier circuits such as the Class B Amplifier.

Class-B amplifiers uses "Complementary" or "Matched Pair" (that is one PNP and one NPN connected together) transistors in its output stage or in reversible H-Bridge motor control circuits where we want to control the flow of current evenly through the motor in both directions at different times for forward and reverse motion.

A pair of corresponding NPN and PNP transistors with near identical characteristics to each other are called Complementary Transistors for example, a TIP3055 (NPN transistor) and the TIP2955 (PNP transistor) are good examples of complementary or matched pair silicon power transistors. They both have a DC current gain, Beta, (Ic/Ib) matched to within 10% and high Collector current of about 15A making them ideal for general motor control or robotic applications.

Also, class B amplifiers use complementary NPN and PNP in their power output stage design. The NPN transistor conducts for only the positive half of the signal while the PNP transistor conducts for negative half of the signal.

This allows the amplifier to drive the required power through the load loudspeaker in both directions at the stated nominal impedance and power resulting in an output current which is likely to be in the order of several amps shared evenly between the two complementary transistors.

Identifying the PNP Transistor

We can use this analogy to determine whether a transistor is of the PNP type or NPN type by testing its Resistance between the three different leads, Emitter, Base and Collector. By testing each pair of transistor leads in both directions with a multimeter will result in six tests in total with the expected resistance values in Ohm's given below.

- Emitter-Base Terminals – The Emitter to Base should act like a normal diode and conduct one way only.
- Collector-Base Terminals – The Collector-Base junction should act like a normal diode and conduct one way only.
- Emitter-Collector Terminals – The Emitter-Collector should not conduct in either direction.

Terminal Resistance Values for PNP and NPN Transistors

Between Transistor Terminals		PNP	NPN
Collector	Emitter	R_{HIGH}	R_{HIGH}
Collector	Base	R_{LOW}	R_{HIGH}
Emitter	Collector	R_{HIGH}	R_{HIGH}
Emitter	Base	R_{LOW}	R_{HIGH}
Base	Collector	R_{HIGH}	R_{LOW}
Base	Emitter	R_{HIGH}	R_{LOW}

Then we can define a PNP Transistor as being normally "OFF" but a small output current and negative voltage at its Base (B) relative to its Emitter (E) will turn it "ON" allowing a much large Emitter-Collector current to flow. PNP transistors conduct when Ve is much greater than Vc.

In other words, a Bipolar PNP Transistor will ONLY conduct if both the Base and Collector terminals are negative with respect to the Emitter

Transistors instead of using the transistor as an amplifying device, we will look at the operation of the transistor in its saturation and cut-off regions when used as a solid-state switch. Bipolar transistor switches are used in many applications to switch a DC current "ON" or "OFF", from LED's which require only a few milliamps of switching current at low DC voltages, or motors and relays which may require higher currents at higher voltages.

Difference between PNP and NPN

One of the major difference between the NPN and PNP transistor is that in the NPN transistor the current flow between collector to emitter when the positive supply is given to the base, whereas in PNP transistor the charge carrier flows from the emitter to collector when negative supply is given to the base. The NPN and PNP transistor are differentiated below in the comparison chart by considering the various other factors.

The NPN and PNP both are the bipolar junction transistor. It is the current controlling devices and mainly used for switching and amplifying the signal. Mostly, the NPN transistor is used in the circuit because in NPN transistor the conduction current is mainly by electrons while in the PNP transistor the conduction current is because of the holes. As the electrons are more mobile the NPN has high conduction.

The letter PNP and NPN show the voltage requires by the emitter, collector and base of the junction transistor. The NPN and PNP transistor, both are made up of different material due to which the current develops in them also differs. Sometimes when the voltage applied across the emitter the electrons cross the base junction and reach the collector region. This happens because the base of the NPN and PNP transistor is very thin and lightly doped.

Comparison Chart

Basis For Comparison	NPN Transistor	PNP Transistor
Definition	Transistor in which two n-type layer are separated by one P-type layer	Two blocks of p-types semiconductors are separated by one thin block of n-type semiconductor.
Symbol	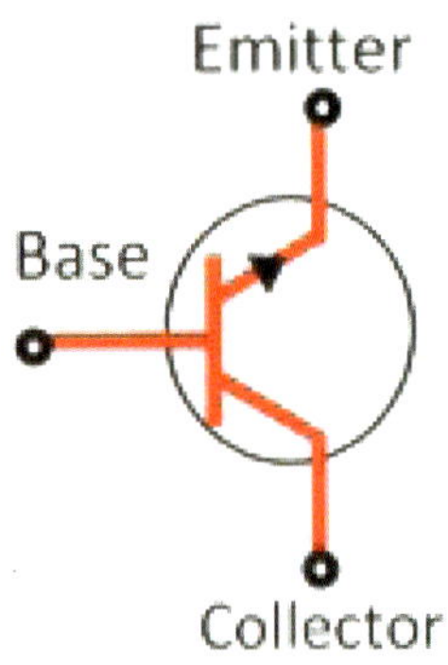	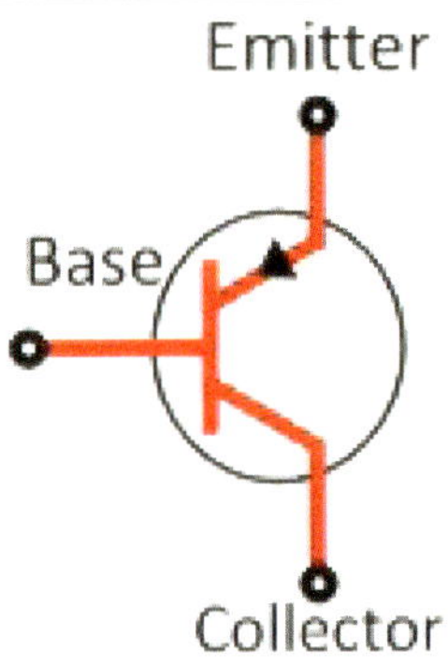
Full Form	Negative Positive and Negative	Positive Negative and Positive
Direction of Current	Collector to Emitter	Emitter to Collector
Turn-on	When electrons enters into the base.	When holes enter into the base.
Inside Current	Develop because of varying position of electrons.	Originate because of varying position of holes.
Outside Current	Current develop because of the flow of holes.	Current develop because of the flow of electrons.
Majority Charge Carrier	Electron	Hole

Switching Time	Faster	Slower
Minority Charge Carrier	Hole	Electron
Positive Voltage	Collector Terminal	Emitter Terminal
Forward Biased	Emitter Base Junction	Emitter Base Junction
Reverse Biased	Collector Base Junction	Collector Base Junction
Small current	Flows from emitter-to-base	Base to emitter
Ground Signal	Low	High

Key Differences between NPN and PNP Transistor

1. The NPN transistor has two blocks of n-type semiconductor materials and one block of p types semiconductor materials whereas the PNP transistor has one thin layer of p-type material and two thick layers of N-type material.
2. The symbol of the NPN and the PNP transistor are almost same the only difference between them is the direction of the arrow which is based on the emitter. In NPN transistor the head of the arrow move outward to the base and in PNP the arrow moves inward.
3. In NPN transistor, the current flow from collector to emitter because positive supply is given to the base, whereas, in PNP transistor the current flows from the emitter to the collector.
4. The NPN transistor turn-on when the electron enters into the base while the PNP transistor is turned on when holes enter into the base.
5. The inside current in NPN transistor constitutes because of the varying position of the electrons whereas in PNP transistor the inside current is because of the varying position of the holes.
6. In NPN transistor the output current exists because flows of the holes and in PNP it is constituted because of the flows of the electrons.
7. In NPN transistor, the majority charge carrier is electron whereas in PNP transistor the majority hole is the majority charge carrier.
8. The minority charge carrier of the NPN transistor is the hole and in PNP transistor it is electrons.
9. The switching time of NPN transistor is more as compared to PNP transistor because the majority charge carrier of NPN transistor is an electron.
10. The emitter-base junction of both the NPN and PNP transistor is connected in forward biased.
11. Note: The forward base junction means the p terminal of the diode is connected to the positive terminal of the supply and the n-type material is connected to the negative terminal of the supply
12. The collector base junction of the NPN and PNP transistor is connected in reversed biased.
13. Note: The reversed bias means the negative region is connected to the positive terminal of the supply and the p-region is connected to the positive terminal of the supply.

14. The NPN transistor turns on when the small current flows from emitter to base, whereas for turning on the PNP transistor the small current flows from base to emitter.
15. The ground signal of the PNP transistor is kept low, whereas, in PNP transistor the ground signal is high.

Field Effect Transistor (FET)

The field effect transistor, FET is a key active device for the electronics industry.

The FET used in many circuits constructed from discrete components in areas from RF technology to power control and switching to general amplification.

However the major use for the field effect transistor, FET is within integrated circuits. In this application FET circuits are able to consume only very small levels of power, and this enables the huge very large scale integrated circuits to operate. If bipolar technology was used the power consumption would be orders of magnitude greater and the power generated far too large to accommodate within a single integrated circuit.

Typical field effect transistors

Field Effect Transistor, FET history

Before the first FETs were introduced into the market, the concept had been known for a number of years. There had been many difficulties in realising a device and making it work.

Some of the early concepts for the field effect transistor were outlined in a paper by Lilienfield in 1926, and in another paper by Heil in 1935.

The next foundations were set in place during the 1940s at Bell Laboratories where the semiconductor research group was set up. This group investigated a number of areas pertaining to semiconductors and semiconductor technology, one of which was a device that would modulate the current flowing in a semiconductor channel buy placing an electric field close to it.

During these early experiments, the researchers were unable to make the idea work, turning their ideas to another idea and ultimately inventing another form of semiconductor electronics component: the bipolar transistor.

After this much of the semiconductor research was focussed on improving the bipolar transistor, and the idea for a field effect transistor was not fully investigated for some while. Now FETs are very widely used, providing the main active element in many integrated circuits. Without them electronics technology would be very different to what it is now.

Field Effect Transistor – the Basics

The concept of the field effect transistor is based around the concept that charge on a nearby object can attract charges within a semiconductor channel. It essentially operates using an electric field effect - hence the name.

The FET consists of a semiconductor channel with electrodes at either end referred to as the drain and the source.

A control electrode called the gate is placed in very close proximity to the channel so that its electric charge is able to affect the channel.

In this way, the gate of the FET controls the flow of carriers (electrons or holes) flowing from the source to drain. It does this by controlling the size and shape of the conductive channel.

The semiconductor channel where the current flow occurs may be either P-type or N-type. This gives rise to two types or categories of FET known as P-Channel and N-Channel FETs.

In addition to this, there are two further categories. Increasing the voltage on the gate can either deplete or enhance the number of charge carriers available in the channel. As a result there are enhancement mode FET and depletion mode FETs.

Junction FET circuit symbol

As it is only the electric field that controls the current flowing in the channel, the device is said to be voltage operated and it has a high input impedance, usually many megohms. This can be a distinct advantage over the bipolar transistor that is current operated and has a much lower input impedance.

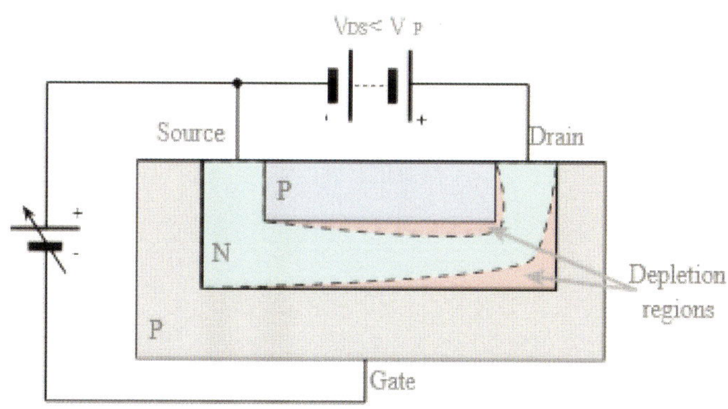

Junction FET working below saturation

Types

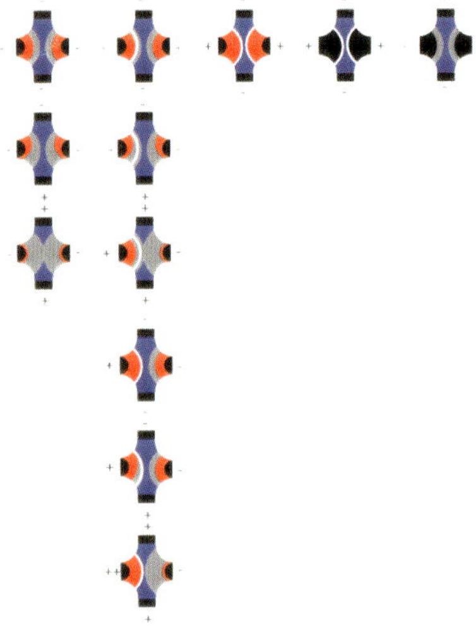

Depletion-type FETs under typical voltages: JFET, poly-silicon MOSFET, double-gate MOSFET, metal-gate MOSFET, MESFET.

Depletion Electrons Holes Metal Insulator

Top: bottom: drain, left: gate, right: bulk. Voltages that lead to channel formation are not shown.

The channel of a FET is doped to produce either an n-type semiconductor or a p-type semiconductor. The drain and source may be doped of opposite type to the channel, in the case of enhancement mode FETs, or doped of similar type to the channel as in depletion mode FETs. Field-effect transistors are also distinguished by the method of insulation between channel and gate. Types of FETs include:

- The JFET (junction field-effect transistor) uses a reverse biased p–n junction to separate the gate from the body.
- The MOSFET (metal–oxide–semiconductor field-effect transistor) utilizes an insulator (typically SiO_2) between the gate and the body.
- The MNOS metal–nitride–oxide–semiconductor transistor utilizes an nitride-oxide layer insulator between the gate and the body.
- The DGMOSFET (dual-gate MOSFET), a FET with two insulated gates.
- The DEPFET is a FET formed in a fully depleted substrate and acts as a sensor, amplifier and memory node at the same time. It can be used as an image (photon) sensor.
- The FREDFET (fast-reverse or fast-recovery epitaxial diode FET) is a specialized FET designed to provide a very fast recovery (turn-off) of the body diode.
- The HIGFET (heterostructure insulated-gate field-effect transistor) is now used mainly in research.
- The MODFET (modulation-doped field-effect transistor) is a high-electron-mobility transistor using a quantum well structure formed by graded doping of the active region.
- The TFET (tunnel field-effect transistor) is based on band-to-band tunneling.

- The IGBT (insulated-gate bipolar transistor) is a device for power control. It has a structure akin to a MOSFET coupled with a bipolar-like main conduction channel. These are commonly used for the 200–3000 V drain-to-source voltage range of operation. Power MOSFETs are still the device of choice for drain-to-source voltages of 1 to 200 V.
- The HEMT (high-electron-mobility transistor), also called a HFET (heterostructure FET), can be made using bandgap engineering in a ternary semiconductor such as AlGaAs. The fully depleted wide-band-gap material forms the isolation between gate and body.
- The ISFET (ion-sensitive field-effect transistor) can be used to measure ion concentrations in a solution; when the ion concentration (such as H^+) changes, the current through the transistor will change accordingly.
- The BioFET (Biologically sensitive field-effect transistor) is a class of sensors/biosensors based on ISFET technology which are utilized to detect charged molecules; when a charged molecule is present, changes in the electrostatic field at the BioFET surface result in a measurable change in current through the transistor. These include EnFETs, ImmunoFETs, GenFETs, DNAFETs, CPFETs, BeetleFETs, and FETs based on ion-channels/protein binding.
- The MESFET (metal–semiconductor field-effect transistor) substitutes the p–n junction of the JFET with a Schottky barrier; and is used in GaAs and other III-V semiconductor materials.
- The NOMFET is a nanoparticle organic memory field-effect transistor.
- The GNRFET (graphene nanoribbon field-effect transistor) uses a graphene nanoribbon for its channel.
- The VeSFET (vertical-slit field-effect transistor) is a square-shaped junctionless FET with a narrow slit connecting the source and drain at opposite corners. Two gates occupy the other corners, and control the current through the slit.
- The CNTFET (carbon nanotube field-effect transistor).
- The OFET (organic field-effect transistor) uses an organic semiconductor in its channel.
- The DNAFET (DNA field-effect transistor) is a specialized FET that acts as a biosensor, by using a gate made of single-strand DNA molecules to detect matching DNA strands.
- The QFET (quantum field effect transistor) takes advantage of quantum tunneling to greatly increase the speed of transistor operation by eliminating the traditional transistor's area of electron conduction.
- The SB-FET (Schottky-barrier field-effect transistor) is a field-effect transistor with metallic source and drain contact electrodes, which create Schottky barriers at both the source-channel and drain-channel interfaces.
- The GFET is a highly sensitive graphene-based field effect transistor used as biosensors and chemical sensors. Due to the 2 dimensional structure of graphene, along with its physical properties, GFETs offer increased sensitivity, and reduced instances of 'false positives' in sensing applications.

Advantages

One advantage of the FET is its high gate to main current resistance, on the order of 100 MΩ or more, thus providing a high degree of isolation between control and flow. Because base current noise will increase with shaping time, a FET typically produces less noise than a bipolar junction transistor (BJT), and is thus found in noise sensitive electronics such as tuners and low-noise amplifiers for VHF and satellite receivers. It is relatively immune to radiation. It exhibits no offset voltage at zero drain current and hence makes an excellent signal chopper. It typically has better thermal stability than a BJT. Because they are controlled by gate charge, once the gate is closed or opened, there is no additional power draw, as there would be with a bipolar junction transistor or with non-latching relays in some states. This allows extremely low-power switching, which in turn allows greater miniaturization of circuits because heat dissipation needs are reduced compared to other types of switches.

Disadvantages

It has a relatively low gain–bandwidth product compared to a BJT. The MOSFET has a drawback of being very susceptible to overload voltages, thus requiring special handling during installation. The fragile insulating layer of the MOSFET between the gate and channel makes it vulnerable to electrostatic discharge or changes to threshold voltage during handling. This is not usually a problem after the device has been installed in a properly designed circuit.

FETs often have a very low "on" resistance and have a high "off" resistance. However, the intermediate resistances are significant, and so FETs can dissipate large amounts of power while switching. Thus efficiency can put a premium on switching quickly, but this can cause transients that can excite stray inductances and generate significant voltages that can couple to the gate and cause unintentional switching. FET circuits can therefore require very careful layout and can involve trades between switching speed and power dissipation. There is also a trade-off between voltage rating and "on" resistance, so high-voltage FETs have a relatively high "on" resistance and hence conduction losses.

Failure Modes

FETs are relatively robust, especially when operated within the temperature and electrical limitations defined by the manufacturer (proper derating). However, modern FET devices can often incorporate a body diode as part of the overall functionality. If the characteristics of the body diode are not taken into consideration, the FET can experience slow body diode behavior, where a parasitic transistor will turn on and allow high current to be drawn from drain to source when the FET is off.

Uses

The most commonly used FET is the MOSFET. The CMOS (complementary metal oxide semiconductor) process technology is the basis for modern digital integrated circuits. This process technology uses an arrangement where the (usually "enhancement-mode") p-channel MOSFET and n-channel MOSFET are connected in series such that when one is on, the other is off.

In FETs, electrons can flow in either direction through the channel when operated in the linear mode. The naming convention of drain terminal and source terminal is somewhat arbitrary, as the devices are typically (but not always) built symmetrically from source to drain. This makes FETs suitable for switching analog signals between paths (multiplexing). With this concept, one can construct a solid-state mixing board, for example.

A common use of the FET is as an amplifier. For example, due to its large input resistance and low output resistance, it is effective as a buffer in common-drain (source follower) configuration.

IGBTs are used in switching internal combustion engine ignition coils, where fast switching and voltage blocking capabilities are important.

Source-gated Transistor

Source-gated transistors are more robust to manufacturing and environmental issues in large-area electronics such as display screens, but are slower in operation than FETs.

MOSFET

MOSFET (Metal Oxide Semiconductor Field Effect Transistor is a unipolar transistor acting as a voltage controlled device where current between the two electrodes 'Drain' and 'Source' is controlled by the applied electric field at another electrode (Gate). Here metal Gate is isolated from the semiconductor by a very thin oxide layer. The characters of metal oxide silicon field effect transistors, or MOSFETs are quite similar to those of JFETs, although there is a difference in their mechanism of production. A replacement of gate-channel p-n junction of JFET is done by a metal oxide silicon capacitors that influences the number of carriers in the channel. The carrier density in the channel is modulated by the applied gate voltage, whereas the width of the channel is hardly affected. Comparison of MOSFTE and JFET is given below:

MOSFET	JFET
It can be operated in both depletion mode and enhancement mode	It can be operated only in depletion mode
Have high input impedance, thus small input current	Relatively low input impedance
MOSFETs have higher speed of operation i.e. they respond instantaneously with a small change in gate voltage	JFETs are relatively slower

Type of MOSFET

TYPES OF MOSFETS (Depending on type of inversion layer)

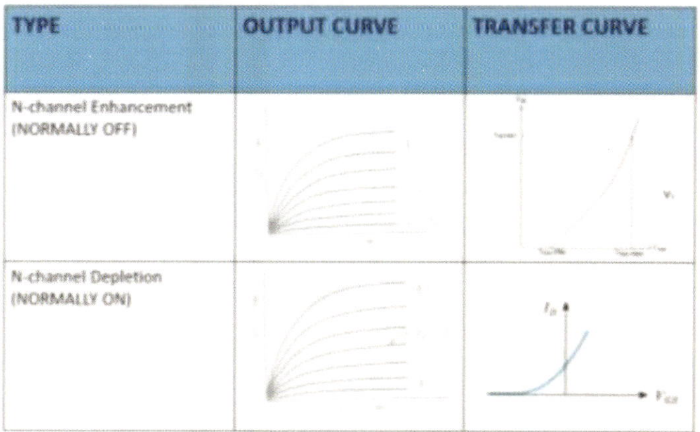

Composition

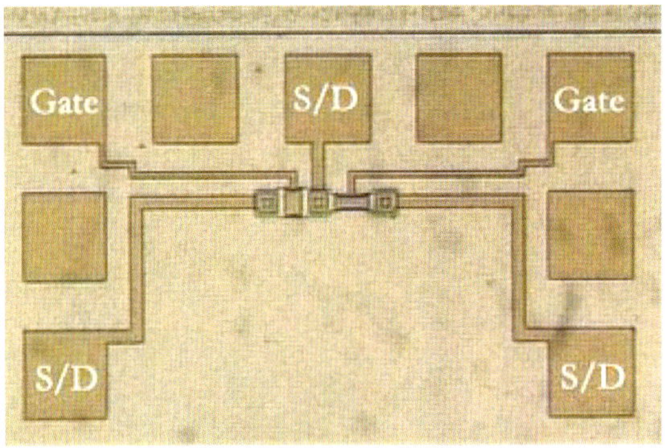

Photomicrograph of two metal-gate MOSFETs in a test pattern. Probe pads for two gates and three source/drain nodes are labeled.

Usually the semiconductor of choice is silicon, processed using local oxidation methods (LOCOS). Recently, some chip manufacturers, most notably IBM and Intel, have started using a chemical compound of silicon and germanium (SiGe) in MOSFET channels. Unfortunately, many semiconductors with better electrical properties than silicon, such as gallium arsenide, do not form good semiconductor-to-insulator interfaces, and thus are not suitable for MOSFETs. Research continues on creating insulators with acceptable electrical characteristics on other semiconductor materials.

To overcome the increase in power consumption due to gate current leakage, a high-κ dielectric is used instead of silicon dioxide for the gate insulator, while polysilicon is replaced by metal gates.

The gate is separated from the channel by a thin insulating layer, traditionally of silicon dioxide and later of silicon oxynitride. Some companies have started to introduce a high-κ dielectric and metal gate combination in the 45 nanometer node.

When a voltage is applied between the gate and body terminals, the electric field generated penetrates through the oxide and creates an *inversion layer* or *channel* at the semiconductor-insulator interface. The inversion layer provides a channel through which current can pass between source and drain terminals. Varying the voltage between the gate and body modulates the conductivity of this layer and thereby controls the current flow between drain and source. This is known as enhancement mode.

Operation

Metal-oxide-semiconductor Structure

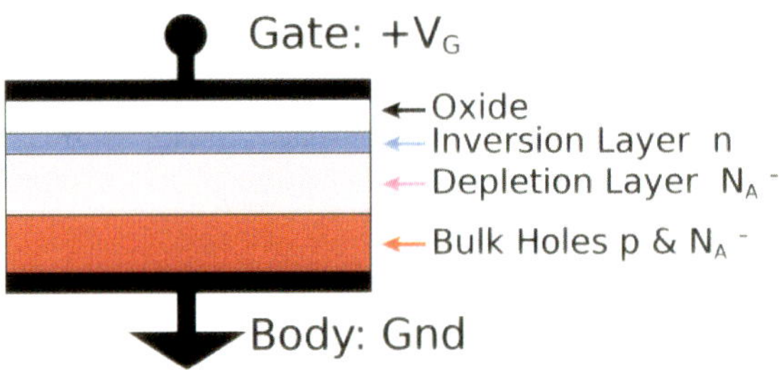

Metal-oxide-semiconductor structure on p-type silicon

The traditional metal-oxide-semiconductor (MOS) structure is obtained by growing a layer of silicon dioxide (SiO_2) on top of a silicon substrate and depositing a layer of metal or polycrystalline silicon (the latter is commonly used). As the silicon dioxide is a dielectric material, its structure is equivalent to a planar capacitor, with one of the electrodes replaced by a semiconductor.

When a voltage is applied across a MOS structure, it modifies the distribution of charges in the semiconductor. If we consider a p-type semiconductor (with N_A the density of acceptors, p the density of holes; $p = N_A$ in neutral bulk), a positive voltage, V_{GB}, from gate to body creates a depletion layer by forcing the positively charged holes away from the gate-insulator/semiconductor interface, leaving exposed a carrier-free region of immobile, negatively charged acceptor ions. If V_{GB} is high enough, a high concentration of negative charge carriers forms in an *inversion layer* located in a thin layer next to the interface between the semiconductor and the insulator.

Conventionally, the gate voltage at which the volume density of electrons in the inversion layer is the same as the volume density of holes in the body is called the threshold voltage. When the voltage between transistor gate and source (V_{GS}) exceeds the threshold voltage (V_{th}), the difference is known as overdrive voltage.

This structure with p-type body is the basis of the n-type MOSFET, which requires the addition of n-type source and drain regions.

Structure and Channel Formation

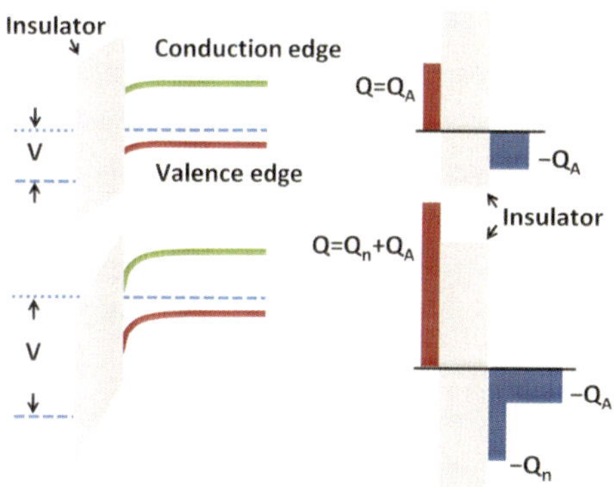

Channel formation in nMOS MOSFET shown as band diagram

Top panels: An applied gate voltage bends bands, depleting holes from surface (left). The charge inducing the bending is balanced by a layer of negative acceptor-ion charge (right). Bottom panel: A larger applied voltage further depletes holes but conduction band lowers enough in energy to populate a conducting channel

A MOSFET is based on the modulation of charge concentration by a MOS capacitance between a *body* electrode and a *gate* electrode located above the body and insulated from all other device regions by a gate dielectric layer. If dielectrics other than an oxide are employed, the device may be referred to as a metal-insulator-semiconductor FET (MISFET). Compared to the MOS capacitor, the MOSFET includes two additional terminals (*source* and *drain*), each connected to individual highly doped regions that are separated by the body region. These regions can be either p or n type, but they must both be of the same type, and of opposite type to the body region. The source and drain (unlike the body) are highly doped as signified by a "+" sign after the type of doping.

If the MOSFET is an n-channel or nMOS FET, then the source and drain are *n+* regions and the body is a *p* region. If the MOSFET is a p-channel or pMOS FET, then the source and drain are *p+* regions and the body is a *n* region. The source is so named because it is the source of the charge carriers (electrons for n-channel, holes for p-channel) that flow through the channel; similarly, the drain is where the charge carriers leave the channel.

The occupancy of the energy bands in a semiconductor is set by the position of the Fermi level relative to the semiconductor energy-band edges.

With sufficient gate voltage, the valence band edge is driven far from the Fermi level, and holes from the body are driven away from the gate.

At larger gate bias still, near the semiconductor surface the conduction band edge is brought close to the Fermi level, populating the surface with electrons in an *inversion layer* or *n-channel* at the interface between the p region and the oxide. This conducting channel extends between the source and the drain, and current is conducted through it when a voltage is applied between the two electrodes. Increasing the voltage on the gate leads to a higher electron density in the inversion layer and therefore increases the current flow between the source and drain. For gate voltages below the threshold value, the channel is lightly populated, and only a very small subthreshold leakage current can flow between the source and the drain.

When a negative gate-source voltage (positive source-gate) is applied, it creates a *p-channel* at the surface of the n region, analogous to the n-channel case, but with opposite polarities of charges and voltages. When a voltage less negative than the threshold value (a negative voltage for the p-channel) is applied between gate and source, the channel disappears and only a very small subthreshold current can flow between the source and the drain. The device may comprise a silicon on insulator device in which a buried oxide is formed below a thin semiconductor layer. If the channel region between the gate dielectric and the buried oxide region is very thin, the channel is referred to as an ultrathin channel region with the source and drain regions formed on either side in or above the thin semiconductor layer. Other semiconductor materials may be employed. When the source and drain regions are formed above the channel in whole or in part, they are referred to as raised source/drain regions.

Comparison of n- and p-type MOSFETs

Parameter		nMOSFET	pMOSFET
Source/drain type		n-type	p-type
Channel type (MOS capacitor)		n-type	p-type
Gate type	Polysilicon	n+	p+
	Metal	φ_m ~ Si conduction band	φ_m ~ Si valence band
Well type		p-type	n-type
Threshold voltage, V_{th}		• Positive (enhancement) • Negative (depletion)	• Negative (enhancement) • Positive (depletion)
Band-bending		Downwards	Upwards
Inversion layer carriers		Electrons	Holes
Substrate type		p-type	n-type

Modes of Operation

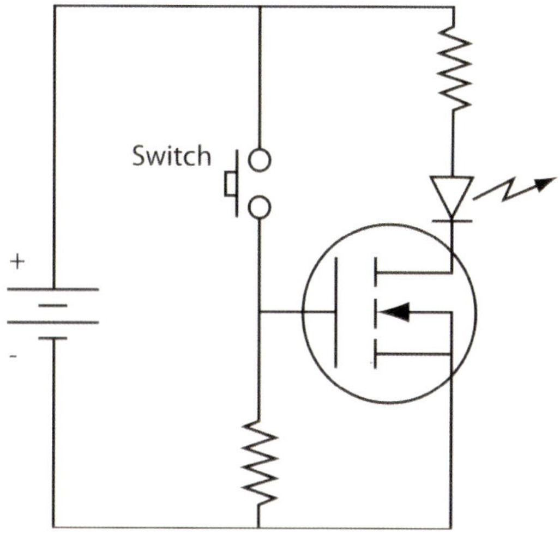

Example application of an n-channel MOSFET. When the switch is pushed, the LED lights up.

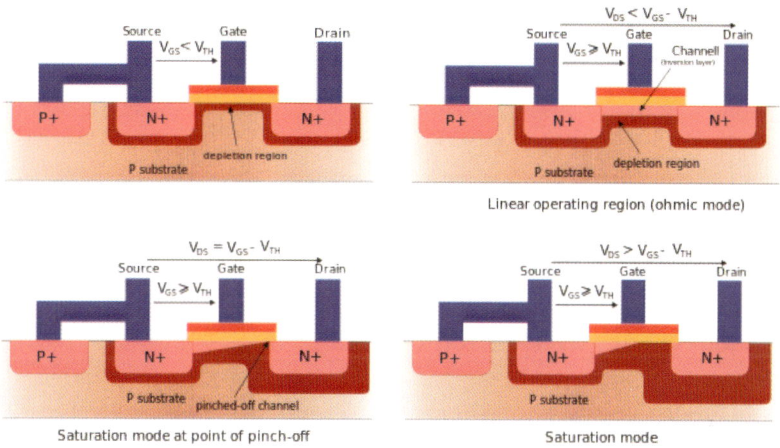

Ohmic contact to body to ensure no body bias; top left: Subthreshold, top right: Ohmic mode, bottom left: Active mode at onset of pinch-off, bottom right: Active mode well into pinch-off – channel length modulation evident

The operation of a MOSFET can be separated into three different modes, depending on the voltages at the terminals. In the following discussion, a simplified algebraic model is used. Modern MOSFET characteristics are more complex than the algebraic model presented here.

For an *enhancement-mode, n-channel MOSFET*, the three operational modes are

Cutoff, subthreshold, and weak-inversion mode

When $V_{GS} < V_{th}$:

where V_{GS} is gate-to-source bias and V_{th} is the threshold voltage of the device.

According to the basic threshold model, the transistor is turned off, and there is no conduction between drain and source. A more accurate model considers the effect of thermal energy on the Fermi–Dirac distribution of electron energies which allow some of the more energetic electrons at the source to enter the channel and flow to the drain. This results in a subthreshold current that is an exponential function of gate-source voltage. While the current between drain and source should ideally be zero when the transistor is being used as a turned-off switch, there is a weak-inversion current, sometimes called subthreshold leakage.

In weak inversion where the source is tied to bulk, the current varies exponentially with V_{GS} as given approximately by:

$$I_D \approx I_{D0} e^{\frac{V_{GS}-V_{th}}{nV_T}},$$

where I_{D0} = current at $V_{GS} = V_{th}$, the thermal voltage $V_T = kT/q$ and the slope factor n is given by:

$$n = 1 + \frac{C_{dep}}{C_{ox}},$$

with C_{dep} = capacitance of the depletion layer and C_{ox} = capacitance of the oxide layer. This equation is generally used, but is only an adequate approximation for the source tied to the bulk. For the source not tied to the bulk, the subthreshold equation for drain current in saturation is

$$I_D \approx I_{D0} e^{\frac{\kappa(V_G-V_{th})-V_S}{V_T}},$$

where the κ is the channel divider that is given by:

$$\kappa = \frac{C_{ox}}{C_{ox}+C_D},$$

with C_D = capacitance of the depletion layer and C_{ox} = capacitance of the oxide layer. In a long-channel device, there is no drain voltage dependence of the current once $V_{DS} \gg V_T$, but as channel length is reduced drain-induced barrier lowering introduces drain voltage dependence that depends in a complex way upon the device geometry (for example, the channel doping, the junction doping and so on). Frequently, threshold voltage V_{th} for this mode is defined as the gate voltage at which a selected value of current I_{Do} occurs, for example, I_{Do} = 1 μA, which may not be the same V_{th}-value used in the equations for the following modes.

Some micropower analog circuits are designed to take advantage of subthreshold conduction. By working in the weak-inversion region, the MOSFETs in these circuits deliver the highest possible transconductance-to-current ratio, namely: $g_m / I_D = 1/(nV_T)$, almost that of a bipolar transistor.

The subthreshold *I–V curve* depends exponentially upon threshold voltage, introducing a strong dependence on any manufacturing variation that affects threshold voltage; for example: variations in oxide thickness, junction depth, or body doping that change the degree of drain-induced barrier lowering. The resulting sensitivity to fabricational variations complicates optimization for leakage and performance.

Triode mode or linear region (also known as the ohmic mode)

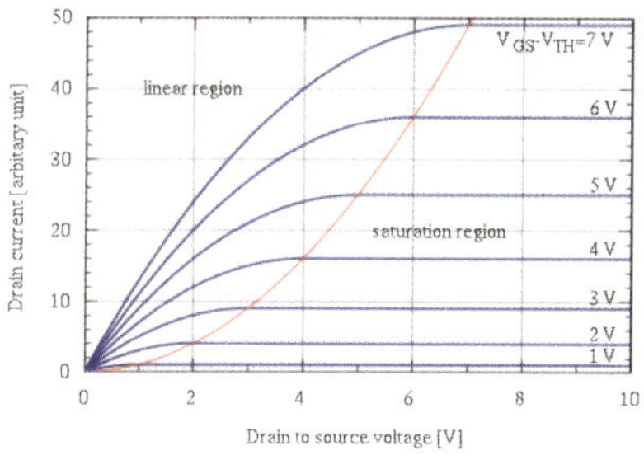

MOSFET drain current vs. drain-to-source voltage for several values of $V_{GS} - V_{th}$; the boundary between linear (Ohmic) and saturation (active) modes is indicated by the upward curving parabola

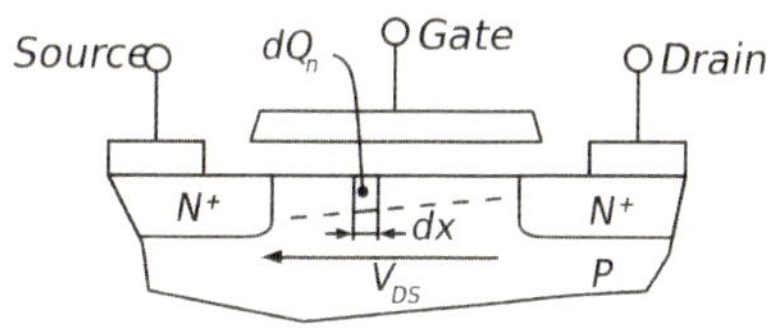

Cross section of a MOSFET operating in the linear (Ohmic) region; strong inversion region present even near drain

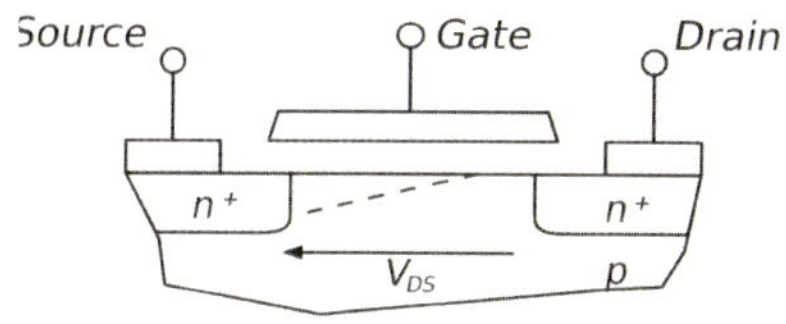

Cross section of a MOSFET operating in the saturation (active) region; channel exhibits channel pinching near drain

When $V_{GS} > V_{th}$ and $V_{DS} < V_{GS} - V_{th}$:

The transistor is turned on, and a channel has been created which allows current between the drain and the source. The MOSFET operates like a resistor, controlled by the gate voltage relative to both the source and drain voltages. The current from drain to source is modeled as:

$$I_D = \mu_n C_{ox} \frac{W}{L}\left((V_{GS} - V_{th}) V_{DS} - \frac{V_{DS}{}^2}{2} \right)$$

where μ_n is the charge-carrier effective mobility, W is the gate width, L is the gate length and C_{ox} is the gate oxide capacitance per unit area. The transition from the exponential subthreshold region to the triode region is not as sharp as the equations suggest.

Saturation or Active Mode

When $V_{GS} > V_{th}$ and $V_{DS} \geq (V_{GS} - V_{th})$:

The switch is turned on, and a channel has been created, which allows current between the drain and source. Since the drain voltage is higher than the source voltage, the electrons spread out, and conduction is not through a narrow channel but through a broader, two- or three-dimensional current distribution extending away from the interface and deeper in the substrate. The onset of this region is also known as pinch-off to indicate the lack of channel region near the drain. Although the channel does not extend the full length of the device, the electric field between the drain and the channel is very high, and conduction continues. The drain current is now weakly dependent upon drain voltage and controlled primarily by the gate-source voltage, and modeled approximately as:

$$I_D = \frac{\mu_n C_{ox}}{2} \frac{W}{L} [V_{GS} - V_{th}]^2 [1 + \lambda (V_{DS} - V_{DSsat})].$$

The additional factor involving λ, the channel-length modulation parameter, models current dependence on drain voltage due to the Early effect, or channel length modulation. According to this equation, a key design parameter, the MOSFET transconductance is:

$$g_m = \frac{\partial I_D}{\partial V_{GS}} = \frac{2I_D}{V_{GS} - V_{th}} = \frac{2I_D}{V_{ov}},$$

where the combination $V_{ov} = V_{GS} - V_{th}$ is called the overdrive voltage, and where $V_{DSsat} = V_{GS} - V_{th}$ accounts for a small discontinuity in I_D which would otherwise appear at the transition between the triode and saturation regions.

Another key design parameter is the MOSFET output resistance r_{out} given by:

$$r_{out} = \frac{1}{\lambda I_D}.$$

r_{out} is the inverse of g_{DS} where $g_{DS} = \frac{\partial I_{DS}}{\partial V_{DS}}$. I_D is the expression in saturation region.

If λ is taken as zero, an infinite output resistance of the device results that leads to unrealistic circuit predictions, particularly in analog circuits.

As the channel length becomes very short, these equations become quite inaccurate. New physical effects arise. For example, carrier transport in the active mode may become limited by velocity saturation. When velocity saturation dominates, the saturation drain current is more nearly linear than quadratic in V_{GS}. At even shorter lengths, carriers transport with near zero scattering, known as quasi-ballistic transport. In the ballistic regime, the carriers travel at an injection velocity that may exceed the saturation velocity and approaches the Fermi velocity at high inversion charge density. In addition, drain-induced barrier lowering increases off-state (cutoff) current and requires an increase in threshold voltage to compensate, which in turn reduces the saturation current.

Body Effect

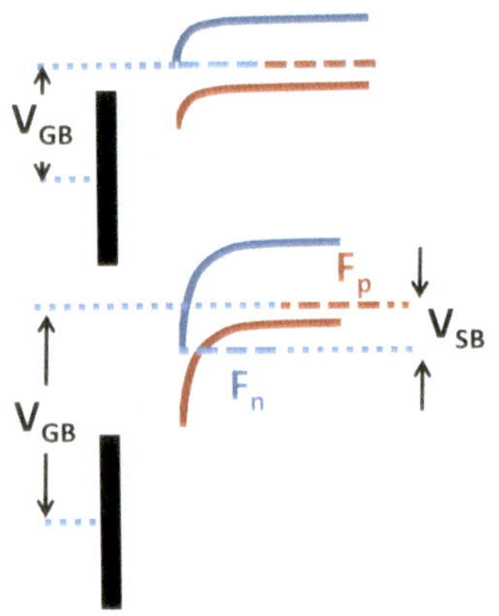

Band diagram showing body effect. V_{SB} splits Fermi levels F_n for electrons and F_p for holes, requiring larger V_{GB} to populate the conduction band in an nMOS MOSFET

The occupancy of the energy bands in a semiconductor is set by the position of the Fermi level relative to the semiconductor energy-band edges. Application of a source-to-substrate reverse bias of the source-body pn-junction introduces a split between the Fermi levels for electrons and holes, moving the Fermi level for the channel further from the band edge, lowering the occupancy of the channel. The effect is to increase the gate voltage necessary to establish the channel, as seen in the figure. This change in channel strength by application of reverse bias is called the 'body effect'.

Simply put, using an nMOS example, the gate-to-body bias V_{GB} positions the conduction-band energy levels, while the source-to-body bias V_{SB} positions the electron Fermi level near the interface, deciding occupancy of these levels near the interface, and hence the strength of the inversion layer or channel.

The body effect upon the channel can be described using a modification of the threshold voltage, approximated by the following equation:

$$V_{TB} = V_{TO} + \gamma\left(\sqrt{V_{SB} + 2\varphi_B} - \sqrt{2\varphi_B}\right),$$

where V_{TB} is the threshold voltage with substrate bias present, and V_{To} is the zero-V_{SB} value of threshold voltage, γ is the body effect parameter, and $2\varphi_B$ is the approximate potential drop between surface and bulk across the depletion layer when $V_{SB} = 0$ and gate bias is sufficient to ensure that a channel is present. As this equation shows, a reverse bias $V_{SB} > 0$ causes an increase in threshold voltage V_{TB} and therefore demands a larger gate voltage before the channel populates.

The body can be operated as a second gate, and is sometimes referred to as the "back gate"; the body effect is sometimes called the "back-gate effect".

Circuit Symbols

A variety of symbols are used for the MOSFET. The basic design is generally a line for the channel with the source and drain leaving it at right angles and then bending back at right angles into the same direction as the channel. Sometimes three line segments are used for enhancement mode and a solid line for depletion mode. Another line is drawn parallel to the channel for the gate.

The *bulk* or *body* connection, if shown, is shown connected to the back of the channel with an arrow indicating pMOS or nMOS. Arrows always point from P to N, so an NMOS (N-channel in P-well or P-substrate) has the arrow pointing in (from the bulk to the channel). If the bulk is connected to the source (as is generally the case with discrete devices) it is sometimes angled to meet up with the source leaving the transistor. If the bulk is not shown (as is often the case in IC design as they are generally common bulk) an inversion symbol is sometimes used to indicate PMOS, alternatively an arrow on the source may be used in the same way as for bipolar transistors (out for nMOS, in for pMOS).

Comparison of enhancement-mode and depletion-mode MOSFET symbols, along with JFET symbols. The orientation of the symbols, (most significantly the position of source relative to drain) is such that more positive voltages appear higher on the page than less positive voltages, implying current flowing "down" the page:

P-channel	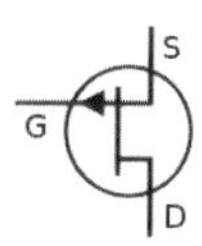	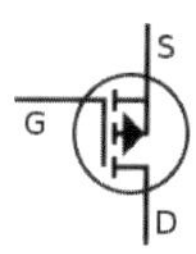	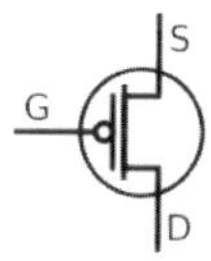	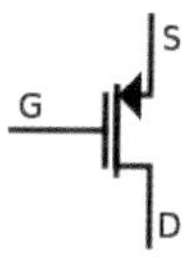	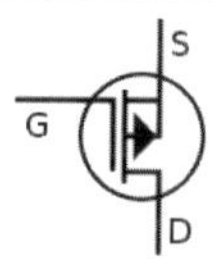
N-channel	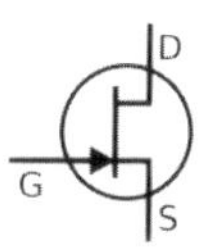	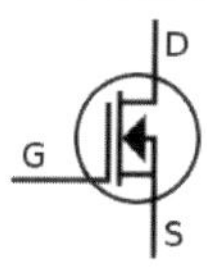	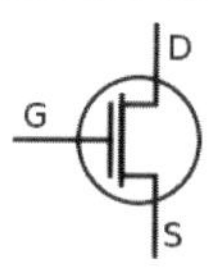	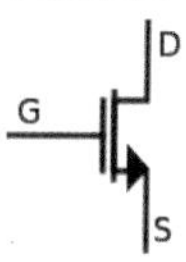	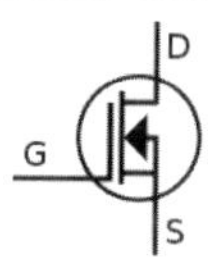
	JFET	MOSFET enh.	MOSFET enh. (no bulk)		MOSFET dep.

In schematics where G, S, D are not labeled, the detailed features of the symbol indicate which terminal is source and which is drain. For enhancement-mode and depletion-mode MOSFET symbols (in columns two and five), the source terminal is the one connected to the triangle. Additionally, in this diagram, the gate is shown as an "L" shape, whose input leg is closer to S than D, also indicating which is which. However, these symbols are often drawn with a "T" shaped gate, so it is the triangle which must be relied upon to indicate the source terminal.

For the symbols in which the bulk, or body, terminal is shown, it is here shown internally connected to the source (i.e., the black triangles in the diagrams in columns 2 and 5). This is a typical configuration, but by no means the only important configuration. In general, the MOSFET is a four-terminal device, and in integrated circuits many of the MOSFETs share a body connection, not necessarily connected to the source terminals of all the transistors.

Applications

Digital integrated circuits such as microprocessors and memory devices contain thousands to millions of integrated MOSFET transistors on each device, providing the basic switching functions required to implement logic gates and data storage. Discrete devices are widely used in applications such as switch mode power supplies, variable-frequency drives and other power electronics applications where each device may be switching thousands of watts. Radio-frequency amplifiers up to the UHF spectrum use MOSFET transistors as analog signal and power amplifiers. Radio systems also use MOSFETs as oscillators, or mixers to convert frequencies. MOSFET devices are also applied in audio-frequency power amplifiers for public address systems, sound reinforcement and home and automobile sound systems.

MOS Integrated Circuits

Following the development of clean rooms to reduce contamination to levels never before thought necessary, and of photolithography and the planar process to allow circuits to be made in very few steps, the Si–SiO_2 system possessed the technical attractions of low cost of production (on a per circuit basis) and ease of integration. Largely because of these two factors, the MOSFET has become the most widely used type of transistor in integrated circuits.

General Microelectronics introduced the first commercial MOS integrated circuit in 1964.

Additionally, the method of coupling two complementary MOSFETS (P-channel and N-channel) into one high/low switch, known as CMOS, means that digital circuits dissipate very little power except when actually switched.

The earliest microprocessors starting in 1970 were all *MOS microprocessors*; i.e., fabricated entirely from PMOS logic or fabricated entirely from NMOS logic. In the 1970s, *MOS microprocessors* were often contrasted with *CMOS microprocessors* and *bipolar bit-slice processors*.

CMOS Circuits

The MOSFET is used in digital complementary metal-oxide-semiconductor (CMOS) logic, which uses p- and n-channel MOSFETs as building blocks. Overheating is a major concern in integrated circuits since ever more transistors are packed into ever smaller chips. CMOS logic reduces power consumption because no current flows (ideally), and thus no power is consumed, except when the inputs to logic gates are being switched. CMOS accomplishes this current reduction by complementing every nMOSFET with a pMOSFET and connecting both gates and both drains together. A high voltage on the gates will cause the nMOSFET to conduct and the pMOSFET not to conduct and a low voltage on the gates causes the reverse. During the switching time as the voltage goes from one state to another, both MOSFETs will conduct briefly. This arrangement greatly reduces power consumption and heat generation.

Digital

The growth of digital technologies like the microprocessor has provided the motivation to advance MOSFET technology faster than any other type of silicon-based transistor. A big advantage of MOSFETs for digital switching is that the oxide layer between the gate and the channel prevents DC current from flowing through the gate, further reducing power consumption and giving a very large input impedance. The insulating oxide between the gate and channel effectively isolates a MOSFET in one logic stage from earlier and later stages, which allows a single MOSFET output to drive a considerable number of MOSFET inputs. Bipolar transistor-based logic (such as TTL) does not have such a high fanout capacity. This isolation also makes it easier for the designers to ignore to some extent loading effects between logic stages independently. That extent is defined by the operating frequency: as frequencies increase, the input impedance of the MOSFETs decreases.

Analog

The MOSFET's advantages in digital circuits do not translate into supremacy in all analog circuits. The two types of circuit draw upon different features of transistor behavior. Digital circuits switch, spending most of their time either fully on or fully off. The transition from one to the other is only of concern with regards to speed and charge required. Analog circuits depend on operation in the transition region where small changes to V_{gs} can modulate the output (drain) current. The MOSFET and bipolar junction transistor (BJT) is preferred for accurate matching (of adjacent devices in integrated circuits), higher transconductance and certain temperature characteristics which simplify keeping performance predictable as circuit temperature varies.

Nevertheless, MOSFETs are widely used in many types of analog circuits because of their own advantages (zero gate current, high and adjustable output impedance and improved robustness vs. BJTs which can be permanently degraded by even lightly breaking down the emitter-base). The characteristics and performance of many analog circuits can be scaled up or down by changing the sizes (length and width) of the MOSFETs used. By comparison, in bipolar transistors the size of the device does not significantly affect its performance. MOSFETs' ideal characteristics regarding gate current (zero) and drain-source offset voltage (zero) also make them nearly ideal switch elements, and also make switched capacitor analog circuits practical. In their linear region, MOSFETs can be used as precision resistors, which can have a much higher controlled resistance than BJTs. In high power circuits, MOSFETs sometimes have the advantage of not suffering from thermal runaway as BJTs do. Also, MOSFETs can be configured to perform as capacitors and gyrator circuits which allow op-amps made from them to appear as inductors, thereby allowing all of the normal analog devices on a chip (except for diodes, which can be made smaller than a MOSFET anyway) to be built entirely out of MOSFETs. This means that complete analog circuits can be made on a silicon chip in a much smaller space and with simpler fabrication techniques. MOSFETS are ideally suited to switch inductive loads because of tolerance to inductive kickback.

Some ICs combine analog and digital MOSFET circuitry on a single mixed-signal integrated circuit, making the needed board space even smaller. This creates a need to isolate the analog circuits from the digital circuits on a chip level, leading to the use of isolation rings and silicon on insulator (SOI). Since MOSFETs require more space to handle a given amount of power than a BJT, fabrication processes can incorporate BJTs and MOSFETs into a single device. Mixed-transistor devices are called bi-FETs (bipolar FETs) if they contain just one BJT-FET and BiCMOS (bipolar-CMOS) if they contain complementary BJT-FETs. Such devices have the advantages of both insulated gates and higher current density.

Analog Switches

MOSFET analog switches use the MOSFET to pass analog signals when on, and as a high impedance when off. Signals flow in both directions across a MOSFET switch. In this application, the drain and source of a MOSFET exchange places depending on the relative voltages of the source/drain electrodes. The source is the more negative side for an N-MOS or the more positive side for a P-MOS. All of these switches are limited on what signals they can pass or stop by their gate-source, gate-drain and source–drain voltages; exceeding the voltage, current, or power limits will potentially damage the switch.

Single-type

This analog switch uses a four-terminal simple MOSFET of either P or N type.

In the case of an n-type switch, the body is connected to the most negative supply (usually GND) and the gate is used as the switch control. Whenever the gate voltage exceeds the source voltage by at least a threshold voltage, the MOSFET conducts. The higher the voltage, the more the MOSFET can conduct. An N-MOS switch passes all voltages less than $V_{gate} - V_{tn}$. When the switch is conducting, it typically operates in the linear (or ohmic) mode of operation, since the source and drain voltages will typically be nearly equal.

In the case of a P-MOS, the body is connected to the most positive voltage, and the gate is brought to a lower potential to turn the switch on. The P-MOS switch passes all voltages higher than $V_{gate} - V_{tp}$ (threshold voltage V_{tp} is negative in the case of enhancement-mode P-MOS).

Dual-type (CMOS)

This "complementary" or CMOS type of switch uses one P-MOS and one N-MOS FET to counteract the limitations of the single-type switch. The FETs have their drains and sources connected in parallel, the body of the P-MOS is connected to the high potential (V_{DD}) and the body of the N-MOS is connected to the low potential (*gnd*). To turn the switch on, the gate of the P-MOS is driven to the low potential and the gate of the N-MOS is driven to the high potential. For voltages between $V_{DD} - V_{tn}$ and $gnd - V_{tp}$, both FETs conduct the signal; for voltages less than $gnd - V_{tp}$, the N-MOS conducts alone; and for voltages greater than $V_{DD} - V_{tn}$, the P-MOS conducts alone.

The voltage limits for this switch are the gate-source, gate-drain and source-drain voltage limits for both FETs. Also, the P-MOS is typically two to three times wider than the N-MOS, so the switch will be balanced for speed in the two directions.

Tri-state circuitry sometimes incorporates a CMOS MOSFET switch on its output to provide for a low-ohmic, full-range output when on, and a high-ohmic, mid-level signal when off.

Construction

Gate Material

The primary criterion for the gate material is that it is a good conductor. Highly doped polycrystalline silicon is an acceptable but certainly not ideal conductor, and also suffers from some more technical deficiencies in its role as the standard gate material. Nevertheless, there are several reasons favoring use of polysilicon:

1. The threshold voltage (and consequently the drain to source on-current) is modified by the work function difference between the gate material and channel material. Because polysilicon is a semiconductor, its work function can be modulated by adjusting the type and level of doping. Furthermore, because polysilicon has the same bandgap as the underlying silicon channel, it is quite straightforward to tune the work function to achieve low threshold voltages for both NMOS and PMOS devices. By contrast, the work functions of metals are not easily modulated, so tuning the work function to obtain low threshold voltages (LVT) becomes a significant challenge. Additionally, obtaining low-threshold devices on both PMOS and NMOS devices sometimes requires the use of different metals for each device type. While bimetallic integrated circuits (i.e., one type of metal for gate electrodes of NFETS and a second type of metal for gate electrodes of PFETS) are not common, they are known in patent literature and provide some benefit in terms of tuning electrical circuits' overall electrical performance.
2. The silicon-SiO2 interface has been well studied and is known to have relatively few defects. By contrast many metal-insulator interfaces contain significant levels of defects which can lead to Fermi level pinning, charging, or other phenomena that ultimately degrade device performance.
3. In the MOSFET IC fabrication process, it is preferable to deposit the gate material prior to certain high-temperature steps in order to make better-performing transistors. Such high temperature steps would melt some metals, limiting the types of metal that can be used in a metal-gate-based process.

While polysilicon gates have been the de facto standard for the last twenty years, they do have some disadvantages which have led to their likely future replacement by metal gates. These disadvantages include:

- Polysilicon is not a great conductor (approximately 1000 times more resistive than metals) which reduces the signal propagation speed through the material. The resistivity can be lowered by increasing the level of doping, but even highly doped polysilicon is not as conductive as most metals. To improve conductivity further, sometimes a high-temperature metal such as tungsten, titanium, cobalt, and more recently nickel is alloyed with the top layers of the polysilicon. Such a blended material is called silicide. The silicide-polysilicon combination has better electrical properties than polysilicon alone and still does not melt in subsequent processing. Also the threshold voltage is not significantly higher than with polysilicon alone, because the silicide material is not near the channel. The process in which silicide is formed on both the gate electrode and the source and drain regions is sometimes called salicide, self-aligned silicide.

- When the transistors are extremely scaled down, it is necessary to make the gate dielectric layer very thin, around 1 nm in state-of-the-art technologies. A phenomenon observed here is the so-called poly depletion, where a depletion layer is formed in the gate polysilicon layer next to the gate dielectric when the transistor is in the inversion. To avoid this problem, a metal gate is desired. A variety of metal gates such as tantalum, tungsten, tantalum nitride, and titanium nitride are used, usually in conjunction with high-κ dielectrics. An alternative is to use fully silicided polysilicon gates, a process known as FUSI.

Present high performance CPUs use metal gate technology, together with high-κ dielectrics, a combination known as *high-κ, metal gate* (HKMG). The disadvantages of metal gates are overcome by a few techniques:

1. The threshold voltage is tuned by including a thin "work function metal" layer between the high-κ dielectric and the main metal. This layer is thin enough that the total work function of the gate is influenced by both the main metal and thin metal work functions (either due to alloying during annealing, or simply due to the incomplete screening by the thin metal). The threshold voltage thus can be tuned by the thickness of the thin metal layer.
2. High-κ dielectrics are now well studied, and their defects are understood.
3. HKMG processes exist that do not require the metals to experience high temperature anneals; other processes select metals that can survive the annealing step.

Insulator

As devices are made smaller, insulating layers are made thinner, often through steps of thermal oxidation or localised oxidation of silicon (LOCOS). For nano-scaled devices, at some point tunneling of carriers through the insulator from the channel to the gate electrode takes place. To reduce the resulting leakage current, the insulator can be made thinner by choosing a material with a higher dielectric constant. To see how thickness and dielectric constant are related, note that Gauss's law connects field to charge as:

$$Q = \kappa\epsilon_0 E,$$

with Q = charge density, κ = dielectric constant, ε_0 = permittivity of empty space and E = electric field. From this law it appears the same charge can be maintained in the channel at a lower field provided κ is increased. The voltage on the gate is given by:

$$V_G = V_{ch} + Et_{ins} = V_{ch} + \frac{Qt_{ins}}{\kappa\epsilon_0},$$

with V_G = gate voltage, V_{ch} = voltage at channel side of insulator, and t_{ins} = insulator thickness. This equation shows the gate voltage will not increase when the insulator thickness increases, provided κ increases to keep t_{ins} / κ = constant.

The insulator in a MOSFET is a dielectric which can in any event be silicon oxide, formed by LOCOS but many other dielectric materials are employed. The generic term for the dielectric is gate dielectric since the dielectric lies directly below the gate electrode and above the channel of the MOSFET.

Junction Design

The source-to-body and drain-to-body junctions are the object of much attention because of three major factors: their design affects the current-voltage (I-V) characteristics of the device, lowering output resistance, and also the speed of the device through the loading effect of the junction capacitances, and finally, the component of stand-by power dissipation due to junction leakage.

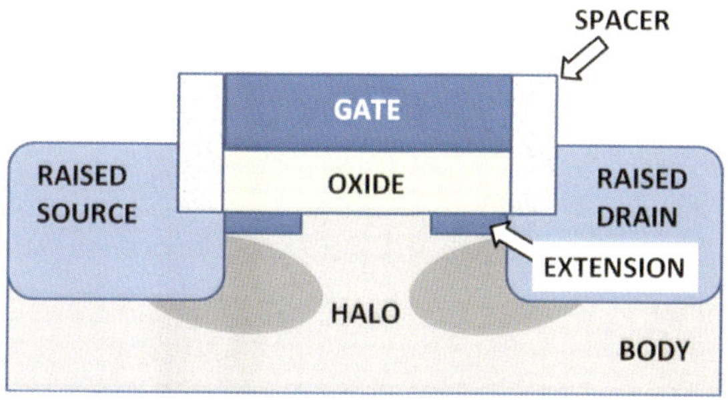

MOSFET showing shallow junction extensions, raised source and drain and halo implant. Raised source and drain separated from gate by oxide spacers

The drain induced barrier lowering of the threshold voltage and channel length modulation effects upon *I-V* curves are reduced by using shallow junction extensions. In addition, *halo* doping can be used, that is, the addition of very thin heavily doped regions of the same doping type as the body tight against the junction walls to limit the extent of depletion regions.

The capacitive effects are limited by using raised source and drain geometries that make most of the contact area border thick dielectric instead of silicon.

These various features of junction design are shown (with artistic license) in the figure.

Scaling

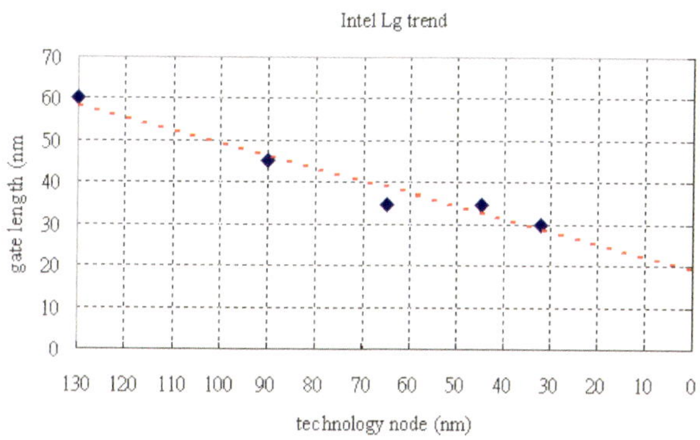

Trend of Intel CPU transistor gate length

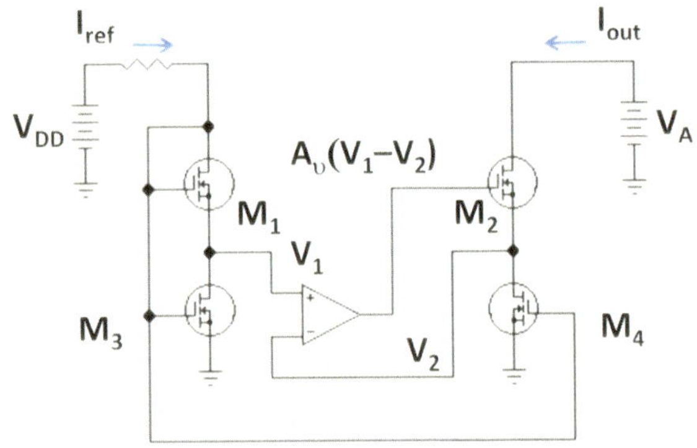

MOSFET version of gain-boosted current mirror; M_1 and M_2 are in active mode, while M_3 and M_4 are in Ohmic mode, and act like resistors. The operational amplifier provides feedback that maintains a high output resistance.

Over the past decades, the MOSFET (as used for digital logic) has continually been scaled down in size; typical MOSFET channel lengths were once several micrometres, but modern integrated circuits are incorporating MOSFETs with channel lengths of tens of nanometers. Robert Dennard's work on scaling theory was pivotal in recognising that this ongoing reduction was possible. Intel began production of a process featuring a 32 nm feature size (with the channel being even shorter) in late 2009. The semiconductor industry maintains a "roadmap", the ITRS, which sets the pace for MOSFET development. Historically, the difficulties with decreasing the size of the MOSFET have been associated with the semiconductor device fabrication process, the need to use very low voltages, and with poorer electrical performance necessitating circuit redesign and innovation (small MOSFETs exhibit higher leakage currents and lower output resistance).

Smaller MOSFETs are desirable for several reasons. The main reason to make transistors smaller is to pack more and more devices in a given chip area. This results in a chip with the same functionality in a smaller area, or chips with more functionality in the same area. Since fabrication costs for a semiconductor wafer are relatively fixed, the cost per integrated circuits is mainly related to the number of chips that can be produced per wafer. Hence, smaller ICs allow more chips per wafer, reducing the price per chip. In fact, over the past 30 years the number of transistors per chip has been doubled every 2–3 years once a new technology node is introduced. For example, the number of MOSFETs in a microprocessor fabricated in a 45 nm technology can well be twice as many as in a 65 nm chip. This doubling of transistor density was first observed by Gordon Moore in 1965 and is commonly referred to as Moore's law. It is also expected that smaller transistors switch faster. For example, one approach to size reduction is a scaling of the MOSFET that requires all device dimensions to reduce proportionally. The main device dimensions are the channel length, channel width, and oxide thickness. When they are scaled down by equal factors, the transistor channel resistance does not change, while gate capacitance is cut by that factor. Hence, the RC delay of the transistor scales with a similar factor. While this has been traditionally the case for the older technologies, for the state-of-the-art MOSFETs reduction of the transistor dimensions does not necessarily translate to higher chip speed because the delay due to interconnections is more significant.

Producing MOSFETs with channel lengths much smaller than a micrometre is a challenge, and the difficulties of semiconductor device fabrication are always a limiting factor in advancing integrated circuit technology. Though processes such as ALD have improved fabrication for small components, the small size of the MOSFET (less than a few tens of nanometers) has created operational problems:

Higher subthreshold conduction

As MOSFET geometries shrink, the voltage that can be applied to the gate must be reduced to maintain reliability. To maintain performance, the threshold voltage of the MOSFET has to be reduced as well. As threshold voltage is reduced, the transistor cannot be switched from complete turn-off to complete turn-on with the limited voltage swing available; the circuit design is a compromise between strong current in the *on* case and low current in the *off* case, and the application determines whether to favor one over the other. Subthreshold leakage (including subthreshold conduction, gate-oxide leakage and reverse-biased junction leakage), which was ignored in the past, now can consume upwards of half of the total power consumption of modern high-performance VLSI chips.

Increased gate-oxide leakage

The gate oxide, which serves as insulator between the gate and channel, should be made as thin as possible to increase the channel conductivity and performance when the transistor is on and to reduce subthreshold leakage when the transistor is off. However, with current gate oxides with a thickness of around 1.2 nm (which in silicon is ~5 atoms thick) the quantum mechanical phenomenon of electron tunneling occurs between the gate and channel, leading to increased power consumption. Silicon dioxide has traditionally been used as the gate insulator. Silicon dioxide however has a modest dielectric constant. Increasing the dielectric constant of the gate dielectric allows a thicker layer while maintaining a high capacitance (capacitance is proportional to dielectric constant and inversely proportional to dielectric thickness). All else equal, a higher dielectric thickness reduces the quantum tunneling current through the dielectric between the gate and the channel. Insulators that have a larger dielectric constant than silicon dioxide (referred to as high-κ dielectrics), such as group IVb metal silicates e.g. hafnium and zirconium silicates and oxides are being used to reduce the gate leakage from the 45 nanometer technology node onwards. On the other hand, the barrier height of the new gate insulator is an important consideration; the difference in conduction band energy between the semiconductor and the dielectric (and the corresponding difference in valence band energy) also affects leakage current level. For the traditional gate oxide, silicon dioxide, the former barrier is approximately 8 eV. For many alternative dielectrics the value is significantly lower, tending to increase the tunneling current, somewhat negating the advantage of higher dielectric constant. The maximum gate-source voltage is determined by the strength of the electric field able to be sustained by the gate dielectric before significant leakage occurs. As the insulating dielectric is made thinner, the electric field strength within it goes up for a fixed voltage. This necessitates using lower voltages with the thinner dielectric.

Increased junction leakage

To make devices smaller, junction design has become more complex, leading to higher doping levels, shallower junctions, "halo" doping and so forth, all to decrease drain-induced barrier lowering. To keep these complex junctions in place, the annealing steps formerly used to remove damage and electrically active defects must be curtailed increasing junction leakage. Heavier doping is also associated with thinner depletion layers and more recombination centers that result in increased leakage current, even without lattice damage.

Drain-induced barrier lowering (DIBL) and V_T roll off

Because of the short-channel effect, channel formation is not entirely done by the gate, but now the drain and source also affect the channel formation. As the channel length decreases, the depletion regions of the source and drain come closer together and make the threshold voltage (V_T) a function of the length of the channel. This is called V_T roll-off. V_T also becomes function of drain to source voltage V_{DS}. As we increase the V_{DS}, the depletion regions increase in size, and a considerable amount of charge is depleted by the V_{DS}. The gate voltage required to form the channel is then lowered, and thus, the V_T decreases with an increase in V_{DS}. This effect is called drain induced barrier lowering (DIBL).

Lower output resistance

For analog operation, good gain requires a high MOSFET output impedance, which is to say, the MOSFET current should vary only slightly with the applied drain-to-source voltage. As devices are made smaller, the influence of the drain competes more successfully with that of the gate due to the growing proximity of these two electrodes, increasing the sensitivity of the MOSFET current to the drain voltage. To counteract the resulting decrease in output resistance, circuits are made more complex, either by requiring more devices, for example the cascode and cascade amplifiers, or by feedback circuitry using operational amplifiers, for example a circuit like that in the adjacent figure.

Lower transconductance

The transconductance of the MOSFET decides its gain and is proportional to hole or electron mobility (depending on device type), at least for low drain voltages. As MOSFET size is reduced, the fields in the channel increase and the dopant impurity levels increase. Both changes reduce the carrier mobility, and hence the transconductance. As channel lengths are reduced without proportional reduction in drain voltage, raising the electric field in the channel, the result is velocity saturation of the carriers, limiting the current and the transconductance.

Interconnect capacitance

Traditionally, switching time was roughly proportional to the gate capacitance of gates. However, with transistors becoming smaller and more transistors being placed on the chip, interconnect capacitance (the capacitance of the metal-layer connections between different parts of the chip) is becoming a large percentage of capacitance. Signals have to travel through the interconnect, which leads to increased delay and lower performance.

Heat production

The ever-increasing density of MOSFETs on an integrated circuit creates problems of substantial localized heat generation that can impair circuit operation. Circuits operate more slowly at high temperatures, and have reduced reliability and shorter lifetimes. Heat sinks and other cooling devices and methods are now required for many integrated circuits including microprocessors. Power MOSFETs are at risk of thermal runaway. As their on-state resistance rises with temperature, if the load is approximately a constant-current load then the power loss rises correspondingly, generating further heat. When the heatsink is not able to keep the temperature low enough, the junction temperature may rise quickly and uncontrollably, resulting in destruction of the device.

Process variations

With MOSFETs becoming smaller, the number of atoms in the silicon that produce many of the transistor's properties is becoming fewer, with the result that control of dopant numbers and placement is more erratic. During chip manufacturing, random process variations affect all transistor dimensions: length, width, junction depths, oxide thickness *etc.*, and become a greater percentage of overall transistor size as the transistor shrinks. The transistor characteristics become less certain, more statistical. The random nature of manufacture means we do not know which particular example MOSFETs actually will end up in a particular instance of the circuit. This uncertainty forces a less optimal design because the design must work for a great variety of possible component MOSFETs.

Modeling challenges

Modern ICs are computer-simulated with the goal of obtaining working circuits from the very first manufactured lot. As devices are miniaturized, the complexity of the processing makes it difficult to predict exactly what the final devices look like, and modeling of physical processes becomes more challenging as well. In addition, microscopic variations in structure due simply to the probabilistic nature of atomic processes require statistical (not just deterministic) predictions. These factors combine to make adequate simulation and "right the first time" manufacture difficult.

Other Types

Dual-gate

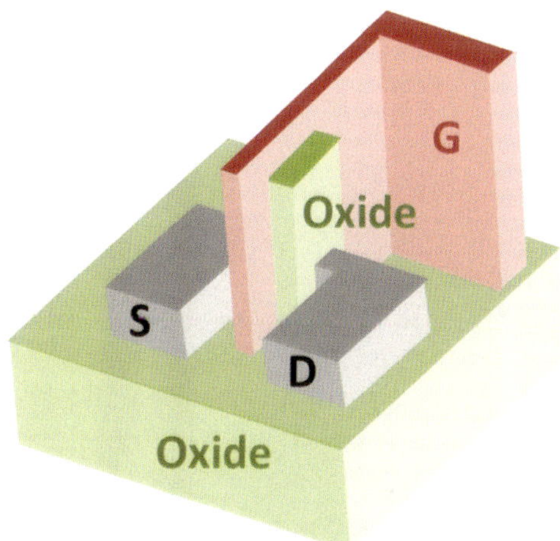

A FinFET

The dual-gate MOSFET has a tetrode configuration, where both gates control the current in the device. It is commonly used for small-signal devices in radio frequency applications where biasing the drain-side gate at constant potential reduces the gain loss caused by Miller effect, replacing two separate transistors in cascode configuration. Other common uses in RF circuits include gain control and mixing (frequency conversion). The *tetrode* description, though accurate, does not replicate the vacuum-tube tetrode. Vacuum-tube tetrodes, using a screen grid, exhibit much lower grid-plate capacitance and much higher output impedance and voltage gains than triode vacuum tubes. These improvements are commonly an order of magnitude (10 times) or considerably more. Tetrode transistors (whether bipolar junction or field-effect) do not exhibit improvements of such a great degree.

The FinFET is a double-gate silicon-on-insulator device, one of a number of geometries being introduced to mitigate the effects of short channels and reduce drain-induced barrier lowering. The *fin* refers to the narrow channel between source and drain. A thin insulating oxide layer on either side of the fin separates it from the gate. SOI FinFETs with a thick oxide on top of the fin are called *double-gate* and those with a thin oxide on top as well as on the sides are called *triple-gate* FinFETs.

Depletion-mode

There are *depletion-mode* MOSFET devices, which are less commonly used than the standard *enhancement-mode* devices already described. These are MOSFET devices that are doped so that a channel exists even with zero voltage from gate to source. To control the channel, a negative voltage is applied to the gate (for an n-channel device), depleting the channel, which reduces the current flow through the device. In essence, the depletion-mode device is equivalent to a normally closed (on) switch, while the enhancement-mode device is equivalent to a normally open (off) switch.

Due to their low noise figure in the RF region, and better gain, these devices are often preferred to bipolars in RF front-ends such as in TV sets.

Depletion-mode MOSFET families include BF960 by Siemens and Telefunken, and the BF980 in the 1980s by Philips (later to become NXP Semiconductors), whose derivatives are still used in AGC and RF mixer front-ends.

Metal-insulator-semiconductor field-effect Transistor (MISFET)

Metal-insulator-semiconductor field-effect-transistor, or *MISFET*, is a more general term than *MOSFET* and a synonym to *insulated-gate field-effect transistor* (IGFET). All MOSFETs are MISFETs, but not all MISFETs are MOSFETs.

The gate dielectric insulator in a MISFET is silicon dioxide in a MOSFET, but other materials can also be employed. The gate dielectric lies directly below the gate electrode and above the channel of the MISFET. The term *metal* is historically used for the gate material, even though now it is usually highly doped polysilicon or some other non-metal.

Insulator types may be:

- Silicon dioxide, in MOSFETs
- Organic insulators (e.g., undoped trans-polyacetylene; cyanoethyl pullulan, CEP), for organic-based FETs.

NMOS Logic

For devices of equal current driving capability, n-channel MOSFETs can be made smaller than p-channel MOSFETs, due to p-channel charge carriers (holes) having lower mobility than do n-channel charge carriers (electrons), and producing only one type of MOSFET on a silicon substrate is cheaper and technically simpler. These were the driving principles in the design of NMOS logic which uses n-channel MOSFETs exclusively. However, neglecting leakage current, unlike CMOS logic, NMOS logic consumes power even when no switching is taking place. With advances in technology, CMOS logic displaced NMOS logic in the mid-1980s to become the preferred process for digital chips.

Power MOSFET

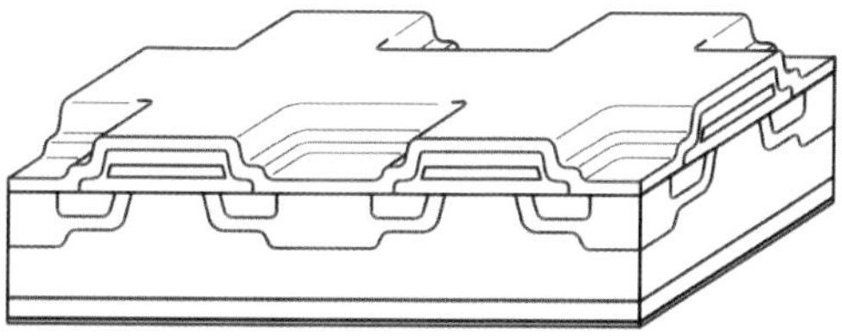

Cross section of a power MOSFET, with square cells. A typical transistor is constituted of several thousand cells

Power MOSFETs have a different structure. As with most power devices, the structure is vertical and not planar. Using a vertical structure, it is possible for the transistor to sustain both high blocking voltage and high current. The voltage rating of the transistor is a function of the doping and thickness of the N-epitaxial layer, while the current rating is a function of the channel width (the wider the channel, the higher the current). In a planar structure, the current and breakdown voltage ratings are both a function of the channel dimensions (respectively width and length of the channel), resulting in inefficient use of the "silicon estate". With the vertical structure, the component area is roughly proportional to the current it can sustain, and the component thickness (actually the N-epitaxial layer thickness) is proportional to the breakdown voltage.

Power MOSFETs with lateral structure are mainly used in high-end audio amplifiers and high-power PA systems. Their advantage is a better behaviour in the saturated region (corresponding to the linear region of a bipolar transistor) than the vertical MOSFETs. Vertical MOSFETs are designed for switching applications.

Double-diffused Metal-oxide-semiconductor (DMOS)

There are LDMOS (lateral double-diffused metal oxide semiconductor) and *VDMOS* (vertical double-diffused metal oxide semiconductor). Most power MOSFETs are made using this technology.

Radiation-hardened-by-design (RHBD)

Semiconductor sub-micrometer and nanometer electronic circuits are the primary concern for operating within the normal tolerance in harsh radiation environments like outer space. One of the design approaches for making a radiation-hardened-by-design (RHBD) device is enclosed-layout-transistor (ELT). Normally, the gate of the MOSFET surrounds the drain, which is placed in the center of the ELT. The source of the MOSFET surrounds the gate. Another RHBD MOSFET is called H-Gate. Both of these transistors have very low leakage current with respect to radiation. However, they are large in size and take more space on silicon than a standard MOSFET. In older STI (shallow trench isolation) designs, radiation strikes near the silicon oxide region cause the channel inversion at the corners of the standard MOSFET due to accumulation of radiation induced trapped charges. If the charges are large enough, the accumulated charges affect STI surface edges along the channel near the channel interface (gate) of the standard MOSFET. Thus the device channel inversion occurs along the channel edges and the device creates off-state leakage path, causing device to turn on. So the reliability of circuits degrades severely. The ELT offers many advantages. These advantages include improvement of reliability by reducing unwanted surface inversion at the gate edges that occurs in the standard MOSFET. Since the gate edges are enclosed in ELT, there is no gate oxide edge (STI at gate interface), and thus the transistor off-state leakage is reduced very much. Low-power microelectronic circuits including computers, communication devices and monitoring systems in space shuttle and satellites are very different from what we use on earth. They are radiation (high-speed atomic particles like proton and neutron, solar flare magnetic energy dissipation in Earth's space, energetic cosmic rays like X-ray, gamma ray etc.) tolerant circuits. These special electronics are designed by applying very different techniques using RHBD MOSFETs to ensure the safe space journey and also space-walk of astronauts.

Enhancement Mode (E- MOSFET)

Although DE-MOSFET is useful in special applications, it does not enjoy widespread use. However, it played an important role in history because it was part of the evolution towards the E-mode MOSFET, a device that has revolutionized the electronic industry. E-MOSFET has become enormously important, in digital electronics and. In the absence of E-MOSFET's the personal computers (PCs) that are now so widespread would not exist.

Construction of an EMOSFET

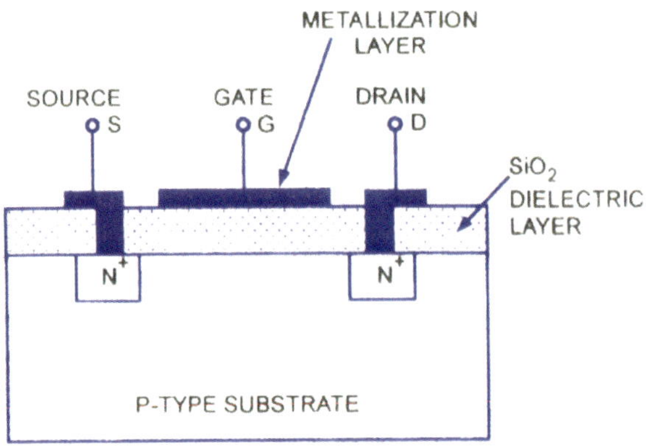

N-Channel E-MOSFET Structure

Construction of EMOSFET

Figure shows the construction of an N-channel E-MOSFET. The main difference between the construction of DE-MOSFET and that of E-MOSFET, as we see from the figures given below the

E-MOSFET substrate extends all the way to the silicon dioxide (SiO_2) and no channels are doped between the source and the drain. Channels are electrically induced in these MOSFETs, when a positive gate-source voltage V_{GS} is applied to it.

Operation of an EMOSFET

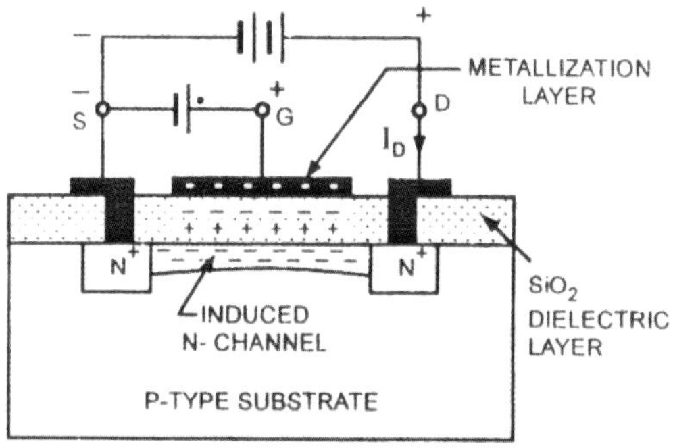

Operation of N-Channel E-MOSFET

Working of an EMOSFET

As its name indicates, this MOSFET operates only in the *enhancement mode* and has no depletion mode. It operates with large positive gate voltage only. It does not conduct when the gate-source voltage $V_{GS} = 0$. This is the reason that it is called normally-off MOSFET. In these MOSFET's drain current I_D flows only when V_{GS} exceeds V_{GST}.

When drain is applied with positive voltage with respect to source and no potential is applied to the gate two N-regions and one P-substrate from two P-N junctions connected back to back with a resistance of the P-substrate. So a very small drain current that is, reverse leakage current flows. If the P-type substrate is now connected to the source terminal, there is zero voltage across the source substrate junction, and the drain-substrate junction remains reverse biased.

When the gate is made positive with respect to the source and the substrate, negative (i.e. minority) charge carriers within the substrate are attracted to the positive gate and accumulate close to the-surface of the substrate. As the gate voltage is increased, more and more electrons accumulate under the gate. Since these electrons can not flow across the insulated layer of silicon dioxide to the gate, so they accumulate at the surface of the substrate just below the gate. These accumulated minority charge carriers N -type channel stretching from drain to source. When this occurs, a channel is induced by forming what is termed an *inversion layer* (N-type). Now a drain current start flowing. The strength of the drain current depends upon the channel resistance which, in turn, depends upon the number of charge carriers attracted to the positive gate. Thus drain current is controlled by the gate potential.

Since the conductivity of the channel is enhanced by the positive bias on the gate so this device is also called the *enhancement MOSFET* or E- MOSFET.

The minimum value of gate-to-source voltage V_{GS} that is required to form the inversion layer (N-type) is termed the *gate-to-source threshold voltage* V_{GST}. For V_{GS} below V_{GST}, the drain current $I_D = 0$. But for V_{GS} exceeding V_{GST} an N-type inversion layer connects the source to drain and the drain current I_D is large. Depending upon the device being used, V_{GST} may vary from less than 1 V to more than 5 V.

JFETs and DE-MOSFETs are classified as the depletion-mode devices because their conductivity depends on the action of depletion layers. E-MOSFET is classified as an enhancement-mode device because its conductivity depends on the action of the inversion layer. Depletion-mode devices are normally ON when the gate-source voltage $V_{GS} = 0$, whereas the enhancement-mode devices are normally OFF when $V_{GS} = 0$.

Characteristics of an EMOSFET

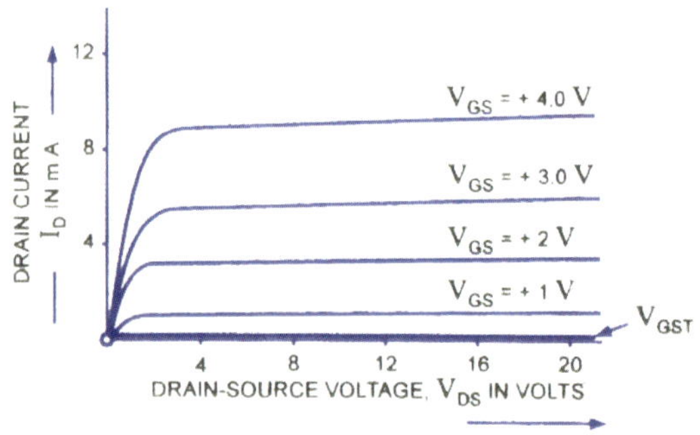

Drain Characteristics

Drain Characteristics-EMOSFET

Drain characteristics of an N-channel E-MOSFET are shown in figure. The lowest curve is the V_{GST} curve. When V_{GS} is lesser than V_{GST}, I_D is approximately zero. When V_{GS} is greater than V_{GST}, the device turns- on and the drain current I_D is controlled by the gate voltage. The characteristic curves have almost vertical and almost horizontal parts. The almost vertical components of the curves correspond to the ohmic region, and the horizontal components correspond to the constant current region. Thus E-MOSFET can be operated in either of these regions *i.e.* it can be used as a variable-voltage resistor (WR) or as a constant current source.

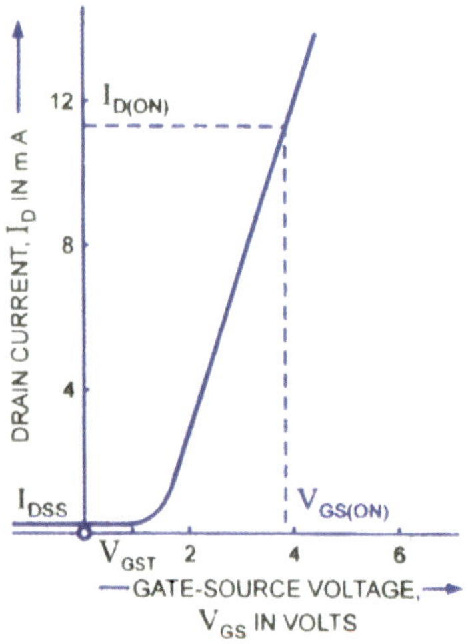

Transfer Characteristic

EMOSFET-Transfer Characteristics

Figure shows a typical transconductance curve. The current IDSS at VGS <=0 is very small, being of the order of a few nano-amperes. When the V_{GS} is made positive, the drain current I_D increases slowly at first, and then much more rapidly with an increase in V_{GS}. The manufacturer sometimes indicates the *gate-source threshold voltage* V_{GST} at which the drain current I_D attains some defined small value, say 10 u A. A current $I_{D\ (oN)}$, corresponding approximately to the maximum value given on the drain characteristics and the values of V_{GS} required to give this current $V_{Gs\ QN}$ are also usually given on the manufacturers data sheet.

The equation for the transfer characteristic does not obey equation. However it does follow a similar "square law type" of relationship. The equation for the transfer characteristic of E-MOSFETs is given as:

$$I_D = K\left(V_{GS} - V_{GST}\right)^2$$

Schematic Symbols of EMOSFET.

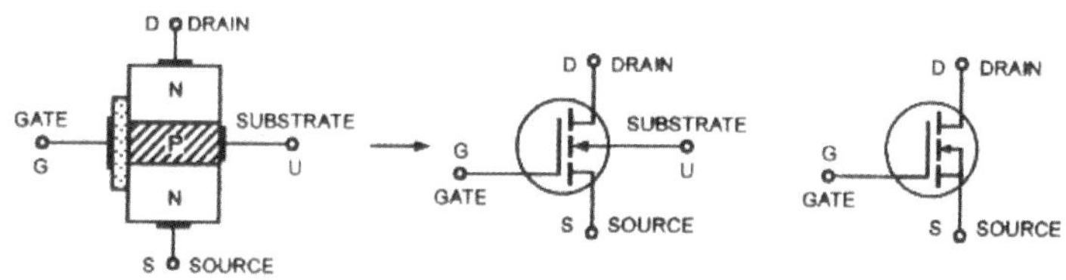

N-Channel E-MOSFET

P-Channel E-MOSFET
Schematic Symbols For E-MOSFETs
EMOSFET-Schematic symbols

Schematic symbols for an N-channel E-MOSFET are shown in figure. For zero value of V_{GS}, the E-MOSFET is OFF because there is no conducting channel between source and drain. Each of schematic symbols shown in figures, has broken channel line to indicate this normally OFF condition. As we know that for V_{GS} exceeding the threshold voltage V_{GST}, an N-type inversion layer, connecting the source to drain, is created. In each of the schematic symbols, the arrow points to this inversion layer, which acts like an N-channel when the device is conducting. In each case, the fact that the device has an insulated gate is indicated by the gate not making direct contact with the channel. The schematic symbol shown in figure shows the source and substrate internally connected, while the other symbol shown in figure shows the substrate connection brought out separately from the source.

The schematic symbols for a P-channel E-MOSFET are also shown. In these cases the arrow points outwards.

Schematic of n-channel E-MOSFET

Figure shows a schematic of a n-channel E-MOSFET. It has "Metal" Gate electrode which is electrically insulated from semiconductor (n-type or p-type) by very thin layer of oxide material usually silicon dioxide. A thin layer of silicon dioxide covers the surface of silicon, which performs three functions: it acts as a protective layer for the silicon surface which ensures stable electrical behavior; it acts as an insulator for the deposition of conductors on its surface, without making electrical contact to the semiconductor; it acts as a dielectric for the capacitor formed by the gate electrode and the substrate. Gate terminal is isolated from current carrying channel, so no current flows into gate.

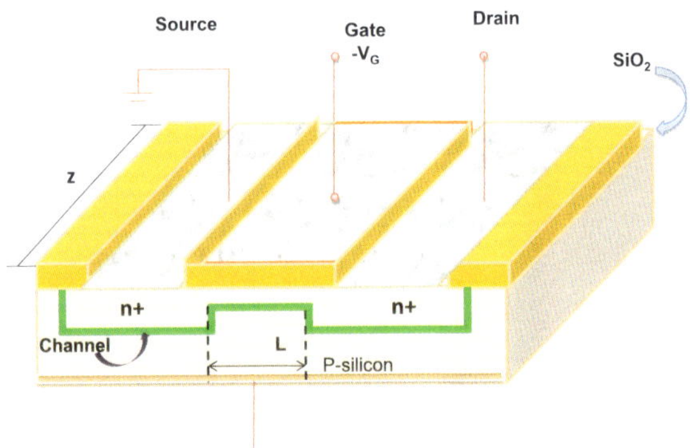

Schematic of n-channel E-MOSFET

It consists of a p- Si substrate into which 2 heavily doped n+ region have been diffused namely Source (S) and Drain (D). Metal contact is deposited on oxide layer that covers gap between two n+ diffusion and serves as Gate (G). Here, L is channel length [distance between two n+ region], Z is channel width, d is thickness of oxide layer, NA is acceptor doping concentration in p-type Si, Central section of device corresponds to MOS diode and Source is used as voltage reference.

Detailed Operation of n-channel E-MOSFET

- Case I: VG = 0

Source-to-drain electrodes corresponds to two n+- p junctions connected back to back. Device is off, when VG = 0 as there is no channel between Source and Drain.

- Case II:VG > VT (Inversion Condition for MOS)

Large positive gate voltage [VG > VT] has to be applied to induce a channel below oxide i.e. a surface inversion layer (accumulated electrons) is created. The electron concentration in channel enhances by increasing VG and hence channel conductance can be modulated by VG.

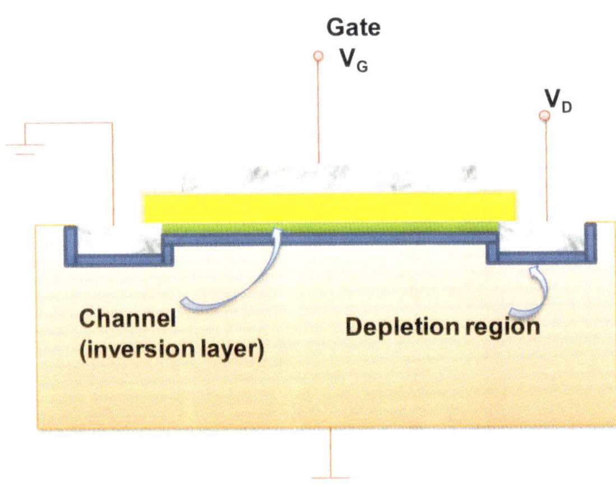

n-channel E-MOSFET with VG > VT

FOR VG > VT

- Case A: For Small Value of +VD

Electron will flow from S to D through conducting channel. Channel acts as a resistance and ID μ VD (linear region). As long as VD is small, the surface potential will remain almost same throughout channel (i.e. near source and drain regions). The electron concentration will remain same and channel behaves as resistor having

$$R = \frac{L}{\sigma A} = \frac{-L}{\mu_n Z Q_n}$$

Where, Qn < 0, is the space charge per unit area of inversion layer (or channel) and Z is the channel width.

- *Case B*: For Moderate Value of +VD

With increase in *VD*, potential will distribute throughout the channel. Potential difference between G and inversion layer will be more near S and very less near D. Thus, electron concentration (Qn) in channel will be relatively less near D, and hence channel resistance increases. ∴ I-V curve will bend downward from initial linear variation.

Case C: For large VD If VD further increases (VDsat), the potential drop across oxide near drain will continue to decrease, and may until it falls below the value required to maintain an inversion layer, i.e. width of inversion layer xi at y = L reduced to zero [known as Pinch off point]. At this condition, Drain gets isolated from conducting channel (inversion layer) by a depletion region. The ID will then saturate to a constant value Idsat at VD = VDsat.

Case D: As VD > VDsat, ID remains as IDsat. The pinch off point P will move towards Source, and effective channel length (L¢) will decrease. However, the potential at P will remain at VDsat, and additional voltage in excess of VDsat will drop across depletion region. Thus, number of carriers arriving at point P from Source and hence current flowing from D to S remains the constant.

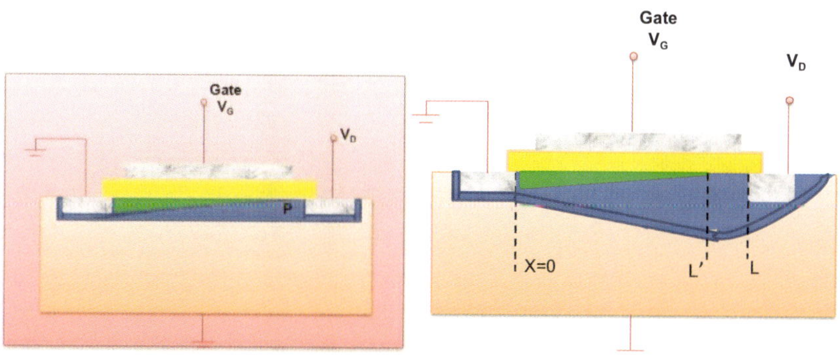

n-channel MOSFET for VD > VDsat

If VG or VGS increases (For Fixed *VD*), leads to increases in the inversion layer and thus charge Qn.

Therefore, channel resistance decreases or ID increases for a given VD. The pinch off will occur at a higher value of VD.

Study of I-V Relationship for n-channel E-MOSFET

Consider ideal condition and following assumptions

1. Ideal MOS structure [no charge in oxide and no difference in work functions of Gate and Semiconductors, f $\phi\ ms = 0$]
2. Current flow in one dimention (i.e. in channel only). No current flows from Channel to Gate.
3. Electron mobility in Inversion layer (channel) is constant and is independent of electric field.
4. Doping in the semiconductor in the channel region is uniform.
5. Reverse-leakage current is negligibly small.
6. Transverse electric field (perpendicular to the current flow) at edge of the depletion region in channel is much larger than the longitudinal electric field (parallel to current flow).

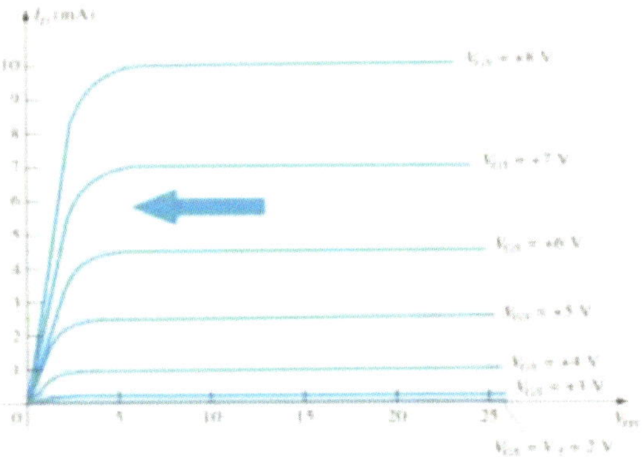

IV characteristics of MOSFET

Depletion Mode MOSFET

We know that when the gate is biased negative with respect to the source in an N-channel JFET, the depletion region widths are increased. Theincrease in the depletion regions reduces the channel thickness, which increases its resistance. The net result is that drain current I_D is reduced.

If the polarity of V_{GG} were reversed so as to apply a positive bias to the gate with respect to source, the P-N junctions between the gate and the channel would then be forward biased. Since a forward bias reduces the width of a depletion region, the thickness of channel would increase with a corresponding decrease in channel resistance. As a result, drain current I_D would increase beyond the JFET's I_{DSS} value.

The normal operation of a JFET is in its depletion mode of operation. However, as discussed above, it is also possible to *enhance* the conductivity of the JFET channel. However, the forward bias of the silicon P-N junction is usually restricted to a maximum of 0.5 V (more conservative limit is 0.2 V) so as to limit the gate current.

As we have seen that, the greater the I_D is compared to I_{DSS} the greater the transconductance g_m will be. We have seen before that the voltage gain is directly proportional to g_m. So, in general, the higher the g_m, the better it is. This is one of the advantages of being able to enhance the channel.

As its name suggests, the depletion-enhancement MOSFET (DE-MOSFET)-was developed to be used in either or both the depletion and enhancement modes.

Construction of a DEMOSFET

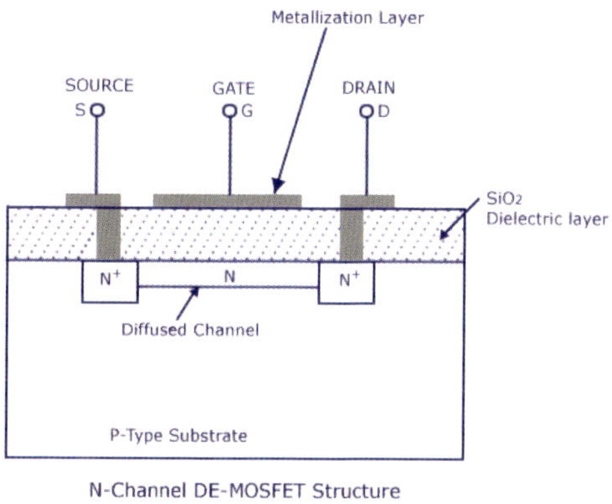

Construction of DEMOSFET

Figure shows the construction of an N-channel depletion MOSFET. It consists of a highly doped P-type substrate into which two blocks of heavily doped N-type material are diffused forming the source and drain. An N-channel is formed by diffusion between the source and drain. The type of impurity for the channel is the same as for the source and drain. Now a thin layer of SiO_2 dielectric is grown over the entire surface and holes are cut through the SiO_2 (silicon-dioxide) layer to make contact with the N-type blocks (Source and Drain). Metal is deposited through the holes to provide drain and source terminals, and on the surface area between drain and source, a metal plate isde-posited. This layer constitutes the gate. Sio_2 layer results in an extremely high input impedance of the order of 10^{10} to 10^{15} Q for this area. The chip area of a MOSFET is typically 0.003 um^2 or less which is about only 5% of the area required by a BJT. A P-channel DE-MOSFET is constructed like an N-channel DE-MOSFET, starting with an N-type substrate and diffusing P-type drain and source blocks and connecting them internally by a P-doped channel region.

Operation of DEMOSFET

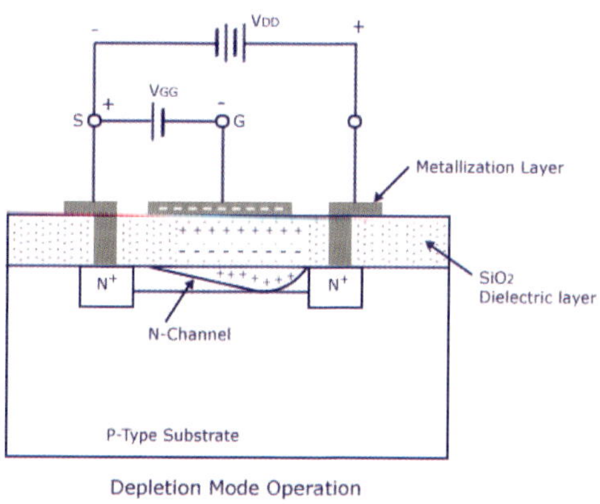

DEMOSFET-Operation

DE-MOSFET can be operated with either a positive or a negative gate. When gate is positive with respect to the source it operates in the enhancement—or E-mode and when the gate is negative with respect to the source, as illustrated in figure, it operates in depletion-mode.

When the drain is made positive with respect to source, a drain current will flow, even with zero gate potential and the MOSFET is said to be operating in E-mode. In this mode of operation gate attracts the negative charge carriers from the P-substrate to the N-channel and thus reduces the channel resistance and increases the drain-current. The more positive the gate is made, the more drain current flows.

On the other hand when the gate is made negative with respect to the substrate, the gate repels some of the negative charge carriers out of the N-channel. This creates a depletion region in the channel, as illustrated in figure, and, therefore, increases the channel resistance and reduces the drain current. The more negative the gate, the less the drain current. In this mode of operation the device is referred to as a *depletion-mode MOSFET*. Here too much negative gate voltage can pinch-off the channel. Thus operation is similar to that of JFET.

Characteristics of DEMOSFET

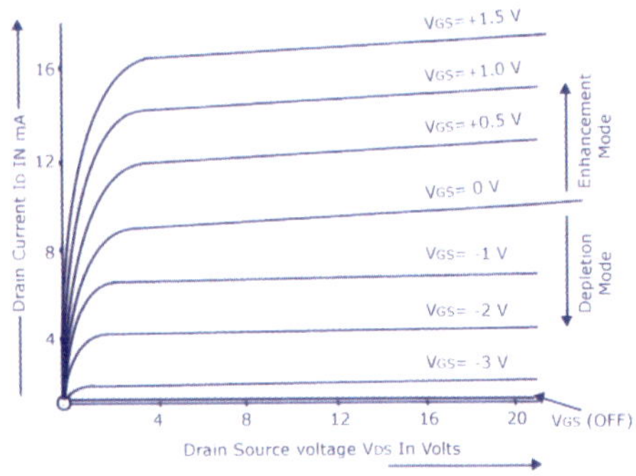

Drain Charecteristics

Drain characteristics

Typical *drain characteristics,* for various levels of gate-source voltage, of an N-channel MOSFET are shown in figure. The upper curves are for positive V_{GS} and the lower curves are for negative V_{GS}. The bottom drain curve is for $V_{GS} = V_{GS}$(OFF). For a specified drain-source voltage V_{DS}, V_{GS}(OFF) is the gate-source voltage at which drain current reduces to a certain specified negligibly small value, as shown in figure. This voltage corresponds to the pinch-off voltage V_p of JFET. For V_{GS} between $V_{GS\ (OFF)}$ and zero, the device operates in *depletion-mode* while for V_{GS} exceeding zero the device operates in *enhancement mode.* These drain curves again display an ohmic region, a constant-current source region and a cut-off region. MOSFET has two major applications: a constant current source and a voltage variable resistor.

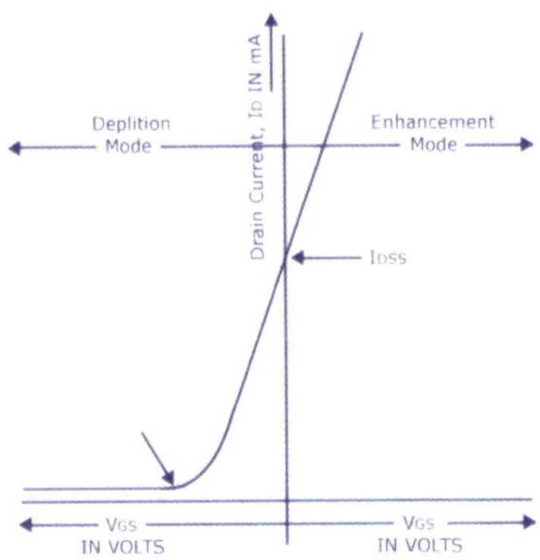

Transfer Characteristics

DEMOSFET-transfer characteristics

The transfer (or transconductance) characteristic for an N-channel DE-MOSFET is shown in figure. I_{DSS} is the drain current with a shorted gate. Since the curve extends to the right of the origin, I_{DSS} is no longer the maximum possibledrain current.

Mathematically, the curve is still part of a parabola and the same square-law relation exists as with a JFET. In fact, the depletion-mode MOSFET has a drain current given by the same transconductance equation as before, equation . Furthermore, it has the same equivalent circuits as a JFET. Because of this, the analysis of a depletion-mode MOSFET circuit is almost identical to that of a JFET circuit. The only difference is the analysis for a positive gate, but even here the same basic formulas are used to determine the drain current I_D, gate-source voltage V_{GS} etc.

The foregoing discussion is applicable in principle also to the P-channel DE-MOSFET. For such a device the sign of all currents and voltages in the characteristics must be reversed.

Schematic Symbols of DEMOSFET

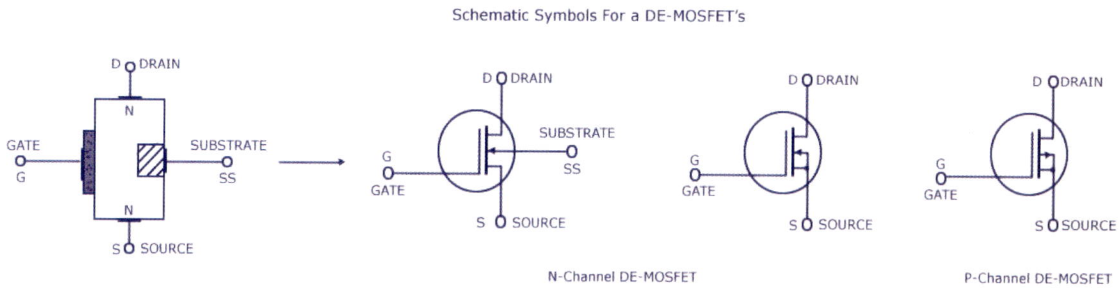

DEMOSFET-Schematic symbols

Figure shows the schematic symbol for a DE-MOSFET. Just to the right of the gate is the thin vertical line representing the channel. The drain lead comes out from the top of the channel and the source lead connects to the bottom. The arrow is on the P-substrate and points to the N-material. In some applications, a voltage can be applied to the substrate for added control of drain current. For this reason, some DE-MOSFETs have four terminal leads. But in most applications, the substrate is connected to the source. Usually the substrate is connected to the source internally by the manufacturer. This results in a three terminal device whose schematic symbol is shown in figure.

Schematic symbol for a three terminal P-channel DE-MOSFET device is shown in figure. The schematic symbol of a P-channel DE-MOSFET is similar to that of an N-channel DE-MOSFET, except that the arrow points outward.

Construction of n and p-channel Depletion Mode MOSFET

Figure shows the n-channel D-MOSFET. Drain (D) and Source (S) leads connect to the to n+-doped regions via an n-channel. n-channel is connected to the Gate (G) via a thin insulating layer of SiO_2. The n-doped material lies on a p-doped substrate that may have an additional terminal connection called SS.

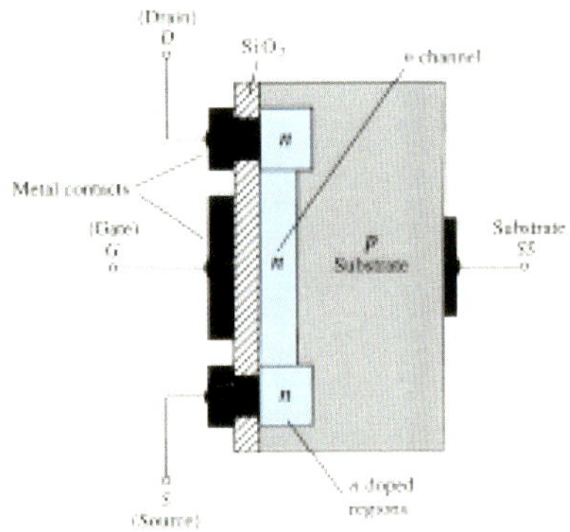

n-channel D-MOSFET

Depletion Mode MOSFET (D-MOSFET) Basic Operation: Figure shows the IV characteristics of n-channel D-MOSFET

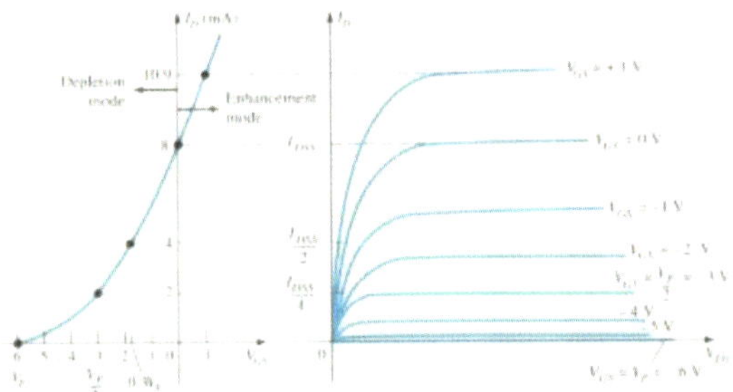

The transfer characteristics are similar to the JFET

Depletion Mode operation: When VGS = oV, ID = IDSS (normally ON condition)

When VGS < oV, ID < IDSS (OFF state)

When VGS > oV, ID > IDSS

p-Channel Depletion Mode MOSFET

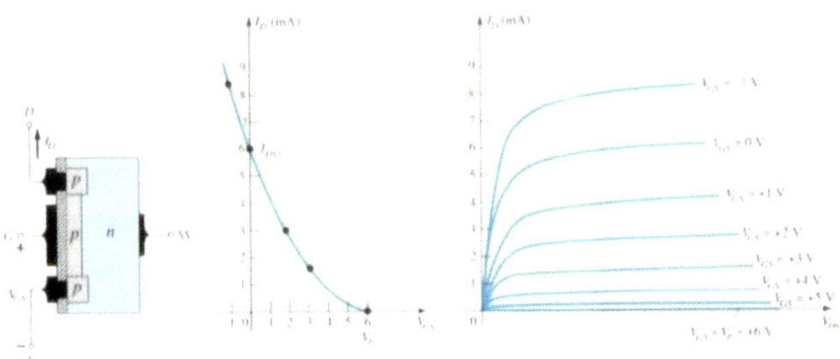

p-channel D-MOSFET

The p-channel Depletion mode MOSFET is similar to the n-channel except that channel is p-type and the voltage polarities and current directions are reversed

Theoretical Characteristics of a MOSFET

The relationship between voltages applied to the MOSFET electrodes and the current that flows from source to the drain, is described in this topic. An ideal model of an n-channel MOSFET is shown in figure. Let source and substrate are connected.

Consider strong inversion condition [No drain current flow in a MOS so system in thermal equilibrium i.e. VD =0], the surface potential.

Consider n-channel MOSFET

Total charge induced in semiconductor at y

$$Q_S(y) = Q_n(y) + Q_{sc}(y)$$

where Q_n is the charge per unit area in inversion region and Q_{sc} is the charge per unit area in depletion region.

Applied VG is partly divided in oxide layer and semiconductor as

$$\begin{aligned} V_G &= V_o + \psi_s(y) \\ &= -\frac{Q_s(y)}{C_o} + \psi_s(y) \end{aligned}$$

where, V_o is voltage drop across oxide layer.

In case of non ideal MOS capacitor, with flat-band voltage V_{FB}, above equation is modified as

$$V_G - V_{FB} = -\frac{Q_s(y)}{C_o} + \psi_s(y)$$

Put value of $Q_s(y)$ from above equation, we have

$$C_o(\psi_s(y) - V_G + V_{FB}) = Q_n(y) + Q_{sc}(y)$$

Solving for $Q_n(y)$, we have

$$Q_n(y) = C_o[\psi_s(y) - V_G + V_{FB}] - Q_{sc}(y)$$

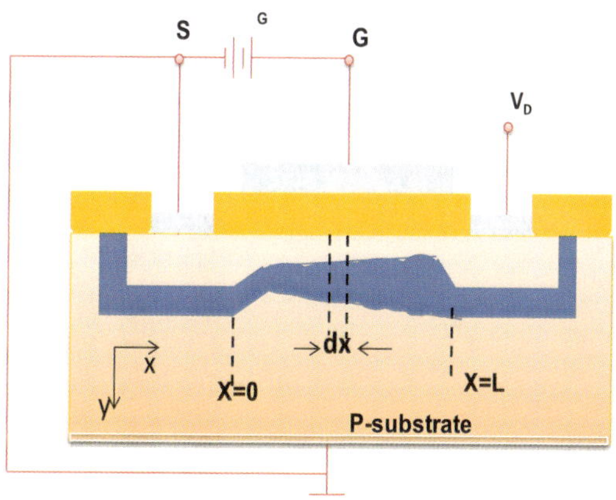

Schematic of n-channel MOSFET

At onset of inversion,

$$\psi_s(y) = 2\,\psi_B$$

Under practical operating condition, space charge region in MOS transistor is not in equilibrium i.e. when VD > 0, the surface potential will increase with y [due to potential distribution in semiconductor]. If V(y) is channel potential at any point y for a drain voltage VD, then to a good approximation the surface potential for strong inversion is

$$\psi_s\big]_{inv.}(y) = 2\,\psi_B + V(y)$$

Also using depletion approximation, the charge per unit area at strong inversion (as calculated while discussing MOS diode) is

$$Q_{sc}(y) = -q\,N_A\,W_m \approx -\sqrt{2\in_s q\,N_A\left[V(y) + 2\,\psi_B\right]}$$

Since V(y) increases with y, $Q_{sc}(y)$ increases with y from S to D. Put $\psi_s(y)$ and $Q_{sc}(y)$ from above equation, we get,

$$Q_n(y) = C_0\left[2\psi_B(y) + V(y) - V_G + \sqrt{2\in_s q\,N_A\left[V(y) + 2\,\psi_B\right]}\right]$$

Consider a small element dy in the channel. For a constant electron mobility (μ_n), the resistance of small element dy of channel is (using above eqn),

$$dR = \frac{-dy}{z\,\mu_n\,Q_n(y)}$$

The voltage drop across element of resistance dR due to drain current (I_D) is

$$dV = I_D\,dR$$

Using above eqn., we have

$$I_D\,dy = -\mu_n\,z\,Q_n(y)\,dV$$

Substitute, $Q_n(y)$ from above equation and integrating from y = 0 to L and V = 0 to V_D

$$I_D\,L = -\mu_n\,z\,C_o\left\{\int_0^{V_D}\left(2\psi_B(y) + V(y) - V_G + V_{FB}\right)dV + \frac{\sqrt{2\in_S q\,N_A}}{C_o}\int_0^{V_D}\sqrt{\left[2\,2\psi_B(y) + V(y)\right]}\,dV\right\}$$

$$I_D = \frac{+\mu_n\,z\,C_o}{L}\left\{\left(V_B - V_{FB} - 2\psi_B - \frac{V_D^2}{2}\right)V_D - \frac{\sqrt{2\in_s q\,N_A}}{C_o}\,\frac{2}{3}\left[(2\,\psi_B + V_D)^{\frac{3}{2}} - (2\,\psi_B)^{\frac{3}{2}}\right]\right\}$$

For ideal MOSFET remove the contribution from VFB in equation (10)

Now, discussing above equation in different conditions

(A) LINEAR REGION

For small VD such that VD << VG - VT or

$$V_D \ll V_G - (V_{FB} + 2\psi_B)$$

Neglecting the contribution from $VD^2/2$

$$I_D = \frac{\mu_n zC_o}{L}\left\{(V_G - V_{FB} - 2\Psi_B)V_D - \frac{\sqrt{2\in_S q N_A}}{C_o}\frac{2}{3}\left[(2\psi_B)^{\frac{3}{2}}\left(1+\frac{V_D}{2\psi_B}\right)^{\frac{3}{2}} - (2\psi_B)^{\frac{3}{2}}\right]\right\}$$

$$\frac{\mu_n zC_o}{L}\left\{(V_G - V_{FB} - 2\Psi_B)V_D - \frac{\sqrt{2\in_S q N_A}}{C_o}\frac{2}{3}\left[(2\Psi_B)^{\frac{3}{2}}\left(1+\frac{3}{2}\frac{V_D}{2\Psi_B}\right) - (2\Psi_B)^{\frac{3}{2}}\right]\right\}$$

$$= \frac{\mu_n zC_o}{L}\left[V_G - V_{FB} - 2\psi_B - \frac{\sqrt{4\in_s q N_A \Psi_B}}{C_o}\right]V_D$$

$$= \frac{\mu_n zC_o}{L}[V_G - V_T]V_D$$

where,

$$V_T = V_{FB} + 2\psi_B + \frac{\sqrt{2\in_s q N_A 2\Psi_B}}{C_o}$$

By plotting I_D versus V_G (for small V_D), V_T can be obtained from linear extrapolation to V_G axis.

For n-channel MOSFET, VFB is small negative and other two terms of equation are positive, => $V_T > 0$ (turn- on voltage).

From equation (12) => $I_D \alpha V_D$ for small V_D. In linear region from above equation we have:

Channel Conductance

$$g_D = \left.\frac{\partial I_D}{\partial V_D}\right]_{V_G = Const} = \frac{\mu_n z C_o}{L}[V_G - V_T]$$

Trans Conductance

$$g_m = \left.\frac{\partial I_D}{\partial V_G}\right]_{V_D = Const} = \frac{\mu_n z C_o}{L} V_D$$

(B) SATURATION REGION

When $V_D > V_{sat}$, then at Drain end the net voltage across MOS is less than or equal to V_T i.e. number of mobile electrons in channel at the Drain are reduced drastically(Pinch off point).

At Source side: $V(y) = 0$, $Q_n(y) = Q_n$ at $y = 0$

At Drain side: $V(y) = V_{Dsat}$, $Q_n(y) = 0$ at $y = L$

Putting source side condition in above equation i.e. for $Q_n(y)$, we have

$$0 = C_o\left[2\Psi_B + V_{Dsat} - V_G + V_{FB}\right] + \sqrt{2\in_s q N_A\left[V_{Dsat} + 2\Psi_B\right]}$$

$$2\Psi_B + V_{Dsat} - V_G + V_{FB} = -\theta\sqrt{V_{Dsat} + 2\Psi_B}$$

where,

$$\theta = \frac{\sqrt{2\in_s q N_A}}{C_o}$$

Find out V_{Dsat} from above equation and substitute $V_D = V_{Dsat}$ in equation, we can obtain expression for I_D in saturation region i.e. I_{Dsat}.

However, when N_A is very low and oxide layer thickness is small (i.e. Co is large), then q << 1. Here voltage drop in oxide is negligibly small compared to $2\Psi_B$.

Thus equation becomes

$$V_{Dsat} = V_G + V_{FB} - 2\Psi_B \approx V_G - V_T$$

Put value of $V_D = V_{Dsat}$ in equation, we have $I_D = I_{Dsat}$ as

$$I_{Dsat} = \frac{\mu_n z C_o}{L}\left\{\left[V_G + V_{FB} - 2\Psi_B - \left(\frac{(V_G - V_T)}{2}\right)\right](V_G - V_T) - \frac{2}{3}\theta\left[(2\Psi_B + V_G - V_T)^{\frac{3}{2}} - (2\Psi_B)^{\frac{3}{2}}\right]\right\}$$

$$= \frac{\mu_n z C_o}{L}\left\{(V_G - V_T)^2 - \frac{(V_G - V_T)^2}{2}\right\}$$

For $\theta \ll 1$, second term is neglected and $V_T = V_{FB} + 2\Psi_B$))

$$= \frac{{}_n z C_o}{}(V_G - V_T)$$

This is square law dependence of I_D on V_G [same as in JFET], and is valid for all $V_D > V_G - V_T$. Here I_D is independent of V_D.

Channel Conductance in Saturation Region:

$$g_D = \frac{\partial I_D}{\partial V_D} = 0$$

Trans Conductance in Saturation Region

$$g_m = \left.\frac{\partial I_D}{\partial V_G}\right]_{V_{D-Const}} = \frac{\mu_n z C_o}{L}(V_G - V_T)$$

Thus Channel conductance (gD) in linear region (eqn.) is same as Trans Conductance (gm) in saturation region (eqn.) for MOSFET.

Frequency Limit for MOSFET

The Small signal ac equivalent circuit for MOSFET is shown in figure.

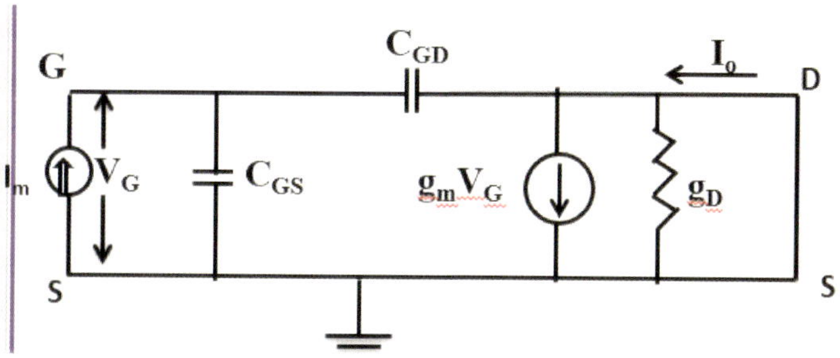

Small signal ac equivalent circuit for MOSFET.

At higher frequency, consider capacitive coupling between device terminals as C_{GS} -MOS capacitance given by dimensions of oxide layer Z,L,C

Where

$$\frac{C_o\, C_j}{C_o \quad C_j}$$

and $C_{GS} = C_o\, z\, L$

C_{GD} - G-D capacitance provide undesirable feedback between I/O.

Consider, f_T = frequency at which ratio of I_{output} to I_{input} becomes unity i.e. device can no longer amplify input signal when output of MOSFET is short circuited

I_m = Input curreent

From the Circuit Shown Above

$$I_m = 2\pi f_T \left[C_{GS} + C_{GD}\right] V_G = 2\pi f_T C_G V_G$$

Where,

$$C_G = C_{GS} + C_{GD}$$

From the Output of Circuit

$$I_0 + wC_{GD} V_G = g_m V_G$$

$$I_0 = g_m V_G - wC_{GD} V_G$$

where I_o=output current

Here, I_m is equal to I_o, thus

$$2\pi f_T C_G V_G = g_m V_G$$

Neglecting contribution from wCGD

$$f_T = \frac{g_m}{2\pi C_G} = \frac{\mu_n z C_o V_D}{L 2\pi C_o z L} = \frac{\mu_n V_D}{2\pi L^2}$$

where Cois the capacitance of oxide and d is thickness of oxide layer. Thus, L should be Low for fast switching.

MESFET

Three-terminal devices such as MESFETs are unipolar, i.e. they use only one type of carrier. This makes optimization easier than if two types of carriers have to be considered. Most of the materials used in three-terminal devices, such as GaA8, InP, etc. have high mobility compared with silicon. A large variety of forms and applications are possible. Monolithic integration of devices on a semi-insulating substrate is easier to accomplish than for two-terminal devices. The availability ofthree terminals means that uni-directional amplification is possible, and circulators are not usually needed A partial list of the type of applications which have been implemented contains amplifiers (low-noise and power-), limiters, harmonic frequency multipliers, mixers, phase-shifters and oscillators. The present-day microwave three-terminal devices of course have as their ancestor the bipolar transistor which was invented in the late 1940's. It took a long time before transistors were fast enough to be useful at microwave frequencies, however. Not until 1965 did germanium bipolar transistors start to be employed at L-band (1 -2 GHz). The idea of a field-effect transistor goes back to Shockley (1952), but took even longer to realize for microwave frequencies. Early field-effect transistors had cut-off frequencies which were inferior to those ofbipolars. For example, thefirst microwave FET using GaAs was reported on in 1968, and had a maximumoscillation frequency (fmax)*= 3 GHz, but already in 1970, Drangeid et al. (1970) had developed a GaAB FET with fmax = 30 GHz, which far surpassed the performance of silicon bipolar devices. Since that time, oscillation has been observed above 100 GH., and noise figures of GaAs. FETs have been shown to be superior at essentially all microwave frequencies. It is now mainly the HFET, a development of the 1980's, which can compete in terms of noise figure.

Functional Architecture

The MESFET differs from the common insulated gate FET or MOSFET in that there is no insulator under the gate over the active switching region. This implies that the MESFET gate should, in transistor mode, be biased such that one does not have a forward-conducting metal–semiconductor diode instead of a reversed-biased depletion zone controlling the underlying channel. While this restriction inhibits certain circuit possibilities as the gate must remain reverse-biased and cannot therefore exceed a certain voltage of forward bias, MESFETs analog and digital devices work reasonably well if kept within the confines of design limits. The most critical aspect of the design is the gate metal extent over the switching region. Generally the narrower the gate modulated carrier channel the better the frequency handling abilities, overall. Spacing of the source and drain with respect to the gate, and the lateral extent of the gate are important though somewhat less critical design parameters. MESFET current handling ability improves as the gate is elongated laterally, keeping the active region constant, however is limited by phase shift along the gate due to the transmission line effect. As a result, most production MESFETs use a built up top layer of low resistance metal on the gate, often producing a mushroom-like profile in cross section.

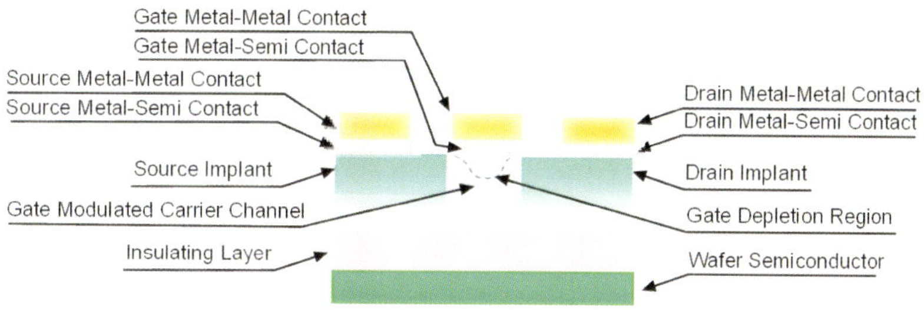

MESFET schematic.

Schematic and Fabrication of MESFET

Metal Semiconductor Field Effect Transistor (MESFET): Figure (a) shows the schematic diagram of MESFET. MESFET consists of a thin layer of n-type semiconductor on a semi-infinite substrate. Ohmic metal contacts are provided at two ends of n-type semiconductor to serve as 'source' and 'drain'. A Schottky metal contact is also with semiconducting layer to act as 'Gate' a shown in figure (a). Carrier flow across the channel (n-type layer) from 'source' to 'drain'. Thickness of the channel is controlled by variation in the depletion region underneath the metal gate this can vary the current between source and drain. Figure (b) shows the process steps used for the fabrication of a MESFET. This figure summarizes the process starting from the initial GaAs wafer and the ending with the final device. An important process of the MESFET fabrication process is the Mask and wafer alignment and the exposure. A special technique used to make sure of the alignment of all of the dies on the wafer is the Split-Field alignment. This technique allows us to view two different sites at a time and align them.

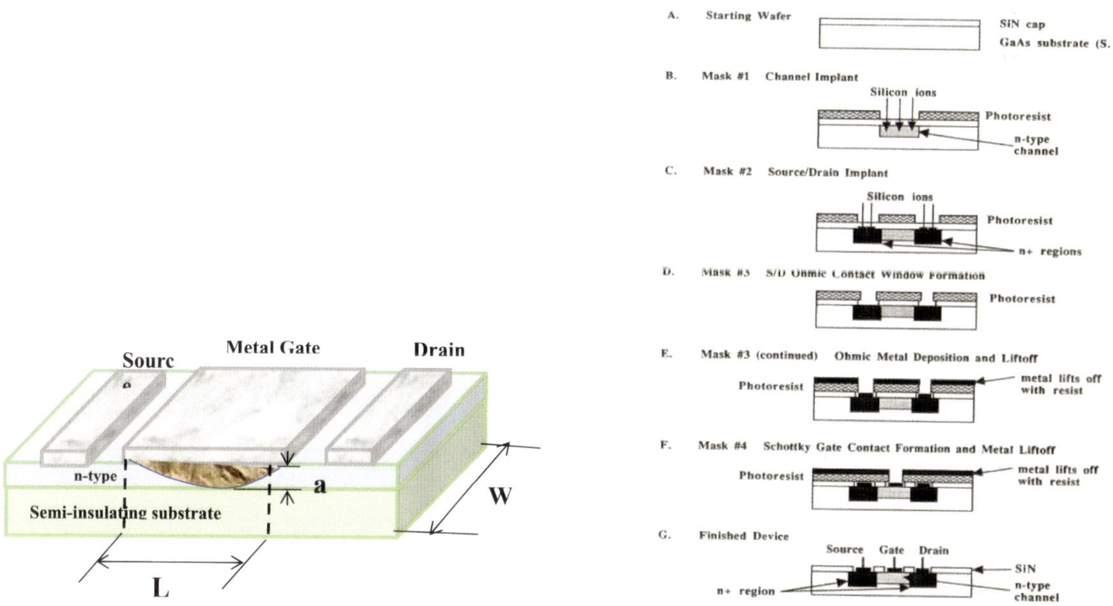

(a) Schematic diagram of MESFET and (b) fabrication steps used for making MESFET

In the case of JFET channel conductance is there with no biasing and 'Source' is not isolated from the 'Drain'. Thus, channel is available for the conduction of electrons and this device is 'Normally On' device or 'Depletion mode' device. But in the case of MESFET, it is possible to design the device with both 'Normally On' and 'Normally Off' features.

Operation of MESFET

ENHANCEMENT MODE: Normally OFF

In enhancement mode, width of depletion region in n-type semiconductor (underneath Schottky contact) is wide is wide enough without applied voltage which can pinch off the channel as shown in Figure. In enhancement mode, the thickness (t) of n-type semiconducting (n-GaAs) layer is less than width (a) of depletion region due to Schottky contact. Thus, MESFET is considered naturally in "OFF" state in the enhancement mode. When some positive voltage bias (V_T>0) is applied across the 'Gate' with respect to 'Source', Schottky diode becomes forward bias, and the width of depletion region starts reducing. This results in the opening of the channel for the flow of charge carriers and makes the channel conducting. Hence, a large drain current (I_D>0) can flow for applied positive bias at drain (V_D) as shown in figure.

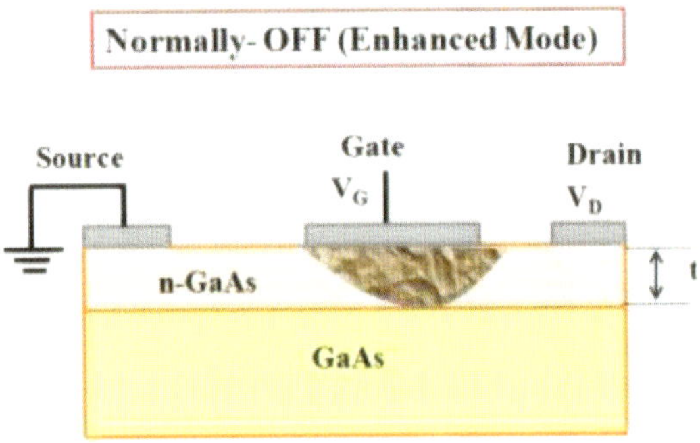

MESFET in Enahcned Mode

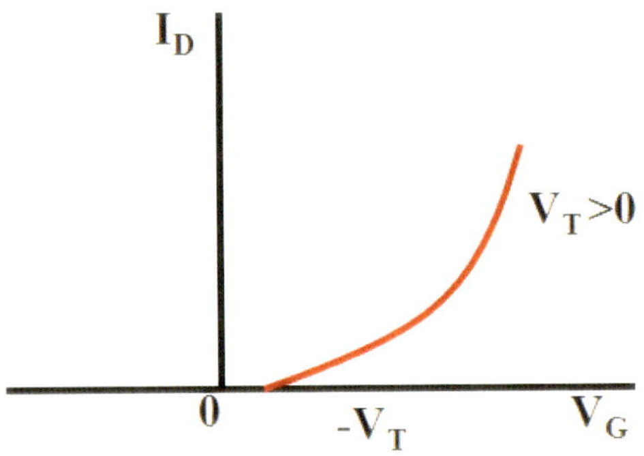

Transfer Characteristics of MESFET in Enahcned Mode

DEPLETION MODE: Normally ON

If the width (a) of depletion region in the n-type semiconductor (underneath Schottky Gate) does not extend to the entire thickness (t) of p-type substrate, the MESFET is in depletion-mode as shown in Figure. Here, depletion width $a < t$ (thickness of n-type layer). Under thermal equilibrium condition, the channel is always available for the flow of charge carriers. Thus, when the 'Gate'-to-'Source' voltage is not applied, the MESFET becomes conductive or in "ON" state.

To switch-off the MESFET, a negative voltage ($V_T < 0$) is required at 'Gate' with respect to the source. The negative applied voltage at 'Gate' makes Schottky diode in reverse bias configuration, and resulting in an increase in the depletion width, such that it 'pinches off' the channel as shown in figure. Thus, drain current (ID) flows when $V_G \geq V_{th}$, and becomes zero for VG = Vth, where $V_T < 0$.

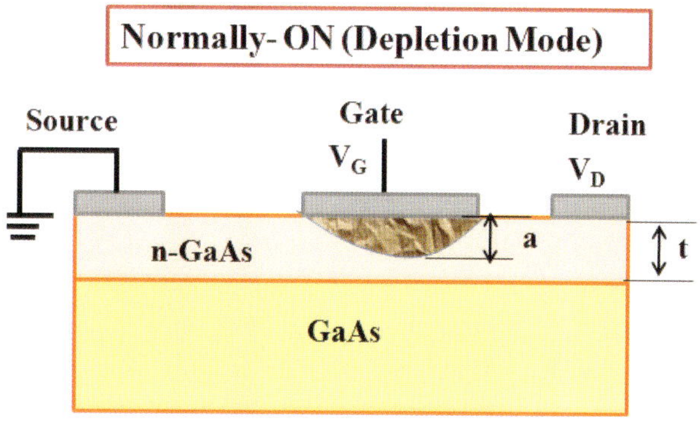

MESFET in Depletion Mode

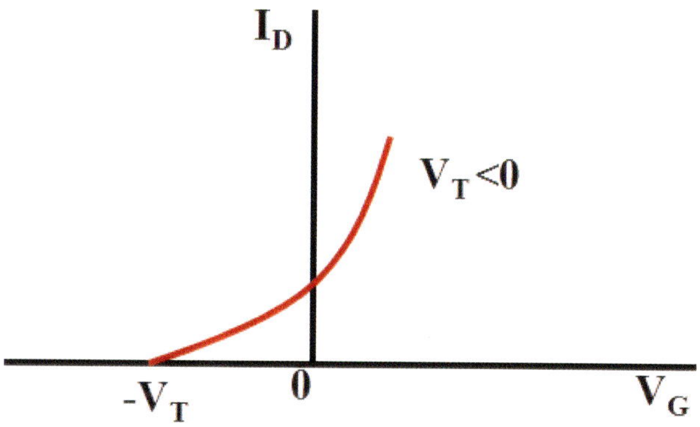

Transfer Characteristics of MESFET in Depeltion Mode

CHARACTERSTICS

The 'Source' is grounded and act as reference voltage (zero) for MESFET device. The Drain is kept at positive voltage with respect to source. The metal 'Gate' which is in reverse biased mode with the ntype semiconductor forms a Schottky diode, so there is no gate current, except for some leakage current in small amount. The applied gate voltage can modulate the depletion width, due to which the conductivity of the channel is modulated.

When the drain (VD) is kept small, the depletion width (χ_{dept}) in the channel can be calculated from the Schottky diode theory as explained below:

$$\chi_{dept} = \sqrt{\frac{2 \in_{sc}}{qNd}(V_i - V_G)}$$

where *Vi* is the potential barrier on the semiconductor side at Schottky diode, Nd is the concentration of donor impurity in n-type semiconductor and ∈*sc* is the permittivity of the semiconductor.

At threshold voltage, $V_G = V_{th}$, the width of depletion region is *xdepl=a*. Using the equation 1, the threshold voltage could be estimated as:

$$V_{Th} = V_i - \frac{qN_d a^2}{2 \in_{sc}}$$

This threshold voltage can be either negative or positive which depends upon the thickness of the n-type semiconducting layer, the doping concentration (*Nd*) or the metal which is used to form the Schottky gate. For positive threshold voltage, the MESFET is in enhancement mode and for negative value of Vth, it is in depletion mode.

Figure shows the variation of Drain current (ID) as a function of Gate voltage (VG) and Drain voltage (VD) in both the Enhancement and Depletion modes. Since the channel provides resistance to the flow of charge carriers from source to drain, it gives rise to a major potential drop along the channel. Consider, VD is the applied Drain voltage with respect to the source, and V(y) is the voltage at any point y (from source) in the channel. The applied voltage is uniformly distributed in the channel and varying from *V(y=0)=VS=0* at the source to *V(y=L)=VD* at the drain.

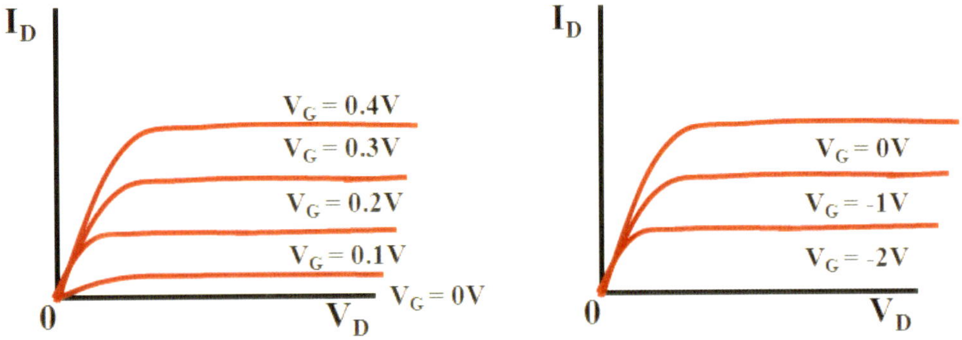

Normally Off (Enhancement mode) Normally On(Depletion mode)

I-V characteristic curve of MESFET in Ehnacement mode and Depletion mode

In every section located vertically at a position '*y*', the reverse bias across the Schottky junction is equal to *VG-V(y)*, due to which the depletion width varies as a function of '*y*'. the depletion width will be

$$\chi_{dept}(y) = \sqrt{\frac{2 \in_{sc}}{qN_d}(V_i - V_G + V(y)}$$

Consider a small element dy of the channel which offer a small resistance 'dR' to the Drain current ID. The Drain current is estimated by integrating Ohm's law from source to drain:

$$dV = I_D dR = \frac{I_D dy}{q\mu N_d w(a - x_{dept}(y)}$$

where, W is the width of device as shown in Figure and dV is the voltage drop across small element 'dy' of the channel.

Replacing *xdepl(y) from equation*:

$$dV = \frac{I_D dy}{q\mu N_d w\left(a - \sqrt{\frac{2 \in_{sc}}{qN_d}(v(y) + V_i - V_G}\right)}$$

Integrate equation and after rearranging we have:

$$q\mu N_d W\int_o^{vD}\left(a-\sqrt{\frac{2\in_{sc}}{qN_d}(v(y)+V_i-V_G}\right)dv = I_D\int_o^l dy$$

$$I_D = g_o\left((V_D - \frac{2\left((V_D+V_i-V_G)^{\frac{3}{2}}-(V_i-V_G)^{\frac{3}{2}}\right)}{3\sqrt{V_p}}\right)$$

where, $g_o = \frac{q\mu N_d wa}{L}$ *and* $V_P = \frac{qN_d a^2}{2\in_{sc}}$

These equations hold if the V_D is small and channel is not pinched-off. It may be noted from equation that I_D increases with increasing V_D for fixed value of V_G (figure). The saturation value of Drain current, $I_D = I_{Dsat}$; occurs when $x_{depl}(L)=a$, at which point $V_D=V_{Dsat}=V_{P-}V_i+V_G$.

Saturation Drain Voltage is $V_{Dsat}=V_P-V_i+V_G=V_{G-}V_{TH}$.

The drain saturation current could be estimated by replacing V_D by the saturation drain voltage V_{Dsat}, in equation:

$$I_{Dsat} = g_o\left(\frac{V_p}{3}+\frac{2(V_i-V_G)^{\frac{3}{2}}}{3\sqrt{V_p}}-V_i+V_G\right)$$

The Drain saturation current (I_{Dsat}) is independent of applied Drain voltage. The transconductance in saturation region is given by:

$$g_m = \left.\frac{\partial I_D}{\partial V_G}\right|_{V_D=constant} = g_o\left(1-\sqrt{\frac{(V_i-V_G)}{V_p}}\right)$$

Advantages and Disadvantages

MESFET has advantage over MOSFET that the carriers have higher mobility in the channel. Their mobility is less than half of the mobility of bulk counter material, as the charge carriers are situated in the inversion layer, which extends into the oxide layer,. However, as the depletion region starts separating the charge carriers from the surface, their mobility starts increasing and approaches to bulk material which leads to increase in current, transconductance and transit frequency of the MESFET. The presence of Schottky metal gate limits the forward bias voltage on the gate to the turn-on voltage of the Schottky diode which is the disadvantage of the MESFET structure. The value of this turn-on voltage for GaAs Schottky diodes is ~0.7 V and the threshold voltage should be lower than the turn-on voltage. It is more difficult to fabricate circuits containing a large number of MESFET in the enhancement-mode. The advantage of higher transit frequency of the MESFET provides a superior microwave amplifier. Moreover, MESFET in depletion mode is used as it provides a larger current and larger transconductance so that threshold control is not a limiting factor. GaAs is most commonly preferred over silicon for fabricating MESFET as the electron mobility at room temperature is more than 5 times larger, while the peak electron velocity is about twice that of silicon. Also GaAs substrate resolves the problem of absorbing microwave power due to free carrier absorption.

Small-Signal Equivalent Circuit Model

The small-signal equivalent circuit of a MESFET is shown in Figure. This model is quite similar to other well-known models of FETs. Measured values for the elements in a circuit model must be obtained by first measuring the small-signal S-parameters of the device, and then fitting these to agree with the equivalent circuit model. and considerable errors may result. Recently introduced microprobes contact the device directly on its. "chip" and allow determination of the equivalent circuit parameters with greaterb accuracy. It is important to note that DC-measurements of especially Rds but also gm often give resultsb which are very noticeably different from measurements at microwave frequencies. Frequency-dispersion of these parameters is observed in the frequency-range of Hz-kHz, and has been correlated with traps, either at the surface, or at the buffer-layer/active layer interface. One thus must be quite careful in using DC data, instead of S-parameter data, for some of the parameters.

g_m - Since the drain current is delayed by a time = the transit-time, with respect to the gate voltage, we must also use a phase-factor, $\exp(-j\omega\tau o)$, to take this into account (not included in the PHS model). We also must distinguish the external $g_{m,e}$ (measured from the terminals) from the intrinsic value, $g_{m,o}$. These are related as follows:

$$g_{m,e} = \frac{g_{m,o}}{1 + g_{m,o}Rs};$$

C_{dg}, and C_{gs} are capacitances from the gate to the channel (source end and drain end, respectively). We of course can not exactly pin-point the channel location of the lower capacitance electrode, but the success of the equivalent circuit model depends on the extent to which measured S-parameter data can be fitted to the predictions based on the model. C_{de} also ends at the source end of the channel. The electric field lines in C_{gs}, are primarily routed through the semi-insulating substrate.

Other series resistances Rs, Rd connect the "actual channel" with the source and drain contacts, respectively. Also, the gate fingers have a resistance, Rg, Rds, is the actual channel resistance.

Cut-Off Frequencies: Several different cut-off frequencies can be defined on the basis of the equivalent circuit model.

MODFET

Modulation doped field effect transistor (MODFET) is a heterojunction device which has the desirable properties of high speed operation and low-power dissipation. MODFET is also called high electron mobility transistor (HEMT) or two-dimensional electron gas field effect transistor or selectively doped heterojunction transistor(SDHT). Frequently, it is referred to by a general name HFET (heterojunction field-effect transistor).

The unique feature of the MODFET is the heterostructure in which the wide energy gap material is doped and carriers diffuse to the undoped narrow energy gap material where the channel is formed. The net result of this modulation doping is that the carriers in the undoped heterointerface are spatially separated from the doped region and have extremely high mobilities because there is no impurity scattering. Carrier transport parallel to the layers of a superlattice was first considered by Esaki and Tsu in 1969. The development of MBE and MOCVD technologies in the 1970s made heterostructures, quantum wells, and superlattices practical. Dingle et al. first demonstrated the enhanced mobility in the AlGaAs-GaAs modulation-doped superlattice in 1978. Stormer et al. subsequently reported a similar effect using a single AlGaAs-GaAs heterojunction in 1979. This effect was applied to field-effect transistors by Mimura et al. in 1980, and later by Delagebeaudeuf et al. in the same year. Since then, the MODFET has been the subject of major research activities, and promises to be an alternative to MESFETs in high-speed circuits.

Structure of MODFET

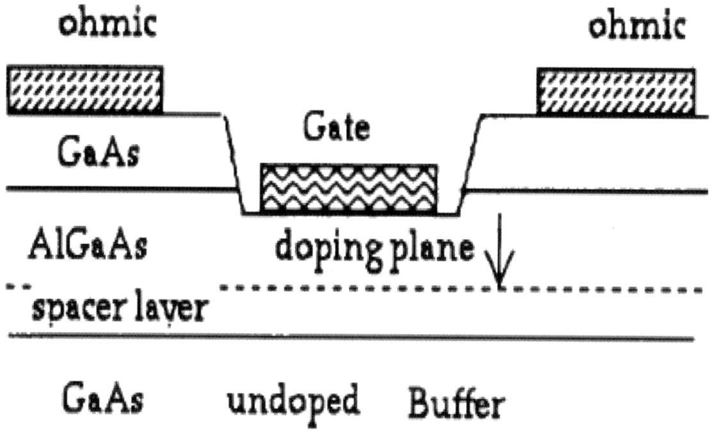

Structure of a typical MODFET device

The structure of a typical MODFET device is shown in figure. There are four layers in this structure: an intrinsic GaAs buffer layer, an undoped Al GaAs spacer layer, a doped Al GaAs donor layer, and the cap layer.There are two ohmic contacts and one gate electrode for providing the electrical connections of the transistor. MODFET works on the principle of modulating the concentration of a two-dimensional carrier gas by a gate potential. MODFET consist of the active region which is generally n-type grown on the top of an undoped buffer layer having low bandgap. This precedes growth of thc high bandgap dopcd material.

Consider the example of AlGaAs/GaAs MODFETs in which the formation of electron gas is due to the diffusion of the electrons which are donated by n-type AlGaAs to GaAs having lower bandgap. Towards aiming in the enhancement of the low temperature electron mobility, the donors are set back away from the interface. A setback layer is formed called as intrinsic layer which leads to weaker Coulombic interaction between the electrons in the donors in AlGaAs and electron gas. Enhancement in electron confinement can be obtained by raising AlAs mole fraction (in AlGaAs) which leads to a large discontinuity in the conduction band between GaAs and AlGaAs. Another way of electron confinement is by introducing an AlGaAs layer between the electron sheet and GaAs buffer. The resultant discontinuity in the conduction band at heterojunction interface resulting in the confinement and localization of those diffused electrons in the material having low bandgap determines the number of electrons that can be sustained at the interface, referred as two-dimensional electron gas (2DEG). This is clearly explained in figure. When the gate voltage in figure is made sufficiently high leading to the state when the source-drain channel is no longer depleted. At this instant, there is conduction by a 2DEG in the undoped material. There is no inhibition on electron mobility and saturation velocity by impurity scattering because undoped material has no added extra donor atoms. Consequently, it is also known that the Hall mobility is limited only by phonon scattering and the conduction mobility is equal to the Hall mobility due to the degeneracy in the GaAs conduction band. The spacer layer initially formed leads to the rise in the channel mobility by shielding the 2DEG from ionized impurities. Now, regarding the electrical contact, two ohmic contacts are provided on the surface of source and the drain. A Schottky gate contact is used to modulate the current flow between these two ohmic contacts.

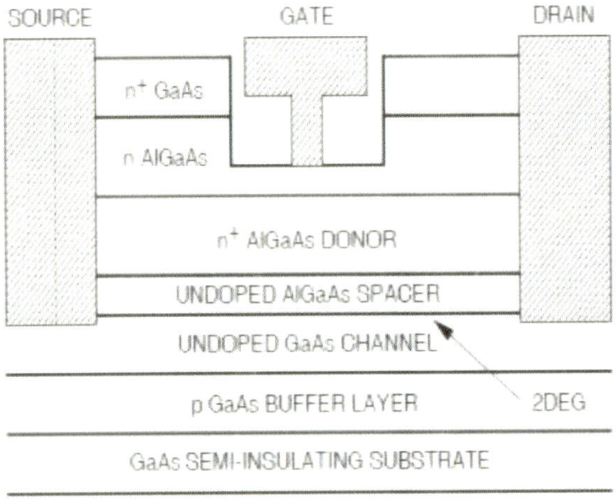

AlGaAs MODFET schematic

Many studies have been done on n-type MODFET devices. The main component of MODFET is the specialized PN junction present inside it and consists of a junction that uses different materials for either side of the junction. From figure, the buffer layer, generally GaAs, is epitaxially grown on the substrate in order to isolate defects from the substrate and to create a smooth surface upon which to grow the active layers of the transistor.

Fabrication

While fabricating MODFET, firstly an intrinsic layer of gallium arsenide of one micron thick is deposited on the semi-insulating gallium arsenide layer. After this, very thin layers of aluminium gallium arsenide of 30 and 60 Angstroms thick were deposited in order to introduce the separation between hetero-junction interface and the doped aluminium gallium arsenide region. The doped layer of aluminium gallium arsenide, about 500 Angstroms thick, is deposited above this layer and its thickness needs to be precise which in turn requires special techniques.

There are two main structures that are generally used i.e. first is self-aligned ion implanted structure and the other is recess gate structure. The electrodes mainly gate, drain and source are made up of metallic contacts in the self-aligned ion implanted structure. Sometimes, source and drain contacts may be made from germanium and gate is generally made from titanium forming a small reverse biased junction similar to that of the GaAsFET.

In the case of recess gate structure, a new layer of n-type gallium arsenide is introduced to fabricate the drain and source contacts and selected areas are etched. The thickness of the layer under the gate is highly important in deciding the threshold voltage of the FET. So, towards aiming this the size of the gate, and the channel is kept to be very small. The typical thickness of the gate is only 0.25 microns or less than this which is beneficial for the device to have a very good high frequency performance.

Conceptual Analysis

HEMTs are heterojunctions. This means that the semiconductors used have dissimilar band gaps. For instance, silicon has a band gap of 1.1 electron volts (eV), while germanium has a band gap of 0.67 eV. When a heterojunction is formed, the conduction band and valence band throughout the material must bend in order to form a continuous level.

HEMTs' exceptional carrier mobility and switching speed come from the following conditions. The wide band element is doped with donor atoms; thus it has excess electrons in its conduction band. These electrons will diffuse to the adjacent narrow band material's conduction band due to the availability of states with lower energy. The movement of electrons will cause a change in potential and thus an electric field between the materials. The electric field will push electrons back to the wide band element's conduction band. The diffusion process continues until electron diffusion and electron drift balance each other, creating a junction at equilibrium similar to a p-n junction. Note that the undoped narrow band gap material now has excess majority charge carriers. The fact that the charge carriers are majority carriers yields high switching speeds, and the fact that the low band gap semiconductor is undoped means that there are no donor atoms to cause scattering and thus yields high mobility.

An important aspect of HEMTs is that the band discontinuities across the conduction and valence bands can be modified separately. This allows the type of carriers in and out of the device to be controlled. As HEMTs require electrons to be the main carriers, a graded doping can be applied in one of the materials making the conduction band discontinuity smaller, and keeping the valence band discontinuity the same. This diffusion of carriers leads to the accumulation of electrons along the boundary of the two regions inside the narrow band gap material. The accumulation of electrons leads to a very high current in these devices. The accumulated electrons are also known as 2DEG or two-dimensional electron gas.

The term "modulation doping" refers to the fact that the dopants are spatially in a different region from the current carrying electrons. This technique was invented by Horst Störmer at Bell Labs.

Explanation

To allow conduction, semiconductors are doped with impurities which donate either mobile electrons or holes. However, these electrons are slowed down through collisions with the impurities (dopants) used to generate them in the first place. HEMTs avoid this through the use of high mobility electrons generated using the heterojunction of a highly doped wide-bandgap n-type donor-supply layer (AlGaAs in our example) and a non-doped narrow-bandgap channel layer with no dopant impurities (GaAs in this case).

The electrons generated in the thin n-type AlGaAs layer drop completely into the GaAs layer to form a depleted AlGaAs layer, because the heterojunction created by different band-gap materials forms a quantum well (a steep canyon) in the conduction band on the GaAs side where the electrons can move quickly without colliding with any impurities because the GaAs layer is undoped, and from which they cannot escape. The effect of this is to create a very thin layer of highly mobile conducting electrons with very high concentration, giving the channel very low resistivity (or to put it another way, "high electron mobility").

Electrostatic Mechanism

Since GaAs has higher electron affinity, free electrons in the AlGaAs layer are transferred to the undoped GaAs layer where they form a two dimensional high mobility electron gas within 100 ångström (10 nm) of the interface. The n-type AlGaAs layer of the HEMT is depleted completely through two depletion mechanisms:

- Trapping of free electrons by surface states causes the surface depletion.
- Transfer of electrons into the undoped GaAs layer brings about the interface depletion.

The Fermi level of the gate metal is matched to the pinning point, which is 1.2 eV below the conduction band. With the reduced AlGaAs layer thickness, the electrons supplied by donors in the AlGaAs layer are insufficient to pin the layer. As a result, band bending is moving upward and the two-dimensional electrons gas does not appear. When a positive voltage greater than the threshold voltage is applied to the gate, electrons accumulate at the interface and form a two-dimensional electron gas.

Versions of HEMT

By Growth Technology: pHEMT and mHEMT

Ideally, the two different materials used for a heterojunction would have the same lattice constant (spacing between the atoms). In practice, the lattice constants are typically slightly different (e.g. AlGaAs on GaAs), resulting in crystal defects. As an analogy, imagine pushing together two plastic combs with a slightly different spacing. At regular intervals, you'll see two teeth clump together. In semiconductors, these discontinuities form deep-level traps and greatly reduce device performance.

A HEMT where this rule is violated is called a pHEMT or pseudomorphic HEMT. This is achieved by using an extremely thin layer of one of the materials – so thin that the crystal lattice simply stretches to fit the other material. This technique allows the construction of transistors with larger bandgap differences than otherwise possible, giving them better performance.

Another way to use materials of different lattice constants is to place a buffer layer between them. This is done in the mHEMT or metamorphic HEMT, an advancement of the pHEMT. The buffer layer is made of AlInAs, with the indium concentration graded so that it can match the lattice constant of both the GaAs substrate and the GaInAs channel. This brings the advantage that practically any Indium concentration in the channel can be realized, so the devices can be optimized for different applications (low indium concentration provides low noise; high indium concentration gives high gain).

By Electrical Behaviour: eHEMT and dHEMT

HEMTs made of nonpolar semiconductors, such as AlGaAs/GaAs, require positive gate voltage or appropriate donor-doping in the AlGaAs barrier to attract the electrons towards the gate, which forms the 2D electron gas and enables conduction of electron currents. This behaviour is similar to that of commonly used field-effect transistors in the enhancement mode, and such a device is called enhancement HEMT, or eHEMT.

When a HEMT is built from AlGaN/GaN, higher power density and breakdown voltage can be achieved. Nitrides also have different crystal structure with lower symmetry, namely the wurtzite one, which has built-in electrical polarisation. Since this polarization differs between the GaN *channel* layer and AlGaN *barrier* layer, a sheet of uncompensated charge in the order of 0.01-0.03 C/mis formed. Due to the crystal orientation typically used for epitaxial growth ("gallium-faced") and the device geometry favorable for fabrication (gate on top), this charge sheet is positive, causing the 2D electron gas to be formed even if there is no doping. Such a transistor is normally on, and will turn off only if the gate is negatively biased - thus this kind of HEMT is known as *depletion HEMT*, or dHEMT. By sufficient doping of the barrier with acceptors (e.g. Mg), the built-in charge can be compensated to restore the more customary eHEMT operation, however high-density p-doping of nitrides is technologically challenging due to dopant diffusion into the channel.

Induced HEMT

In contrast to a modulation-doped HEMT, an induced high electron mobility transistor provides the flexibility to tune different electron densities with a top gate, since the charge carriers are "induced" to the 2DEG plane rather than created by dopants. The absence of a doped layer enhances the electron mobility significantly when compared to their modulation-doped counterparts. This level of cleanliness provides opportunities to perform research into the field of Quantum Billiard for quantum chaos studies, or applications in ultra stable and ultra sensitive electronic devices.

Operation

The improved performance of MODFET over other standard junctions or MOSFETs in microwave radio applications making it slightly different from other types of FET. There is a movement of electrons from the ntype region through the crystal lattice and many of them remain close to the heterojunction. These electrons for one layer thick forming a two dimensional electron gas, where the electrons are able to move freely. Here the electrons have very high mobility because there are no extra donor electrons or other species with which the electrons will collide. Now, MODFET works by modulating the number of electrons in the conducting channel formed by 2-D electron gas with the application of a voltage bias applied to the gate. This leads to the formation of a schottky barrier diode which in turn controls the conductivity of the device. Hence, MODFET can be linked with other traditional types of FET where the applied bias is used to modulate the width of the channel.

Characteristics

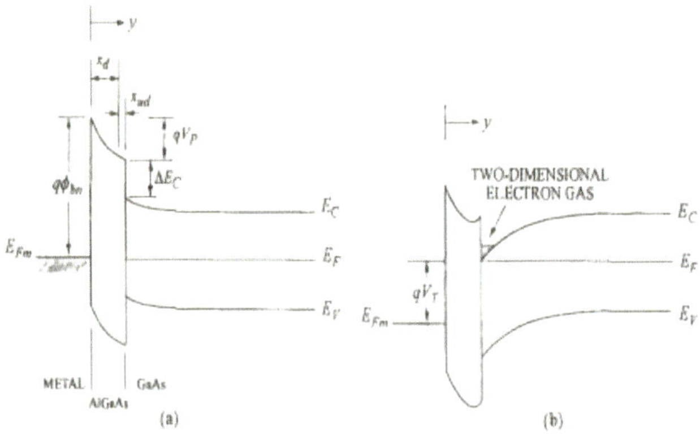

Energy band diagram of an enhancement mode MODFET (a) equilibrium (b) onset of threshold.

From the above diagram we can conclude that the gate bias at which the channel starts to form between the source and drain occurs when *EF* at the GaAs surface coincides with *Ec* and is given as

$$V_T = \phi_{bn} - V_p - \frac{\Delta E_C}{q}$$

Where *Vp* is the pinch off voltage for the AlGaAs given by

$$V_p = \frac{qN_D x_d^2}{2 \epsilon_s}$$

By varying, can be varied ,between positive and negative values.With the gate voltage larger than the threshold the charge sheet induced by the gate is capacitively coupled and given by:

$$n_s(x) = \frac{C_O [V_G - V_T - \psi(x)]}{q}$$

Where, $C_O = \frac{\epsilon_s}{x_d + x_{ud} + \Delta d}$ where*xd* and *xud* are the doped and undopedAlGaAs thickness and

Δd is the channel thickness of the two dimensional electron gas,estimated to be ≈ 80Å .ψ is the channel potential with respect to the source.It varies along the channel from 0 to the drain bias *VD*.The drift current at any point along the channel is given by

$$I = W \mu_n q n_s \epsilon_x$$

$$= W \mu_n C_O [[V_G - V_T - \psi(x)]] \left(\frac{d\psi(x)}{dx} \right)$$

Since the current is constant along the channel, integrating this equation from source to drain gives,

$$I = \frac{W \mu_n C_O}{L} \left[(V_G - V_T) V_D - \frac{V_D^2}{2} \right]$$

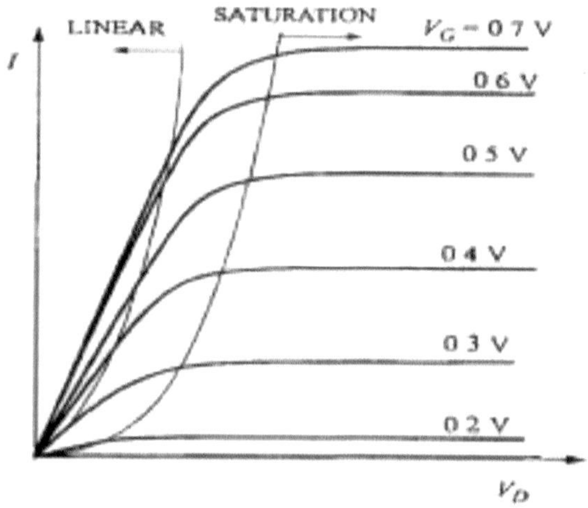

IV characteristics of MODFET

In the linear region above equation reduced to an ohmic equation

$$I_{lin} = \frac{\mu_n C_O W (V_G - V_T) V_D}{L}$$

And the transconductance is

$$g_{m,lin} = \frac{dI_{lin}}{dV_G} = \frac{\mu_n C_0 W V_D}{L}$$

At high $V_D n_s$ at the drain is reduced to zero corresponding to the pinchoff conditionand current saturates with V_D.

$$V_{D,sat} = V_G - V_T$$

$$I_{sat} = \frac{\mu_n C_0 W}{2L}(V_G - V_T)^2$$

$$g_{m,sat} = \frac{dI_{sat}}{dV_G} = \frac{\mu_n C_0 W (V_G - V_T)}{L}$$

Reliability

The reliability of MODFETs is an important parameter which is affected by the epitaxial structure, device fabrication, and device geometry. There is one drawback of using $Al_xGa_{1-x}As$ in the material structure is the occurrence of traps, called DX centers, for AlxAs mole fractions around x = 0.26. These traps are deep donor levels which can cause reduction in drain current, enhancement in low frequency noise and photoconductivity, and creates problems particularly at low temperature. Further, the creation of DX centers increases with higher doping of the $Al_xGa_{1-x}As$. This formation of DX centers can be controlled by keeping the AlxAs content below x = 0.24 for n-type doped $Al_xGa_{1-x}As$. A second possible reliability problem that can occur is the deconfinement of the 2DEG under high-temperature conditions. Thermally accelerated testing has shown that the Al can migrate laterally into the gate, resulting in a change in the conduction band discontinuity.